Engineering, Planning and Management of Groundwater Resources

Engineering, Planning and Management of Groundwater Resources

Edited by **William Sobol**

SYRAWOOD
PUBLISHING HOUSE
New York

Published by Syrawood Publishing House,
750 Third Avenue, 9th Floor,
New York, NY 10017, USA
www.syrawoodpublishinghouse.com

Engineering, Planning and Management of Groundwater Resources
Edited by William Sobol

© 2016 Syrawood Publishing House

International Standard Book Number: 978-1-68286-084-7 (Hardback)

Printed in the United States of America.

Contents

Preface

The main aim of this book is to educate learners and enhance their research focus by presenting diverse topics covering this vast field. This is an advanced book which compiles significant studies by distinguished experts. This book addresses successive solutions to the challenges arising in the area of application, along with it; the book provides scope for future developments.

The rapidly increasing population and urbanization has led to over exploitation of groundwater resources around the world. This book deals with the planning and management of groundwater resources. The chapters discussed in this book encompass the recent studies in the field of groundwater hydrology and discusses concepts like biogeochemical processes, engineering hydrology, global groundwater resources, etc. It attempts to assist those with a goal of delving into the field of groundwater resource management. The book will be a crucial source of knowledge for all the students and academicians engaged in this field.

It was a great honour to edit this book, though there were challenges, as it involved a lot of communication and networking between me and the editorial team. However, the end result was this all-inclusive book covering diverse themes in the field.

Finally, it is important to acknowledge the efforts of the contributors for their excellent chapters, through which a wide variety of issues have been addressed. I would also like to thank my colleagues for their valuable feedback during the making of this book.

Editor

Hydrological hysteresis and its value for assessing process consistency in catchment conceptual models

O. Fovet[1,2], L. Ruiz[1,2], M. Hrachowitz[3], M. Faucheux[1,2], and C. Gascuel-Odoux[1,2]

[1]INRA, UMR1069 SAS, 65 route de Saint Brieuc, 35042 Rennes, France
[2]Agrocampus Ouest, UMR1069 SAS, 65 route de Saint Brieuc, 35042 Rennes, France
[3]Delft University of Technology, Water Resources Section, Faculty of Civil Engineering and Applied Geosciences, Stevinweg 1, 2600 GA Delft, the Netherlands

Correspondence to: O. Fovet (ophelie.fovet@rennes.inra.fr)

Abstract. While most hydrological models reproduce the general flow dynamics, they frequently fail to adequately mimic system-internal processes. In particular, the relationship between storage and discharge, which often follows annual hysteretic patterns in shallow hard-rock aquifers, is rarely considered in modelling studies. One main reason is that catchment storage is difficult to measure, and another one is that objective functions are usually based on individual variables time series (e.g. the discharge). This reduces the ability of classical procedures to assess the relevance of the conceptual hypotheses associated with models.

We analysed the annual hysteric patterns observed between stream flow and water storage both in the saturated and unsaturated zones of the hillslope and the riparian zone of a headwater catchment in French Brittany (Environmental Research Observatory ERO AgrHys (ORE AgrHys)). The saturated-zone storage was estimated using distributed shallow groundwater levels and the unsaturated-zone storage using several moisture profiles. All hysteretic loops were characterized by a hysteresis index. Four conceptual models, previously calibrated and evaluated for the same catchment, were assessed with respect to their ability to reproduce the hysteretic patterns.

The observed relationship between stream flow and saturated, and unsaturated storages led us to identify four hydrological periods and emphasized a clearly distinct behaviour between riparian and hillslope groundwaters. Although all the tested models were able to produce an annual hysteresis loop between discharge and both saturated and unsaturated storage, the integration of a riparian component led to

overall improved hysteretic signatures, even if some misrepresentation remained. Such a system-like approach is likely to improve model selection.

1 Introduction

Rainfall-runoff models are tools that mimic the low-pass filter properties of catchments. Specifically, they aim at reproducing observed stream flow time series by routing time series of meteorological drivers through a sequence of mathematically formalized processes that allow a temporal dispersion of the input signals in a way that is consistent with the modeller's conception of how the system functions. The core of most models, in particular in temperate, humid climates dominated by some type of subsurface flow, is a series of storage–discharge functions that, in the most general terms, express system output (i.e. discharge and evaporation) as a function of the system state (i.e. storage), thereby generating a signal that is attenuated and lagged with respect to the input signal (i.e. precipitation).

However, modelling efforts on the catchment scale typically face the problem that, on that scale, neither integrated internal fluxes nor the integrated storage and the partitioning between different storage components at a given time can be easily observed within limited uncertainty. Indeed, indicators of catchment storage such as groundwater levels and soil water content can be highly variable in space and exhibit heterogeneous spatio-temporal dynamics. While spatial aggregation of storage estimates (e.g. catchment averages) in lumped models may lead to a loss of crucial information and thus to overly simplistic representations of reality, allowing for the explicit incorporation of spatial storage heterogeneity in (semi-)distributed models may prove elusive in the presence of data error and the frequent absence of detailed spatial knowledge of the properties of the flow domain. A time series of groundwater table levels from a single piezometer is not representative of the behaviour of the groundwater, even on the hillslope scale; therefore, it is difficult to link it with either a reservoir volume simulated by a lumped model or an average water table level of a grid point simulated by a fully distributed model. These problems were recently addressed in some studies that intended to assess catchment storage using all available data (Mc-Namara et al., 2011; Tetzlaff et al., 2011) and showing the importance of this storage in thresholds observed in the response of discharge to precipitation in catchments. For example, Spence (2010) argued that the observed nonlinear relationships between stream flow and catchment storage (i.e. no unique storage–discharge relations) are the manifestation of thresholds occurring in catchment runoff generation. Thus, depending on the structure of the system, storage–discharge dynamics can exhibit hysteretic patterns, i.e. the system response depends on the history and the memory of the system (e.g. Everett and Whitton, 1952; Ali et al., 2011; Gabrielli et al., 2012; Haught and van Meerveld, 2011). Andermann et al. (2012) found a hysteretic relationship between precipitation and discharge in both glaciated and unglaciated catchments in the Himalaya Mountains that was shown to be due to groundwater storage rather than to snow or glacier melt. Hrachowitz et al. (2013a), demonstrating the presence of hysteresis in the distribution of water ages, highlighted the importance of an adequate characterization of all system-relevant internal states at a given time to predict the system response within limited uncertainty as flow can be generated from different system components depending on the wetness state of the system.

In catchment-scale rainfall-runoff models, the need for calibration remains inevitable (Beven, 2001) due to the presence of data errors (e.g. Beven, 2013) and to the typically oversimplified process representations (e.g. Gupta et al., 2012). In spite of their comparatively high degrees of freedom, such models are frequently evaluated only against one single observed output variable, e.g. stream flow. Although the calibrated models may then adequately reproduce the output variable, model equifinality (e.g. Savenije, 2001) will lead to many apparently feasible solutions that do not sufficiently well reproduce system-internal dynamics as they are mere artefacts of the mathematical optimization process rather than suitable representations of reality (Gharari et al., 2013; Hrachowitz et al., 2013b; Andréassian et al., 2012; Beven, 2006; Kirchner, 2006). The realisation that there is a need for multivariable and multiobjective model evaluation strategies to identify and discard solutions that do not satisfy all evaluation criteria applied is therefore gaining ground (e.g. Freer et al., 1996; Gupta et al., 1998, 2008, Gascuel-

Odoux et al., 2010) as this will eventually lead to models that are not only capable of reproducing the observed output variables (e.g. stream flow) but that also represent the system-internal dynamics in a more realistic way (Euser et al., 2013). The value of such multivariable and/or multiobjective evaluation strategies has been demonstrated in the past, for example using groundwater levels (e.g. Fenicia et al., 2008; Molénat et al., 2005, Giustolisi and Simeone, 2006; Freer et al., 2004; Seibert, 2000; Lamb et al., 1998), soil moisture (Kampf and Burges, 2007; Parajka et al., 2006), saturated-area extension (Franks et al., 1998), snow cover patterns (e.g. Nester et al., 2012), remotely sensed evaporation, (e.g. Mohamed et al., 2006; Winsemius et al., 2008), stream flow at subcatchment outlets (e.g. Moussa et al., 2007) and even water quality data such as, e.g., chloride concentrations (Hrachowitz et al., 2011), atmospheric tracers (Molénat et al., 2013) or nitrates and sulfate concentrations (Hartmann et al., 2013a) and water isotopes such as $\delta^{18}O$ (Hartmann et al., 2013b). However, most studies using multiple response variables only evaluate them individually to identify Pareto-optimal solutions. This practice may result in the loss of critical information, such as the timing between the multiple variables. In other words it is conceivable that model calibration leads to Pareto-optimal solutions with adequate model performance for all variables while at the same time misrepresenting the dynamics between these variables. Instead, using a synthetic catchment property (Sivapalan et al., 2005) or a hydrological signature (Wagener and Montanari, 2011; Yadav et al., 2007), combining different variables into one function, may potentially serve as a instructive diagnostic tool, a calibration objective or even as a metric for catchment classification (Wagener, 2007).

Hysteretic patterns between hydrological variables are potentially good candidates to build such tools. The objective of this paper is to explore (i) the potential of using annual hysteric patterns observed between stream flow and water storage both in the saturated and unsaturated zones of the hillslope and of the riparian zone for characterizing the hydrological functioning of a small headwater catchment in French Brittany (Environmental Research Observatory ERO AgrHys (ORE AgrHys)), (ii) to which degree a suite of conceptual rainfall-runoff models with increasing complexity, which were calibrated and evaluated for this catchment in previous work using a flexible modelling framework (Hrachowitz et al., 2014), can reproduce the observed storage–discharge hysteresis and (iii) whether the use of the storage–discharge hysteresis can provide additional information for model diagnostics compared to traditional model evaluation metrics.

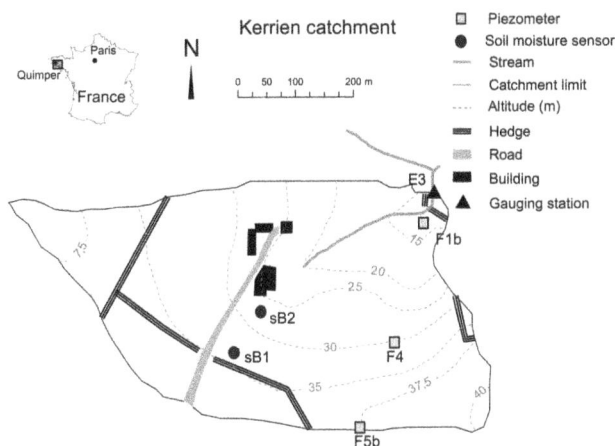

Figure 1. Study site in west Brittany (indicated by the square near Quimper) and location of the monitoring equipments. The weather station is located 500 m north of the catchment.

2 Materials and methods

2.1 Study sites

Kerrien (10.5 ha) is a headwater catchment located in south-western Brittany (47°35′ N, 117°52′ E; see Fig. 1). Elevations range from 14 to 38 m a.s.l.; slopes are less than 8.5 %. The climate is oceanic, with a mean annual temperature of 11.9 °C with a minimum of 5.9 °C in winter and a maximum of 17.9 °C in summer. Mean annual rainfall over the period 1992–2012 is 1113 mm (±20 %) and mean annual Penman potential evapotranspiration (PET) is 700 mm (±4 %). Mean annual drainage is 360 mm (±60 %) at the outlet. There is a high water deficit in the annual budget almost every year due to underflows below the outlet (Ruiz et al., 2002). The catchment lies under granite (leucogranodiorite of Plomelin), the upper part of which is weathered from 1 to more than 20 m deep. Soils are mainly sandy loam with an upper horizon rich in organic matter; depths are between 40 and 90 cm. Soils are well drained except in the bottomlands, which represent 7 % of the total area. Agriculture dominates the land use, with 86 % of the total area covered by grassland, maize and wheat, none of them irrigated. The base flow index is about 80 to 90 %; thus, the hillslope aquifer is the main contributor to stream flow (Molénat et al., 2008; Ruiz et al., 2002). Both stream flow and shallow groundwater tables exhibit a strong annual seasonality in this catchment (Figs. 2 and 3a).

2.2 Data

Meteorological data were recorded in an automatic weather station (CIMEL, Fig. 1) which provides hourly rainfall and variables required to estimate daily Penman PET (net solar radiation, air and soil temperatures, wind speed and direction). Discharge was calculated from water level measurements at the outlet (Fig. 1) using a V-notch weir equipped

with a shaft encoder with integrated data logger (OTT Thalimedes) and recorded every 10 min since 2000 (E3). Groundwater levels have been monitored every 15 min since 2001 in three piezometers – F1b, F4, and F5b (Fig. 1) – using vented pressure probe sensors (OTT Orpheus Mini).

Moisture in the unsaturated zone has been recorded every 30 min since July 2010 at seven depth (25, 55, 85, 125, 165, 215, and 265 cm) and at two locations (sB1 and sB2; Fig. 1), using capacitive probes which provide volumic humidity based on frequency domain reflectometry (Environ-Scan SenteK). Due to technical problems, data are missing in December 2012 and January 2013, so only 2 complete water years were available (2010–2011 and 2011–2012). In summary, stream discharge water table levels were considered for the years 2002–2012, and soil moisture was considered for the years 2010–2012.

2.3 Catchment storage estimates

In order to obtain a proxy for the saturated-zone storage on the catchment scale, the time series of groundwater level were normalized between their minimal and maximal values over the 10 years of records so that the normalized value lies between 0 and 1. The resulting normalized variable exhibited very similar dynamics among all the piezometers (see Fig. 2a). However, the piezometer located in the riparian zone (F1b) exhibited variations at a higher frequency, especially during the winter. Therefore, in the following, we used the average of the normalized level in the two hillslope piezometers (F5b, F4) as a proxy for the hillslope groundwater storage dynamics and the normalized level in the riparian piezometer as a proxy for the riparian groundwater storage dynamics.

In order to obtain a proxy for the unsaturated-zone storage, moisture time series were also normalized using the minimal and maximal values observed in all the sensors of the two profiles over the 2 water years with complete records, setting the minimal value as 0 and the maximal value as 1. As the normalized unsaturated storage variables obtained followed very similar trends and dynamics, we used, in the following, an average of the normalized unsaturated-zone storage among all the measurement points (depths and profiles) (Fig. 2b). The two profiles are located on the upslope and downslope parts of the hillslope. Thus, we assumed that averaging their normalized values will allow us to build a proxy for the dynamics of the unsaturated-zone storage on the whole hillslope.

2.4 Hysteresis indexes

Studies on hysteretic relationships in catchments generally focus on qualitative descriptions of patterns associated with a cross-correlation analysis between the two variables (Frei et al., 2010; Hopmans and Bren, 2007; Jung et al., 2004; Salant et al., 2008; Schwientek et al., 2013; Spence et al., 2010;

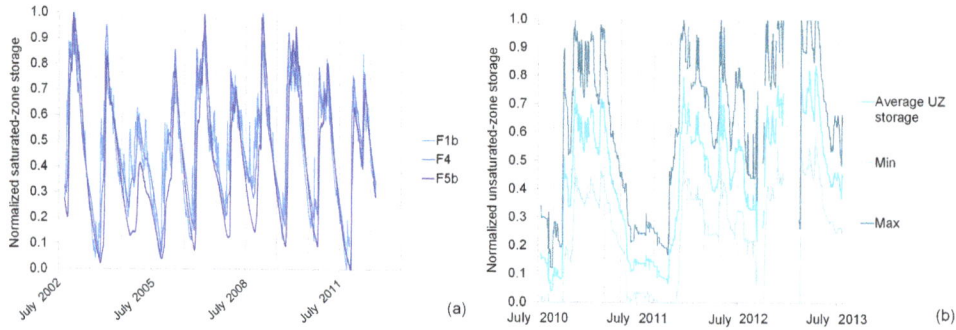

Figure 2. Normalized **(a)** groundwater levels for piezometers in the hillslope (F4 and F5b) and in the riparian zone (F1b) and **(b)** average, maximum and minimum unsaturated-zone storages for all the sensors in the two profiles in the Kerrien catchment.

Velleux et al., 2008). Some authors proposed a typology of hysteretic loops based on their rotational direction, curvature and trend to identify solute controls during storm events (Butturini et al., 2008; Evans and Davies, 1998). For storage–discharge hysteresis on the annual scale, this approach is not sufficient as the same type of hysteretic loop is likely to happen for almost all the years when a strong seasonality exists and its pattern is repeated across years. This is the case in our study, where seasonality of groundwater level and discharge showed a strong unimodal pattern for all years, except 2011–2012, which was bimodal (Figs. 2 and 3a). Moreover, a preliminary cross-correlation analysis revealed that storage and stream flow are strongly correlated, and the cross-correlation value is the greatest for a lag time of 0 days (results not shown).

Quantitative descriptions of the hysteretic loop are also found in the literature, and various ways of computing hysteresis indexes (HIs) have been proposed, for example using the relative difference between extreme concentration values (Butturini et al., 2008) or using the ratio of turbidity values in rising and falling limbs of the storm hydrograph at the midpoint discharge value (Lawler et al., 2006). The latter authors argue that computing HIs by using midpoint discharge usually allows avoiding the small convolutions which are frequently observed at both ends of the hysteretic loop.

In this paper, as the hydrological variables exhibit a strong annual unimodal cycle, we calculated the hysteresis index each year as the difference between water storages at the dates of midpoint discharge in the two phases of the hydrological year – during the recharge period (R) and the recession period (r), i.e. respectively before and after reaching the maximal discharge Q_{max} – as follows:

$$\begin{cases} HI = S\left(t_{R,mid}\right) - S\left(t_{r,mid}\right) \\ Q\left(t_{R,mid}\right) = Q_{mid} \quad \text{and} \quad t_{R,mid} < t_{Q_{max}} \\ Q\left(t_{r,mid}\right) = Q_{mid} \quad \text{and} \quad t_{r,mid} > t_{Q_{max}}, \\ Q\left(t_{Q_{max}}\right) = Q_{max} \end{cases} \quad (1)$$

where $S(t)$ is the storage value at time t and $Q(t)$ the stream flow value at time t. The midpoint discharge Q_{mid} is defined as the mean value of discharge between Q_0, the initial

value at the beginning of the hydrological year (October), and Q_{max}, the maximal value reached during that year:

$$Q_{mid} = \frac{Q_0 + Q_{max}}{2}. \quad (2)$$

In order to reduce the impact of the quick variations of discharge or groundwater level due to individual storm events, we smoothed the time series using 7-day moving averages. The strong seasonal discharge cycle led us to identify two occurrences of Q_{mid} per year only – during the recharge period (t_R) and during the recession period (t_r) – while high and low stream flow values are taken several times per year as explained by Lawler et al. (2006). Computing the HI using the difference in storage was possible here because storage and stream flow values vary among years within a narrow range of magnitude, while Lawler et al. (2006) used the ratio because turbidity can differ by several orders of magnitude from one storm to the other. Computing the HI with the difference between the values of storage and not with their ratio allowed maintaining its sensitivity to the year-to-year variations of the width of the hysteretic loop. The difference in water storage dynamics in the unsaturated and saturated zones were approximated by the difference in normalized soil moisture content and by the difference in normalized groundwater level respectively.

The HI gives two types of information: (i) its sign indicates the direction of the loop (anticlockwise loop induces a negative value of the HI, whereas a clockwise loop leads to a positive value of the HI) and (ii) its absolute value is proportional to the magnitude of the hysteresis (i.e. the width of the hysteretic loop). The HI is a proxy for the importance of lag time response between variations in catchment storages (unsaturated and saturated) and stream discharge; its sign indicates whether storage reacts before or after the stream flow. Therefore, it can be used for comparing the capacity of the different models to reproduce to some extent the observed storage–discharge relationships. The normalization of the observed variables related to the storage (here either groundwater level or soil moisture) has no effect on the sign of the HI; the HI values are only being divided by the maximal amplitude ob-

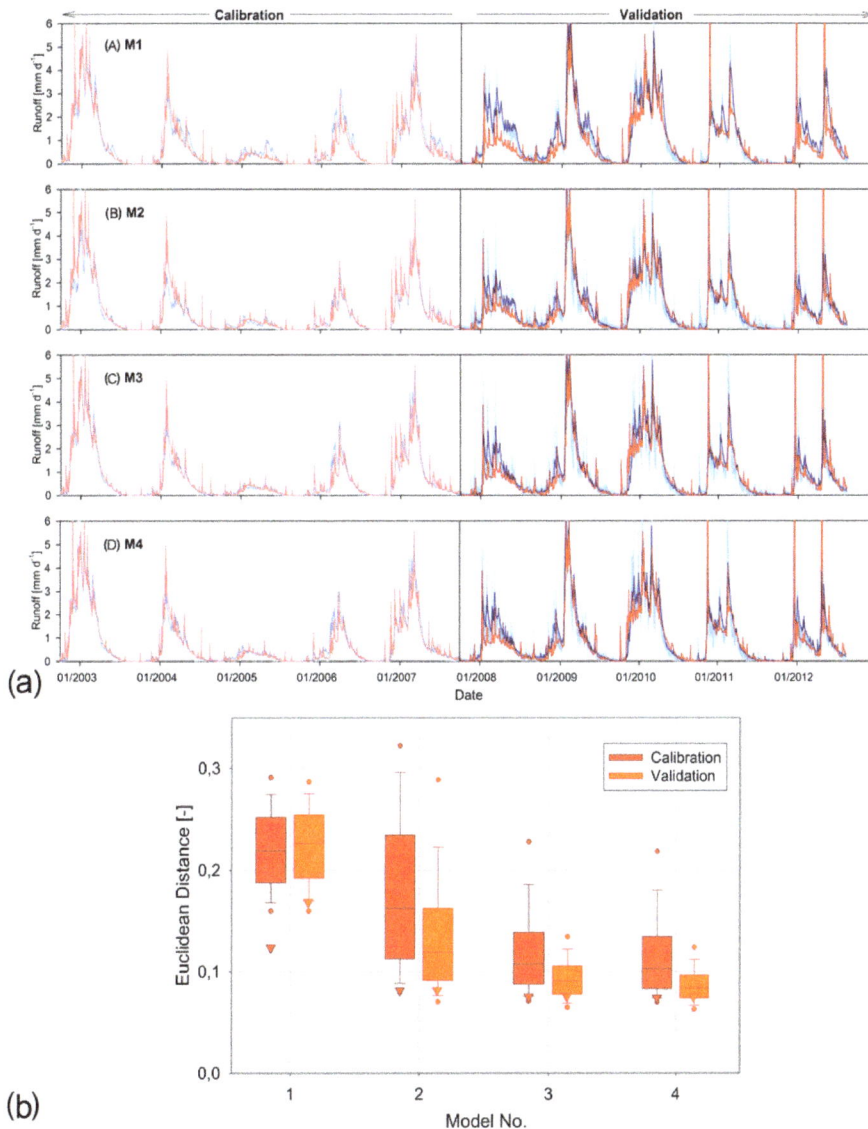

Figure 3. (a) Observed (red line) and modelled runoff for model set-ups **(A)** M1, **(B)** M2, **(C)** M3 and **(D)** M4 in calibration and independent evaluation (validation) periods. Modelled runoff shown as the most balanced solution (dark blue line) and the 5/95th uncertainty bounds (light blue shaded area). Adapted from Hrachowitz et al. (2014). **(b)** Overall model performance for all model set-ups (M1–M4) expressed as Euclidean distance from the "perfect model" computed from all calibration objectives and signatures with respect to calibration and validation periods. Triangles represent the optimal solution, i.e. the solution obtained from the parameter set with the lowest Euclidean distance during calibration. Box plots represent the Euclidean distance for the complete sets of all feasible solutions (the dots indicate 5/95th percentiles, the whiskers 10/90th percentiles and the horizontal central line the median). From Hrachowitz et al. (2014).

served in the storage during the whole period. Therefore, as long as the normalization is applied to the whole period (to all years and to both measurements and simulations), it does not affect the interpretation related to absolute values of the HI.

2.5 Models

In previous work, a range of conceptual models was calibrated and evaluated for the Kerrien catchment in a step-

wise development using a flexible modelling framework (see Hrachowitz et al., 2014). This section aims at summarizing the results of this previous study as they are used as a basis for the present work. In this previous study, adopting a flexible stepwise modelling strategy, 11 models with increasing complexity, i.e. allowing for more process heterogeneity, were calibrated and evaluated for the study catchment. Four of these 11 models (hereafter referred to as M1 to M4; details given in Tables 1 and 2) were selected for the present work as they correspond to the sequence of model

Table 1. Water balance and state and flux equations of the models used.

Process	Water balance	Eq.	Models	Flux and state equations	Eq.	Models
Unsaturated zone	$dS_U/dt = P - E_U - R_F - R_P - R_S$	(1.1)	M1, 2, 3, & 4	$E_U = E_P \text{Min}\left(1, \frac{S_U}{S_{U\text{max,H}}} \frac{1}{L_P}\right)$	(1.2)	M1, 2, 3, & 4
				$R_U = (1 - C_R) P$	(1.3)	M1, 2, 3, & 4
				$R_F = C_R (1 - C_P) P$	(1.4)	M1, 2, 3, & 4
				$R_S = P_{\max}\left(\frac{S_U}{S_{U\text{max,H}}}\right)$	(1.5)	M1, 2, 3, & 4
				$C_R = \frac{1}{1+\exp\left(\frac{-S_U/S_{U\text{max,H}}+0.5}{\beta}\right)}$	(1.6)	M1, 2, 3, & 4
Fast reservoir	$dS_F/dt = R_F - Q_F - E_F$	(2.1)	M1, 2, 3, & 4	$S_{F,\text{in}} = S_F + R_F$	(2.2)	M1, 2, 3, & 4
				$Q_F = S_{F,\text{in}}\left(1 - e^{-k_F t}\right)$	(2.3)	M1, 2, 3, & 4
				$E_F = \text{Min}\left(E_P - E_U, S_{F,\text{in}} - Q_F\right)$	(2.4)	M1, 2, 3, & 4
Slow reservoir	$dS_S/dt = R_S + R_P - Q_S$	(3.1)	M1	$S_{S,\text{in}} = S_S + R_S + R_P$	(3.2)	M1
				$Q_S = S_{S,\text{in}}\left(1 - e^{-k_S t}\right)$	(3.3)	M1
	$dS_{s,a}/dt = \begin{cases} S_{s,a} - \text{Max}\left(0, S_{S,\text{tot,out}}\right), & S_{S,\text{tot,in}} > 0 \\ 0, & S_{S,\text{tot,in}} \leq 0 \end{cases}$	(3.4)	M2, 3 & 4	$Q_S = \text{Max}\left(0, Q_{S,\text{tot}} - Q_{L,\text{cst}}\right)$	(3.7)	M2, 3 & 4
	$dS_{s,p}/dt = \begin{cases} S_{s,p} + \text{Min}\left(0, S_{S,\text{tot,out}}\right), & S_{S,\text{tot,in}} > 0 \\ S_{s,p} + S_{S,\text{tot,in}}, & S_{S,\text{tot,in}} \leq 0 \end{cases}$	(3.5)	M2, 3 & 4	$S_{S,\text{tot,in}} = S_{s,a} + S_{s,p} + R_S + R_P$	(3.8)	M2, 3 & 4
	$dS_s/dt = dS_{s,a}/dt + dS_{s,p}/dt = R_S + R_P - Q_{L,\text{cst}}$	(3.6)	M2, 3 & 4	$S_{S,\text{tot,out}} = \begin{cases} S_{S,\text{tot,in}} e^{-k_S t} - \frac{Q_{L,\text{cst}}}{k_S}\left(1 - e^{-k_S t}\right), & S_{S,\text{tot,in}} > 0 \\ S_{S,\text{tot,in}} - Q_{L,\text{cst}}, & S_{S,\text{tot,in}} \leq 0 \end{cases}$	(3.9)	M2, 3 & 4
				$Q_{L,\text{cst}} = \text{constant}$	(3.10)	M2, 3 & 4
Unsaturated riparian zone	$dS_{U,R}/dt = P - E_{U,R} - R_R$	(4.1)	M3 & 4	$E_{U,R} = E_P \text{Min}\left(1, \frac{S_{U,R}}{S_{U\text{max,R}}} \frac{1}{L_P}\right)$	(4.2)	M3 & 4
				$R_R = C_{R,R} P$	(4.3)	M3 & 4
				$C_{R,R} = \text{Min}\left(1, \frac{S_{U,R}}{S_{U\text{max,R}}}\right)$	(4.4)	M3
				$C_{R,R} = \text{Min}\left(1, \left(\frac{S_{U,R}}{S_{U\text{max,R}}}\right)^{\beta_R}\right)$	(4.5)	M4
Riparian reservoir	$dS_R/dt = R_R - Q_R - E_R$	(5.1)	M3 & 4	$S_{R,\text{in}} = S_R + R_R$	(5.2)	M3 & 4
				$Q_R = S_{R,\text{in}}\left(1 - e^{-k_R t}\right)$	(5.3)	M3 & 4
				$E_R = \text{Min}\left(E_P - E_{U,R}, S_{R,\text{in}} - Q_R\right)$	(5.4)	M3 & 4
Total runoff	$Q_T = Q_F + Q_S$	(6.1)	M1 & 2			
Total evaporative fluxes	$Q_T = (1 - f)(Q_F + Q_S) + f Q_R$	(6.2)	M3 & 4			
	$E_A = E_U + E_F$	(7.1)	M1 & 2			
	$E_A = (1 - f)(E_U + E_F) + f\left(E_{U,R} + E_R\right)$	(7.2)	M3 & 4			

List of symbols: C_P – preferential recharge coefficient [-]; P – total precipitation [LT^{-1}]; S_F – storage in fast reservoir [L]; C_R – hillslope runoff generation coefficient [-]; E_F – transpiration fast responding reservoir [LT^{-1}]; S_R – storage in riparian reservoir [L]; $C_{R,R}$ – riparian runoff generation coefficient [-]; E_P – potential evaporation [LT^{-1}]; S_S – storage in slow reservoir [L]; k_F – storage coefficient of fast reservoir [T^{-1}]; E_R – transpiration from riparian reservoir [LT^{-1}]; $S_{s,a}$ – active storage in slow reservoir [L]; k_S – storage coefficient of slow reservoir [T^{-1}]; E_U – transpiration from unsaturated reservoir [LT^{-1}]; $S_{s,p}$ – passive storage in slow reservoir [L]; k_L – storage coefficient for deep infiltration loss [T^{-1}]; $E_{U,R}$ – transpiration unsaturated riparian reservoir [LT^{-1}]; $S_{S,\text{tot}}$ – total storage in slow reservoir [L]; k_R – storage coefficient of riparian reservoir [T^{-1}]; Q_R – runoff from riparian reservoir [LT^{-1}]; S_U – storage in unsaturated reservoir [L]; f – proportion wetlands in the catchment [-]; Q_S – runoff from slow reservoir [LT^{-1}]; $S_{S,\text{tot,in}}$ – total storage incoming in slow reservoir [L]; L_P – transpiration threshold [-]; Q_F – runoff from fast reservoir [LT^{-1}]; $S_{S,\text{tot,out}}$ – total storage outcoming from slow reservoir [L]; $P_{\max}$ – percolation capacity [LT^{-1}]; $Q_{L,\text{const}}$ – constant deep infiltration loss [LT^{-1}]; $S_{U\text{max,H}}$ – unsaturated hillslope storage capacity [L]; $S_{U\text{max,R}}$ – unsaturated riparian storage capacity [L]; R_P – preferential recharge of slow reservoir [LT^{-1}]; β – hillslope shape parameter for CR [-]; R_R – recharge of riparian reservoir [LT^{-1}]; β_R – riparian shape parameter for $C_{R,R}$ [-]; R_S – recharge of slow reservoir [LT^{-1}]; R_U – infiltration into unsaturated reservoir [LT^{-1}].

architectures that provide the most significant performance improvements among the tested set-ups. As a starting point and benchmark, Model M1 with seven parameters, resembling many frequently used catchment models, such as HBV, was used (e.g. Bergström, 1995). The three boxes represent respectively an unsaturated zone, a slow-responding and a fast-responding reservoir. In Model M2, additional deep infiltration losses are integrated from the slow store to take into account the significant groundwater export to adjacent catchments in this study catchment as indicated by the observed long-term water balance (Ruiz et al., 2002). This is done by adding a second outlet together with a threshold to this storage to allow for continued groundwater export from a storage volume below the stream during zero-flow conditions, i.e. when the stream runs dry. As riparian zones frequently exhibit a distinct hydrological functioning (e.g. Molénat et al., 2005; Seibert et al., 2003), indicated in the study catchment by distinct response dynamics in the riparian piezometers (Martin et al., 2006), Models M3 and M4 additionally integrate a wetland/riparian zone component, composed of an unsaturated-zone store and a fast-responding reservoir, parallel to the other boxes. The riparian unsaturated zone generates flow using a linear function in M3 and a nonlinear func-

tion in M4. The complete set of water balance and constitutive model equations of the four models is listed in Table 1, while the model structures are schematized in Table 2.

2.6 Calibration and evaluation

This section is also a summary of the findings of Hrachowitz et al. (2014) that served as a basis for this study and does not consist of the results of the current study. The models have been calibrated for the period 1 October 2002–30 September 2007 after a 1-year warm-up period, using a multiobjective calibration strategy (e.g. Gupta et al., 1998) based on Monte Carlo sampling (10^7 realizations). The uniform prior parameter distributions used for M1–M4 are provided in Table 3. To reduce parameter and associated predictive uncertainty, the models were calibrated using a total of four calibration objective functions (see Table 4), i.e. the Nash–Sutcliffe efficiencies (Nash and Sutcliffe, 1970) for stream flow ($E_{\text{NS},Q}$), for the logarithm of the stream flow ($E_{\text{NS,log}(Q)}$) and for the flow duration curve ($E_{\text{NS,FDC}}$) as well as the volumetric efficiency for stream flow ($V_{E,Q}$; Criss and Winston, 2008). To facilitate a clearer assessment, the calibration objective functions ($n = 4$) were combined in a

Table 2. Model structures and parameters.

Model structure	Name	Parameters	Equations
	M1	$k_F, k_S, P_{max}, L_P, S_{Umax,H},$ β, C_P	(1.1) to (1.6); (2.1) to (2.4); (3.1) to (3.3); (6.1) & (7.1)
	M2	$k_F, k_S, P_{max}, L_P, S_{Umax,H},$ $\beta, C_P, Q_{L,cst}$	(1.1) to (1.6); (2.1) to (2.4); (3.4) to (3.10); (6.1) & (7.1)
	M3	$k_F, k_S, P_{max}, L_P, S_{Umax,H}, S_{Umax,R}$ $\beta, C_P, Q_{L,cst}, k_R, f,$	(1.1) to (1.6); (2.1) to (2.4); (3.4) to (3.10); (4.1) to (4.4); (5.1) to (5.4); (6.2) & (7.2)
	M4	$k_F, k_S, P_{max}, L_P, S_{Umax,H},$ $\beta, C_P, Q_{L,cst}, f, k_R,$ $S_{Umax,R}, \beta_R$	(1.1) to (1.6); (2.1) to (2.4); (3.4) to (3.10); (4.1) to (4.3); (4.5); (5.1) to (5.4); (6.2) & (7.2)

single calibration metric: the Euclidean distance to the perfect model ($D_{E,cal}$; e.g. Hrachowitz et al., 2013a; Gascuel-Odoux et al., 2010):

$$D_E = \sqrt{\frac{(1-E_{NS,Q})^2 + (1-E_{NS,log(Q)})^2 + (1-E_{V,Q})^2 + (1-E_{NS,FDC})^2}{n}}. \quad (3)$$

As mathematically optimal parameter sets are frequently hydrologically suboptimal, i.e. unrealistic (e.g. Beven, 2006), all parameter sets within the 4-dimensional space spanned by the calibration Pareto fronts, as approximated by the cloud of sample points, were retained as feasible.

The calibrated models were then evaluated against their respective skills to predict the system response with respect to a selection of 13 catchment signatures (described in Table 4) in a multicriteria posterior evaluation strategy. Figure 3 and Table 4 show the global performance D_E of the four models in terms of the Euclidean distance to the perfect model, constructed from all calibration objective functions and evaluation signatures. Model M1 provided good performance in calibration on the objective functions while its validation performances were considerably decreased. Its ability to reproduce the different signatures showed that it failed in particular to reproduce flow in wet periods (such as the evaluation period in Fig. 3a) and groundwater dynamics. Model M2 led to calibration performances slightly lower than model M1 but higher validation performances. The hydrological signatures

simulated by M2 exhibited lower uncertainties both in validation and calibration periods because of a better simulation of low-flow conditions and groundwater dynamics. Model M3 provided similar performances to M2 for calibration and for validation but with clearly reduced uncertainty bounds. Overall signature reproduction was improved because of a clear improvement of low-flow and groundwater-related signatures even if performance in calibration objective functions remained lower than that for model M1. Model M4 exhibited similar performances to the previous models both in calibration and validation periods but a better performance for the whole set of signatures and lower uncertainties.

More details on the model calibration and evaluation with respect to hydrological signatures can be found in Hrachowitz et al. (2014; note that M1, M2 , M3 and M4 presented in this study correspond respectively to M1, M6, M8 and M11 in the original paper). Within the obtained range of parameter uncertainty, the types of simulated hysteresis patterns were not affected by the parameter values but only by the model structures. Note that we restricted the following analysis only to the optimal parameter set in each case, first for the sake of clarity and also because, at this stage, our interest was in assessing the ability of model structures to reproduce the observed general features in hysteresis pat-

Table 3. Prior and posterior distribution of the model parameters.

	C_p [-]	f [-]	k_F [d^{-1}]	k_R [d^{-1}]	k_S [d^{-1}]	L_p [-]	$Q_{L,const}$ [mm d^{-1}]	P_{max} [mm d^{-1}]	$S_{S,p,max}$ [mm]	$S_{Umax,H}$ [mm]	$S_{Umax,R}$ [mm]	β [-]	β_R [-]
Prior distribution	0–1	0.1	0.025–1	0.05–2	0.001–0.05	0–1	0.37	0–4	0–2000	0–1500	0–750	0–100	0–2
Posterior distribution													
M1	0.12/0.63		0.042/0.094		0.03/0.049	0.00/0.07		0.03/0.29		637/1446		10.5/61.5	
M2	0.14/0.55		0.054/0.627		0.041*	0.05/0.34	0.37*	0.27/1.98		722/1461		2.4/36.9	
M3	0.15/0.64		0.054/0.619		0.041*	0.04/0.27	0.37*	0.34/2.29		686/1442	132/725	13.6/69.7	
M4	0.19/0.64	0.1*	0.054/0.466	0.318/1.857	0.041*	0.04/0.27	0.37*	0.29/2.18		683/1444	120/730	13.0/69.2	0.13/1.86

* Fixed parameter values.

terns and not in quantifying their performance in fitting the observations.

In the present work the sensitivity of the hysteresis indexes to parameter uncertainty is investigated by computing the HI values for the all sets of feasible parameters.

3 Results and discussion

3.1 Hysteretic pattern of the groundwater storage–discharge relationship

3.1.1 Observations in hillslope and riparian zones: saturated storage vs. flow

The 2-dimensional observed relationship between saturated storage in the hillslope (HSS) or in the riparian zone (RSS) and stream discharge (Q) for each year was hysteretic, highlighting the nonuniqueness of the response of discharge to storage depending on the initial conditions and a lag time between both variable dynamics, in particular during the recharge period, as illustrated in Fig. 4 for two contrasting water years.

The direction of the hysteretic loop was different depending on the topographic position of the piezometer: loops were always anticlockwise (leading to negative values of the HI) for the piezometer located at the top of the hillslope HSS-F5b(Q), mostly anticlockwise for the midslope piezometer HSS-F4(Q) and mostly clockwise (positive values of the HI) for the piezometer in the riparian zone RSS-F1b(Q) (Fig. 5).

In the riparian zone, storage at Q_{mid} was usually lower in the recession period than in the recharge period, especially in dry years, leading to a positive HI. This is due to the fact that the riparian groundwater level increased early at the beginning of the recharge period, before the stream discharge, due to the limited storage capacity of the narrow unsaturated layer in bottomlands, reinforced by groundwater ridging, which is linked to the extent of the capillary fringe. However, the hysteretic loops were narrow, and, for wet years, the storage value during the recession period occasionally exceeded the value in the recession period without modifying the general direction of the hysteresis when looking at the whole pattern (e.g. in 2003–2004, see Fig. 4a). When this occurred at the time of Q_{mid}, it led to a negative HI although absolute values remained small (Fig. 5)

The hillslope groundwater responded later than the stream, due to the deeper groundwater levels and higher unsaturated storage capacity (Rouxel et al., 2011), both introducing a time lag for the recharge and thus for the groundwater response. This led to negative values of the HI as groundwater levels in recession periods were higher than in recharge periods for the same level of discharge (in particular at Q_{mid}). The loops were also wider in the hillslope, leading to high absolute values of the HI (Fig. 5).

Table 4. Hydrological calibration criteria and evaluation signatures. The performance metrics include the Nash–Sutcliffe efficiency (E_{NS}), the volume error (E_V) and the relative error (E_R). For all variables and signatures, except for Q, Qlow and GW, the long-term averages were used.

$$E_{NS,X} = 1 - \frac{\sum_{i=1:n} \sqrt{(X_{obs,i} - X_{sim,i})^2}}{\sum_{i=1:n} \sqrt{\left(X_{obs,i} - \frac{1}{n}\sum_{i=1:n} X_{obs,i}\right)^2}}$$

$$E_{V,X} = 1 - \frac{\sum_{i=1:n} |X_{obs,i} - X_{sim,i}|}{\sum_{i=1:n} X_{obs,i}}$$

$$E_{R,X} = \frac{X_{obs} - X_{sim}}{X_{obs}}$$

	Variable/signature	ID	Performance metric	Reference
Calibration	Time series of flow	O1	$E_{NS,Q}$	Nash and Sutcliffe (1970)
		O2	$E_{NS,\log(Q)}$	
		O3	$E_{V,Q}$	Criss and Winston (2008)
	Flow duration curve	O4	$E_{NS,FDC}$	Jothityangkoon et al. (2001)
			$D_{E,cal}$	Schoups et al. (2005)
Evaluation	Flow during low-flow period	S1	$E_{NS,Q,low}$	Freer et al. (2003)
	Groundwater dynamics[a]	S2	$E_{NS,GW}$	Fenicia et al. (2008a)
	Flow duration curve low-flow period	S3	$E_{NS,FDC,low}$	Yilmaz et al. (2008)
	Flow duration curve high-flow period	S4	$E_{NS,FDC,high}$	Yilmaz et al. (2008)
	Groundwater duration curve[a]	S5	$E_{NS,GDC}$	–
	Peak distribution	S6	$E_{NS,PD}$	Euser et al. (2013)
	Peak distribution low-flow period	S7	$E_{NS,PD,low}$	Euser et al. (2013)
	Rising-limb density	S8	$E_{R,RLD}$	Shamir et al. (2005)
	Declining-limb density	S9	$E_{R,DLD}$	Sawicz et al. (2011)
	Autocorrelation function of flow[b]	S10	$E_{NS,AC}$	Montanari and Toth (2007)
	Lag-1 autocorrelation of high-flow period	S11	$E_{R,AC1,Q10}$	Euser et al. (2013)
	Lag-1 autocorrelation of low-flow period	S12	$E_{R,AC1,low}$	Euser et al. (2013)
	Runoff coefficient[c]	S13	$E_{R,RC}$	Yadav et al. (2007)
			D_E	Schoups et al. (2005)

[a] Averaged and normalized time series data of the five piezometer were compared to normalized fluctuations in model state variable S_S (see Table 1). [b] Describing the spectral properties of a signal and thus the memory of the system, the observed and modelled autocorrelation functions with lags from 1 to 100 d where compared. [c] Note that in catchments without long-term storage changes and intercatchment groundwater flow, long-term average RC equals the long-term average 1-E_A (Table 1).

The intermediate behaviour of the midslope piezometer (F4), exhibiting varying patterns throughout the years, reflects the fact that the riparian zone extends spatially towards the hillslope and reaches a larger spatial extension during wet years.

Similar observations have been reported by other authors. For example, anticlockwise hysteresis between groundwater tables and discharge are observed by Gabrielli et al. (2012) in the Maimai catchment, while studies on riparian groundwater or river bank groundwater report clockwise hysteresis on the storm event scale (Frei et al., 2010; Jung et al., 2004). Similar patterns were also observed by Jung et al. (2004), who found that in the inner floodplain and in river bank piezometers, the hysteresis curve between the water table and river stage exhibits a synchronous response, while in the hillslope hysteresis, curves are relatively open as the water table is higher during the recession than during the rising limb.

3.1.2 Observations regarding hillslope: saturated and unsaturated storages vs. flow

Figure 6 shows the 3-dimensional relationship between hillslope saturated storage (HSS), unsaturated storage (HUS) and stream flow (Q) for the year 2010–2011. Four main periods can be identified, similar to what was outlined in recent studies (e.g. Heidbuechel et al., 2012; Hrachowitz et al., 2013a): three characterized the recharge period and the last one the recession period. First, stream flow was close or equal to 0 and was almost exclusively sustained by drainage of the saturated storage, while the unsaturated zone exhibited a significant storage deficit and only minor fluctuations due to transpiration and small summer rain events (dry period). As steadier precipitation patterns set in, here typically around November, the unsaturated-zone storage reached its maximal value relatively quickly, rapidly establishing connectivity with fast-responding flow pathways (wetting period). This led to a relatively rapid increase in stream flow while the saturated storage did not change much until the end of this

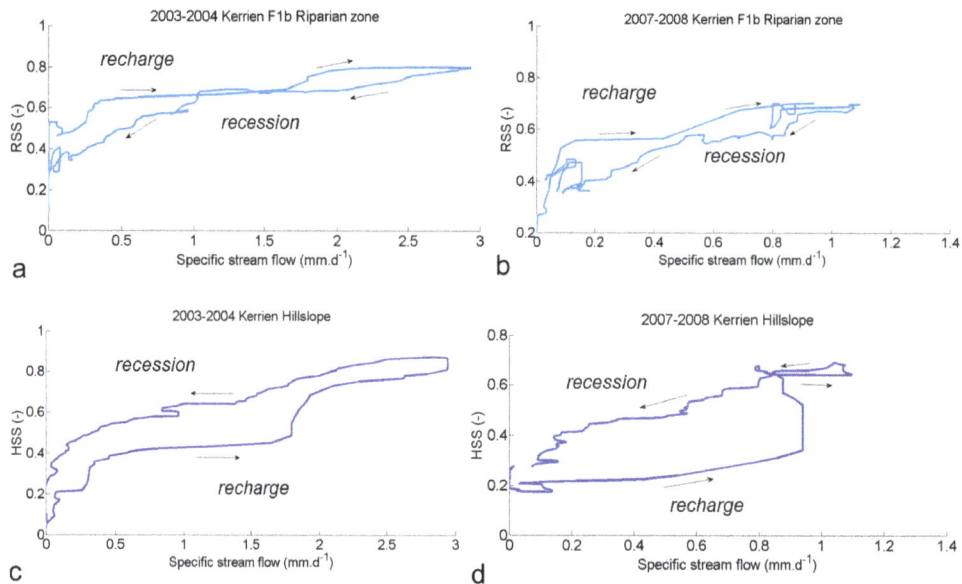

Figure 4. Examples of annual hysteretic loops for saturated-zone storage vs. stream flow which are clockwise in the riparian zone (**a, b**) and anticlockwise in the hillslope (**c, d**) for the wet year 2003–2004 (**a, c**) and the dry year 2007–2008 (**b, d**).

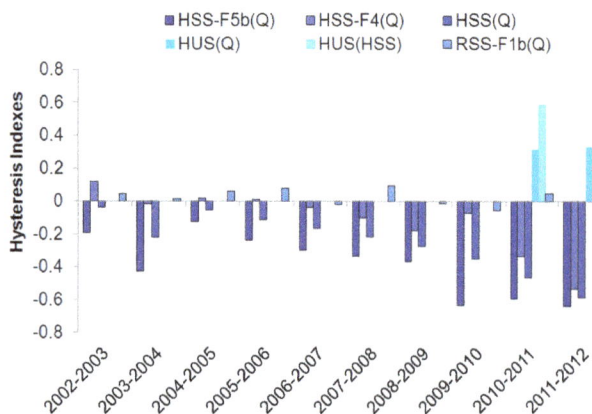

Figure 5. Annual hysteresis indexes (HI) computed for the piezometers in the Kerrien catchment from 2002 to 2012. F5b is located upslope, F4 midslope and F1b downslope in the riparian area. RRS(Q) is the hysteresis between stream flow and riparian saturated-zone storage (measured at F1b). HSS-F5b(Q), HSS-F4(Q) and HSS(Q) are hystereses between stream flow and upslope (at F5b), midslope (at F4) and hillslope (average of F5b and F4) saturated storages respectively. HUS(Q) is the hysteresis between stream flow and hillslope unsaturated storage (HUS) (computed from the average of normalized volumic moisture sensors in profiles sB1 and sB2), and HUS(HSS) is that between the hillslope unsaturated- and saturated-zone storage (average of F5b and F4).

period as incoming precipitation first had to fill the storage deficit in the unsaturated zone before a significant increase in percolation could occur. A further lag was introduced by the time taken for water to percolate and eventually recharge the relatively deep groundwater. As soon as conditions were

wet enough to allow for established percolation, the saturated storage eventually also responded, increasing faster than the stream flow (wet period), while unsaturated storage remained full. During the wet period (or high-flow period), no pattern appeared clearly because all storage elements were almost full and the responses of all the compartments were more directly linked to the short-term dynamics of rain events. Finally during the recession period (drying period), unsaturated storage decreased comparatively quickly by drainage and transpiration, while the saturated storage kept increasing for a while by continued percolation from the unsaturated zone before decreasing through groundwater drainage at a relatively slow rate. A similar pattern was also observed for 2011–2012 (not shown).

The unsaturated-zone storage followed a clockwise hysteresis loop with the stream flow and with the saturated-zone storage. The hysteresis indexes (Fig. 5, years 2010–2011 and 2011–2012) reflected these directions and showed that the hysteresis loops were narrower for unsaturated storage than for saturated storage, inducing smaller absolute values of the hysteresis indexes due to the small size of the unsaturated storage compartment compared to the saturated storage compartment.

3.1.3 Interpretation

There are three main hypotheses generally proposed to interpret storage–discharge hystereses in hydrology. The first one is related to the increase in transmissivity with the groundwater level due to the frequently observed exponential decrease in hydraulic conductivity with depth. However, this would lead to systematic clockwise hysteresis loops and can-

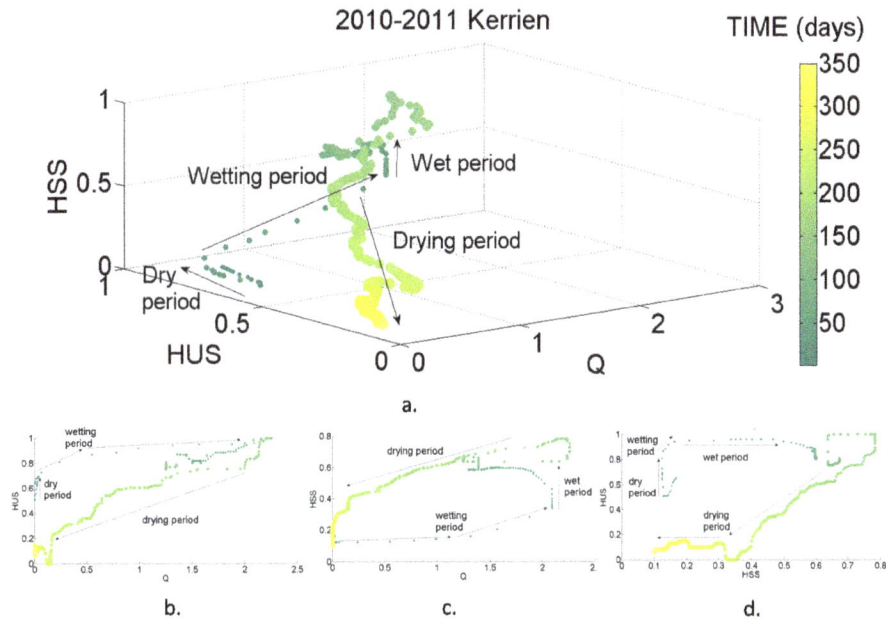

Figure 6. Evolution of stream flow (Q in mm d^{-1}) and normalized hillslope unsaturated storage (HUS) and hillslope saturated storage (HSS) for the water year 2010–2011 (October to September). The size of the dots increases with time. Unsaturated storage (HUS) is computed from the moisture sensors in profiles sB1 and sB2; saturated storage (HSS) is represented using a normalized groundwater table level (computed from two piezometers in the hillslope). (**a**) is the 3-dimensional plot and (**b, c, d**) are the respective 2-dimensional projections of (**a**) on the three plans.

not explain the anticlockwise patterns observed between hillslope saturated storage and stream flow. The second hypothesis proposed by Spence et al., 2010 is that during the recharge period, the groundwater storage not only increases locally (as measured by the piezometric variations), but the spatial extension of connected storage also increases gradually, while during the recession period, the storage decreases homogenously across the entire contribution area. This is likely for riparian groundwater and could explain the clockwise hysteresis observed on this piezometer but cannot explain the anticlockwise hysteresis observed in the hillslope groundwater. The third hypothesis is that dominant hydrological processes are different between recharge and recession periods. For instance, Jung et al. (2004) interpret their clockwise hysteresis in peatlands groundwater as the results of a stepwise filling process during the rising flows (fill and spill mechanism) opposed to a more gradual drainage of the groundwater during the recession combined with the first hypothesis result, similar to what was found by Hrachowitz et al. (2013a). This hypothesis of different hydrological pathways allows an adequate interpretation of the opposite directions of the observed hystereses. The recharge period is characterized by a quick filling of the unsaturated and saturated storages in the riparian zone, which is always close to saturation, while the saturated storage on the hillslope is not yet filling up (wetting period). Thus, the wetting period is characterized by an increase in stream flow, here mainly generated in the riparian zone, and eventual quick flows in the hillslope, while the hill-

slope unsaturated zone reaches the storage capacity volume. At the beginning of the wet period, hillslope saturated storage fills and starts to contribute to the stream, along with riparian and fast flows. During the recession period (drying period), the hillslope saturated zone is the only compartment which continues to sustain stream flow. If this hypothesis is correct, there are three contributions to stream flow in the wet period, while, during the recession period, hillslope groundwater remains the only contributor to stream flow (cf. Hrachowitz et al., 2013a, see Fig. 7). This can explain the difference between storage values in recharge and recession periods. Finally, the hysteretic hydrological signature is not only related to the amount of stored water in the catchment but rather to where it is stored.

These results are consistent with previous studies: the distinction between riparian groundwater and hillslope groundwater components has also been identified in similar catchments (by Molénat et al. (2008) based on nitrate concentration analysis and by Aubert et al. (2013a) based on a range of solutes) and at other site (by Haught and van Meerveld (2011)) using such Q–S relationships and lag time analysis.

3.1.4 Sensitivity of the HI to initial conditions

Sensitivity to antecedent soil moisture conditions is often cited as an explanation for observed storage–discharge hysteresis and its variability between years. The initial levels of each store will obviously influence the time required to fill

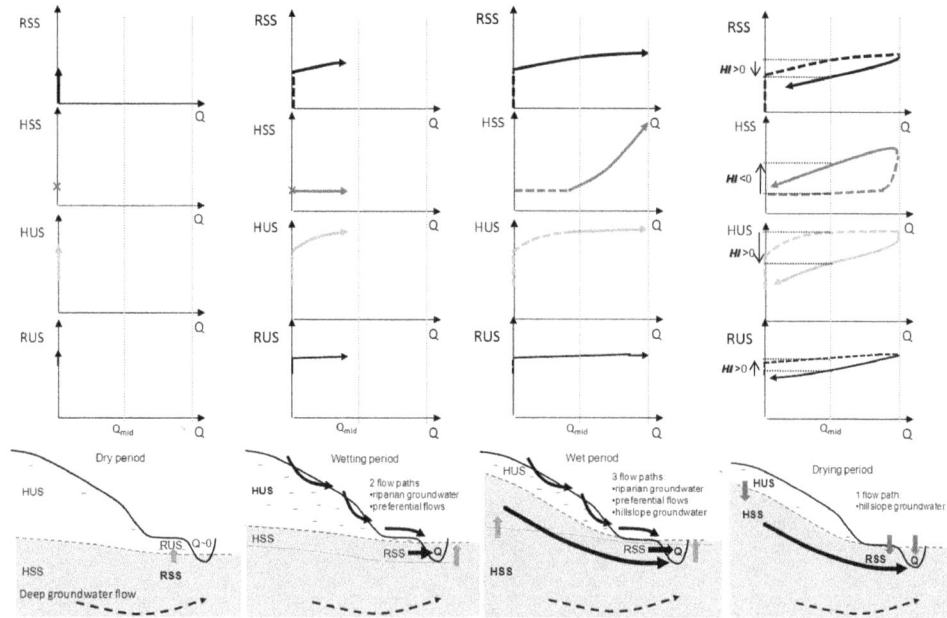

Figure 7. Conceptual scheme of successive mechanisms explaining the annual hysteresis between storages and stream flows. HUS: hillslope unsaturated storage; HSS: hillslope saturated storage; RUS: riparian unsaturated storage; RSS: riparian saturated storage; Q: stream flow. Bold characters indicate compartments with varying storage; grey arrows indicate whether the compartment is filling or emptying; black arrows indicate the water flow paths.

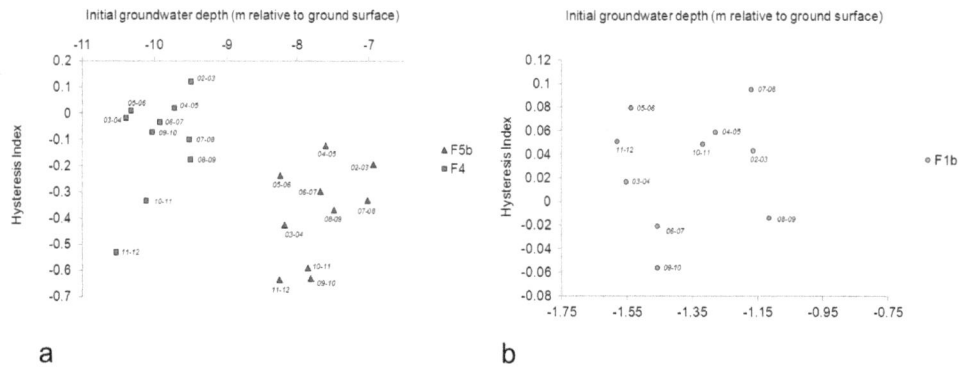

Figure 8. Year-to-year variations, for the 10 monitoring years, of the hysteresis indexes: **(a)** HSS-F5b(Q) and HSS-F4(Q) (HI) versus the initial groundwater table level depth in the corresponding hillslope piezometer (F5b or F4) and **(b)** HSS-F1b(Q) versus the initial groundwater table level depth in the piezometer in the riparian area (F1b).

them and consequently the duration of the successive periods identified in the whole recharge period. As only 2 years of data were available, it was not possible to define a relationship between the initial average soil moisture and the magnitude of the hysteresis indexes. However, the magnitude of the HI was lower for high initial values of average unsaturated-zone storage for both the saturated and unsaturated zones in 2011–2012 (Table 5). The HI for the midslope saturated zone (F4b) seemed to be more sensitive to these initial moisture conditions than the HI for the upslope saturated zone and unsaturated zone. Similarly, the width of the loop (absolute value of the HI) was not very sensitive to initial groundwater levels in the hillslope: although the larger absolute values of

the HI were observed for the lower initial water table levels, no clear correlation was observed (Fig. 8).

3.1.5 Sensitivity of the HI to annual rainfall

For the saturated zone, the observed values of the HI were negatively correlated with the total annual rainfall for both the hillslope and the riparian zone, with a more negative slope for the hillslope (Fig. 9). Wet years (i.e. large values of annual rainfall) are generally associated with large values of annual maximal and midpoint stream flows and also with large values of groundwater table level, leading to larger saturated-storage values during the recession period, while

Table 5. Hysteresis indexes (HIs) and initial hillslope unsaturated-storage values (HUS) at the beginning of the water year.

Year	Initial HUS	Hysteresis index (HI)			
		HSS-F5b(Q)	HSS-F4(Q)	HSS(Q)	RSS-F1b(Q)
2010-2011	0.148	−0.591	−0.334	−0.462	0.590
2011–2012	0.026	−0.635	−0.532	−0.583	0.003

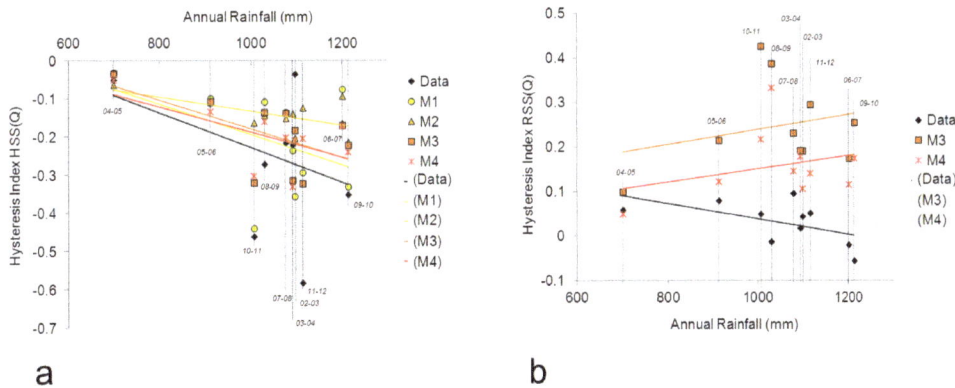

Figure 9. Variations of observed (data) and simulated (M1 to M4) hysteresis index versus annual rainfall for the 10 monitored water years for (**a**) hillslope saturated storage versus discharge HSS(Q) and (**b**) riparian saturated storage vs. discharge RSS(Q). Solid lines indicate the linear regressions.

the storage values during the recharge period do not change much from year to year. Thus, larger storage values at the time of midpoint discharge in the recession period led to smaller values of the HI (i.e larger absolute values for the hillslope, where hystereses are anticlockwise, and smaller absolute value of the HI for the riparian zone, where hystereses are clockwise). In the riparian zone, when rainfall and maximal drainage reached a very high value, it could lead to a saturated-storage value at the time of midpoint discharge in the recession period that was larger than the corresponding value during the recharge period, explaining the inversion of the sign of the HI for RSS(Q) in very wet years.

3.2 Model assessment based on their ability to reproduce the observed hysteresis

3.2.1 Hysteresis simulations

For all years, all models (M1–M4) exhibited a hysteretic relationship between stream flow and storage, as shown in Fig. 10 for the years 2003–2004 and 2007–2008, pertaining to the calibration and validation periods respectively. This means that all tested models introduced a lag time between catchment stores and the stream dynamics. Fig. 11a presents the observed and modelled average and standard deviation of the annual hysteresis indexes, for hillslope saturated storage vs. discharge HSS(Q), hillslope unsaturated storage vs. discharge HUS(Q), hillslope unsaturated storage vs. hillslope saturated storage HUS(HSS) and riparian saturated storage

vs. discharge RSS(Q). As riparian saturated storage (RSS) is not modelled in M1 and M2, simulated RSS(Q) was available only for M3 and M4.

For M1, the shape of the simulated hysteresis showed an overestimation of hillslope saturated storage (HSS) and of flow during dry years (e.g. the year 2007–2008 shown in Fig. 10). This was expected as we have seen that the model was unable to reproduce groundwater dynamics and the low signatures during the validation period (Fig. 3 and supplementary material). Simulated HI values were close to the observed ones for HSS(Q) (Fig. 11a). The simulated hysteresis indexes were small and negative for HUS(Q), while the observed values were large and positive. Simulated HI values for HUS(HSS) were also overestimated. These results show that, in model M1, the overestimation of the hillslope saturated storage was partially compensated by the underestimation of the hillslope unsaturated storage. This reveals the poor consistency of the model and explains why it was able to reach good performance in the calibration period but not in the validation period (Fig. 3).

For the model M2, the shape of the hysteresis loops showed a considerable underestimation of HSS and a large underestimation of stream flow in wet years (Fig. 10). Compared to M1, although the introduction of deep losses in M2 led to higher validation performances and better simulation of hydrological signatures (Fig. 3), the simulated HIs (Fig. 11a) worsened, suggesting a poorer model consistency with respect to internal hydrologic processes.

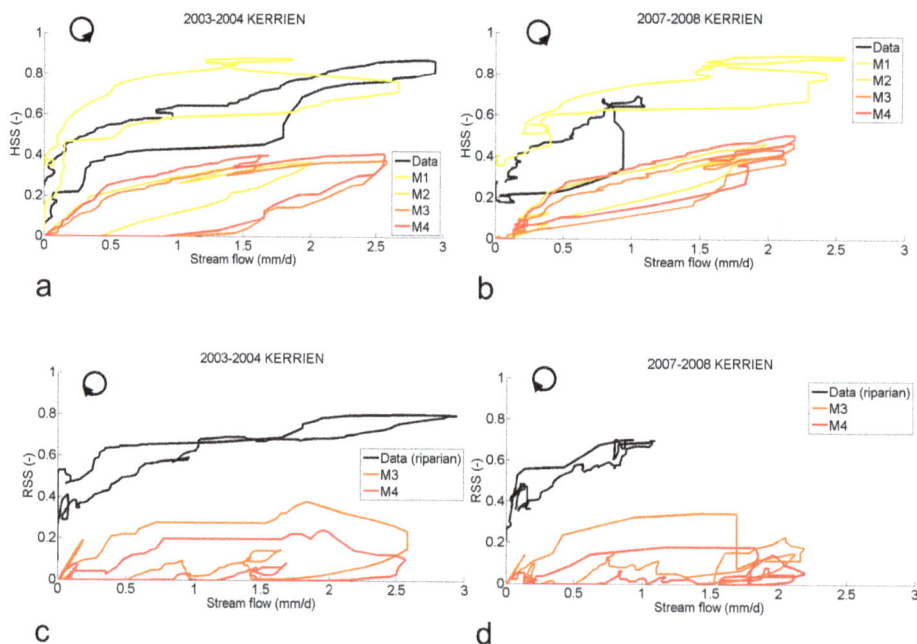

Figure 10. Observed and simulated annual hysteresis between stream flow (Q) and (**a, b**) saturated storage in the hillslope HSS (for observed hysteresis, HSS is the average of F5b and F4) and (**c, d**) saturated storage in the riparian area RSS (for simulated hysteresis, only M3 and M4 represent the riparian area), for the water years (**a, c**) 2003–2004 (wet year, calibration period) and (**b, d**) 2007–2008 (dry year, validation period).

For both models M3 and M4, the introduction of a riparian compartment improved the simulated hysteretic loops, due to a better simulation of stream flow in wet years, but HSS was still largely underestimated (Fig. 10). The mean HI values for HSS(Q) were close to the observed one, but the range of variation was smaller, indicating a reduced sensitivity to climate (Fig. 9). The mean values for HUS(Q) were clearly improved compared to M1 and M2 as the direction of the loop was clockwise as for the observations, although the values were still underestimated. The mean HI values for HUS(HSS) were also greatly improved. The shape of the simulated hysteresis loops between riparian saturated storage (RSS) and stream flow (Q) showed a large underestimation of RSS, especially during the recession period (Fig. 10c, d). This led to simulated HIs for RSS(Q) which are positive, like the observed ones, but also largely overestimated (Fig. 11a). Overall, these results suggest that for models including a riparian component, the underestimation of the hysteresis between HUS and Q was compensated for by an overestimation of the hysteresis between RSS and Q. This highlights that, despite a significant improvement in performances and improved hydrological signature reproduction, these models still involve a certain degree of inconsistency with respect to internal processes. However, M4 provided the most balanced performance considering hysteretic signatures between all storage components and strongly underlines the limitations of overly simplistic model architectures (e.g. M1) and the need for more complete representations

of process heterogeneity. The hysteresis index sensitivity to parameter uncertainty increases with the number of parameters from M1 to M2 and then stays in the same range from M2 to M4 (Fig. 11b). This analysis confirms the importance of considering the hysteresis indexes both between saturated and unsaturated storage (HSS and HUS) to avoid accepting a wrong model. For example, considering only the performance regarding the HSS(Q) relationship could lead one to accept model M1, while its performance on HUS is lower and it is not able to reproduce the Riparian compartment hysteresis. For readability purposes, Fig. 11b only illustrates this sensitivity for the different HIs in the year of 2011–2012 but similar behaviour is observed every year. It shows that best behavioural parameter sets (bbp) lead to modelled HI values closer to the observed values than average modelled HI values. Using an additional calibration criterion related to the hysteresis could reduce the sensitivity of the HI to parameter uncertainty and lead to a narrow range of feasible parameter sets.

3.2.2 Sensitivity of modelled hysteresis indexes to annual rainfall

All models were able to represent the decrease in the hysteresis indexes with annual rainfall on the hillslope, the slope of the correlation getting closer to the observed one from M1 to M4 (Fig. 9). The introduction of deep-groundwater losses (M2) led to smaller saturated storage during recharge periods

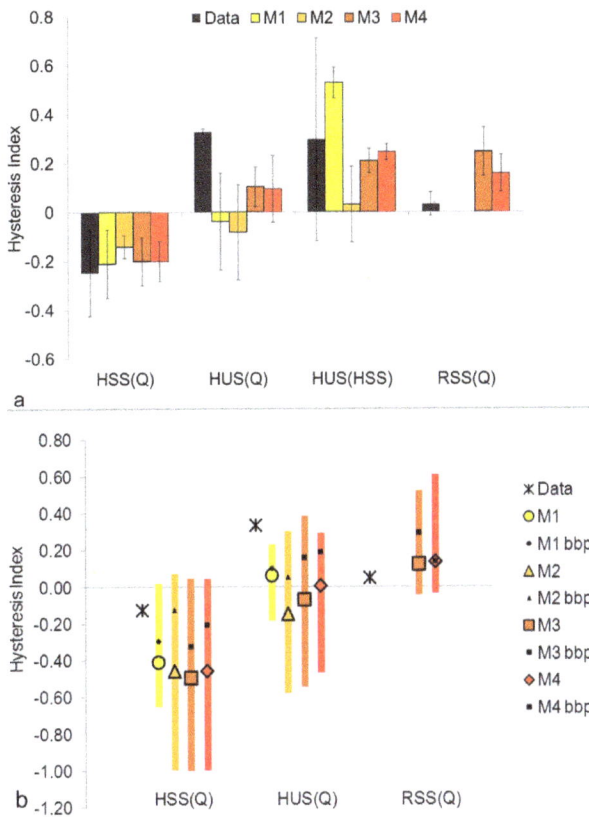

Figure 11. (a) Mean annual hysteresis indexes observed and simulated with the four models M1 to M4 for hillslope saturated storage vs. discharge HSS(Q), hillslope unsaturated storage vs. discharge HUS(Q), hillslope unsaturated storage vs. hillslope saturated storage HUS(HSS) and riparian saturated storage vs. discharge RSS(Q). RSS is only simulated in models M3 and M4. Error bars show the standard deviation for the 10 years for HSS(Q) and RSS(Q) and the values for the 2 available years for HUS(Q) and HUS(HSS). **(b)** Sensitivity of hysteresis index values to parameter uncertainty for the year 2011–2012. Mx bbp indicates the value for best behavioural parameter sets; the circles, triangles, squares and diamonds indicate the mean HI value for the all the behavioural parameter sets, and the corresponding bars indicate its range of variation.

and increased the difference between saturated storage during recharge and recession periods at the time of midpoint discharge. However, as all models tended to overestimate low stream flow values, the slopes of the correlations between annual rainfall and the simulated HI were smaller than for the observed one.

In the riparian zone, the modelled trends were the inverse of the observed one. The modelled recessions were always very sharp (see Hrachowitz et al., 2014), and the simulated riparian storage dried up every year, explaining why saturated storage at the time of midpoint discharge during the recession periods was much greater than during the recharge periods. This led to a general overestimation of HI values, which were even stronger for wet years. This overestimation may be re-

lated to an improper conceptualization of the riparian-zone functioning, which is never connected to the hillslope reservoir in the tested models. In reality, during high-flow periods, the observed hydraulic gradient increased along the hillslope, inducing a connection between riparian and hillslope reservoirs which are disconnected during low-flow periods.

3.2.3 Value of such internal signatures for model evaluation

The use of hydrological hysteretic signatures in model assessments led to conclusions that were consistent with the classical hydrological signatures used in Hrachowitz et al. (2014). However, model M2 was less able to reproduce the different hysteretic signatures, whereas it led to a real improvement regarding to the classical signatures in low flows. Considering only the distance between observed and simulated hysteresis indexes on hillslope saturated storage and stream flow would lead one to select model M1. This highlights the fact that using saturated-storage dynamics alone can be deceptive for understanding the system response behaviour and that it is thus crucial to also consider the hysteretic signatures of unsaturated and riparian zones in a combined approach to develop a more robust understanding of the system. Here, hysteretic signatures of the unsaturated and riparian zones provided valuable additional assessment metrics regarding the performance of models M3 and M4 to represent the riparian zone. It was possible to identify when the model failed to represent processes, which processes were mostly compensating for missing ones and therefore why the model may provide some good performance for the wrong reasons. In this regard, the hysteresis index proved to be a useful proxy of hystereses themselves as it exhibited contrasted patterns sensitive to climate and localization within the catchment.

3.3 Perspectives: toward an integrated hydrological-signature-based modelling?

A general issue in model calibration is that, because of the overparameterization of hydrological models and because the objective functions generally only integrate one variable, such as the stream flow, automatic calibration techniques may lead to parameter sets which compensate for internal model errors. These parameter sets are mathematically correct but wrong from a hydrological point of view. The subsequent model should then be considered nonbehavioural (Beven, 2006). For instance, if storage properties are not taken into account well by the model, this is likely to lead to a wrong simulation of storage dynamics in response to precipitation. Thus, the parameterization using traditional objective functions can lead to compensation of these errors in order to simulate a discharge value close to the observed one while the storage is wrong. In such a case, a model able to represent the internal catchment behaviour will generate a wrong discharge value which is, however, consistent with the storage value and will be rejected in traditional calibration procedures. To handle this issue and in order to select behavioural models, one can use multiple objective functions (Gupta et al., 1998; Seibert and McDonnell, 2002; Freer et al., 2003), including a range of hydrological signatures to be reproduced or additional realism constraints (Kavetski and Fenicia, 2011; Yadav et al., 2007; Yilmaz et al., 2008; Euser et al., 2013; Gharari et al., 2013; Hrachowitz et al., 2014). We argue that, rather than increasing the number of constraints or objective functions which have to be satisfied, an alternative could be to use some objective functions based on a combination of different variables, such as stream flow and the groundwater level, soil moisture or stream concentrations. Among the possible combination of variables, objective functions based on the relative dynamics of storage in different spatial locations, such as riparian versus hillslope, might provide new insights into the catchment-internal processes. We suggest that such combined objective functions would be more constraining for model selection. Therefore, the present study is a first step which aims at highlighting the still underexploited potential of hydrological hysteresis. The next step would be to quantify these relationships through functions or several indexes usable in calibration criteria, such as the hysteresis index proposed in this study. Moreover, such criteria could be used in classification studies. Indeed, some studies in the literature present storage–discharge relationships for different catchments that show patterns that are similar or dissimilar to the ones we observed in the Kerrien catchment (Ali et al., 2011; Gabrielli et al., 2012). This signature may help to classify catchments in terms of dominant processes driving their behaviour.

A remaining difficulty with integrating storage into calibration or evaluation procedure in hydrological modelling is how to measure this storage. McNamara et al. (2011) and Tetzlaff et al. (2011) proposed using all available data from groundwater level monitoring, soil moisture records, water budget, modelling results and so on to estimate the storage in catchments. In this study, we used quite a dense network of piezometers and soil moisture measurements relative to the small size of the catchment. Promising ways to estimate spatial quantification of storage in catchments include remote sensing of soil moisture (Sreelash et al., 2013; Vereecken et al., 2008), gravimetric techniques (Creutzfeldt et al., 2012), geodesy and geophysical methods. The interest in such techniques would be to provide a spatially integrated vision of the catchment water content.

As for the different hydrological variables, the combination of hydrological and chemical variables appears relevant to investigating the hydrochemical behaviour of catchments. Hysteresis patterns between concentration and discharge have been largely documented for storm event characterization (Evans and Davies, 1998; Evans et al., 1999; Taghavi et al., 2011). Some studies also report similar patterns on the annual scale (e.g. Aubert et al., 2013b). Such hysteretic relationships have been observed also between water and chemistry in groundwater (Rouxel et al., 2011; Hrachowitz et al., 2013a), emphasizing a disconnection between water and solute dynamics that simple diffusion or partial mixing processes cannot explain. Stream water chemistry also exhibits particular seasonal cycles with different phasing and with discharge depending on the solutes (Aubert et al., 2013b). This provides extra information on the water pathways within the catchment. These relationships also appear to be powerful in constraining hydrochemical modelling.

4 Conclusions

A method to characterize and partially quantify the relationship between storages in a headwater catchment and stream flow throughout a year has been proposed. It allowed us to then assess the ability of a range of conceptual lumped models to reproduce this catchment-internal signature. Catchment storage has been approximated using a network of piezometric data and several unsaturated-zone moisture profiles to consider the storage in the saturated as well as in the unsaturated zones.

The observations showed that storage–discharge relationships in catchments can be hysteretic, highlighting a successive activation of different hydrological components during the recharge period, while the recession exhibits a fast decrease in unsaturated and riparian storage and a slow decrease in hillslope saturated storage which sustains the stream flow. Four periods have been identified in the hydrological year: (1) first, at the end of the dry period, rainfall starts to refill unsaturated storage; (2) in the wetting period, riparian unsaturated storage is filled and the saturated storage starts to supply the stream while hillslope unsaturated storage is still being replenished; (3) during the wet period, unsaturated storage in the hillslope is also filled and the saturated hillslope storage also feeds the stream. (4) Finally when rainfall declines, flow from the riparian groundwater recedes and, during the recession period, the stream discharge is sustained only by hillslope groundwater. Stream discharge and riparian and hillslope saturated storages exhibited different patterns of hysteresis, with opposite directions of the hysteretic loops.

The tested models were characterized by an increasing degree of complexity and also an increasing consistency, as shown in a previous study using classical hydrologic signatures. In this study, we showed that, if all of the models simulated a hysteretic relationship between storage and discharge, their ability to reproduce the hysteresis index also increased with model complexity. In addition, we suggest that, if classical hydrological signatures help to assess model consistency, the hysteretic signatures also help to identify quickly when and why the models give "right answers for the wrong reasons" and can be used as a descriptor of the internal catchment functioning.

Acknowledgements. The investigations benefited from the support of INRA and CNRS for the Research Observatory ORE AgrHys, and from Allenvi for the SOERE RBV. Data are available at http://geowww.agrocampus-ouest.fr/web/.

Edited by: N. Romano

References

Ali, G. A., L'Heureux, C., Roy, A. G., Turmel, M.-C., and Courchesne, F.: Linking spatial patterns of perched groundwater storage and stormflow generation processes in a headwater forested catchment, Hydrol. Process., 25, 3843–3857, 2011.

Andermann, C., Longuevergne, L., Bonnet, S., Crave, A., Davy, Ph., and Gloaguen, R.: Impact of transient groundwater storage on the discharge of Himalayan rivers, Nat. Geosci., 5, 127–132, 2012.

Andréassian, V., Le Moine, N., Perrin, C., Ramos, M. H., Oudin, L., Mathevet, T., Lerat, J., and Berthet, L.: All that glitters is not gold: the case of calibrating hydrological models, Hydrol. Process., 26, 2206–2210, 2012.

Aubert, A. H., Gascuel-Odoux, C., Gruau, G., Akkal, N., Faucheux, M., Fauvel, Y., Grimaldi, C., Hamon, Y., Jaffrézic, A., Lecoz-Boutnik, M., Molénat, J., Petitjean, P., Ruiz, L., and Merot, P.: Solute transport dynamics in small, shallow groundwater-dominated agricultural catchments: insights from a high-frequency, multisolute 10 yr-long monitoring study, Hydrol. Earth Syst. Sci., 17, 1379–1391, doi:10.5194/hess-17-1379-2013, 2013.

Aubert, A. H., Gascuel-Odoux, C., and Merot, P.: Annual hysteresis of water quality: A method to analyse the effect of intra- and inter-annual climatic conditions, J. Hydrol., 478, 29–39, 2013.

Bergström, S.: The HBV model, in: Computer Models of Watershed Hydrology, edited by: Singh, V. P., Water Resources Publications: Highlands Ranch, CO, 443-476, 1995.

Beven, K.: How far can we go in distributed hydrological modelling?, Hydrol. Earth Syst. Sci., 5, 1–12, doi:10.5194/hess-5-1-2001, 2001.

Beven, K.: A manifesto for the equifinality thesis, J. Hydrol., 320, 18–36, 2006.

Beven, K.: So how much of your error is epistemic? Lessons from Japan and Italy, Hydrol. Process., 27, 1677–1680, 2013.

Butturini, A., Alvarez, M., Bernal, S., Vazquez, E., and Sabater, F.: Diversity and temporal sequences of forms of DOC and NO3-discharge responses in an intermittent stream: Predictable or random succession?, J. Geophys. Res.-Biogeosci., 113, G03016, doi:10.1029/2008JG000721, 2008.

Creutzfeldt, B., Ferré, T., Troch, P., Merz, B., Wziontek, H., and Güntne, A.: Total water storage dynamics in response to climate variability and extremes: Inference from long-term terrestrial gravity measurement, J. Geophys. Res.-Atmos., 117, D08112, doi:10.1029/2011JD016472, 2012.

Criss, R. E. and Winston, W.,E.: Do Nash values have value? Discussion and alternate proposals, Hydrol. Process., 22, 2723–2725, 2008.

Euser, T., Winsemius, H. C., Hrachowitz, M., Fenicia, F., Uhlenbrook, S., and Savenije, H. H. G.: A framework to assess the realism of model structures using hydrological signatures, Hydrol. Earth Syst. Sci., 17, 1893–1912, doi:10.5194/hess-17-1893-2013, 2013.

Evans, C. and Davies, T. D.: Causes of concentration/discharge hysteresis and its potential as a tool for analysis of episode hydrochemistry, Water Resour. Res., 34, 129–137, 1998.

Evans, C., Davies, T. D., and Murdoch, P. S.: Component flow processes at four streams in the Catskill Mountains, New York, analysed using episodic concentration/discharge relationships, Hydrol. Process., 13, 563–575, 1999.

Everett, D. H. and Whitton, W. I.: A general approach of hysteresis, Trans. Faraday Soc., 48, 749–757, 1952.

Fenicia, F., McDonnell, J. J., and Savenije, H. H. G.: Learning from model improvement: On the contribution of complementary data to process understanding, Water Resour. Res., 44, W06419, doi:10.1029/2007WR006386, 2008.

Franks, S. W., Gineste, P., Beven, K. J., and Merot, P.: On constraining the predictions of a distributed moder: The incorporation of fuzzy estimates of saturated areas into the calibration process, Water Resour. Res., 34, 787–797, 1998.

Freer, J., Beven, K., and Ambroise, B.: Bayesian estimation of uncertainty in runoff prediction and the value of data: An application of the GLUE approach, Water Resour. Res., 32, 2161–2173, 1996.

Freer, J., Beven, K., and Peters, N.: Multivariate Seasonal Period Model Rejection Within the Generalised Likelihood Uncertainty Estimation Procedure, in: Calibration of Watershed Models, edited by: Duan, Q., Gupta, H. V., Sorooshian, S., Rousseau, A. N. and Turcotte, R., American Geophysical Union, Washington, D. C., doi:10.1029/WS006p0069, 69–87, 2003.

Freer, J. E., McMillan, H., McDonnell, J. J., and Beven, K. J.: Constraining dynamic TOPMODEL responses for imprecise water table information using fuzzy rule based performance measures, J. Hydrol., 291, 254–277, 2004.

Frei, S., Lischeid, G., and Fleckenstein, J. H.: Effects of microtopography on surface-subsurface exchange and runoff generation in a virtual riparian wetland - A modeling study, Adv. Water Resour., 33, 1388–1401, 2010.

Gabrielli, C. P., McDonnell, J. J., and Jarvis, W. T.: The role of bedrock groundwater in rainfall–runoff response at hillslope and catchment scales, J. Hydrol., 450–45, 117–133, 2012.

Gascuel-Odoux, C., Weiler, M., and Molenat, J.: Effect of the spatial distribution of physical aquifer properties on modelled water table depth and stream discharge in a headwater catchment, Hydrol. Earth Syst. Sci., 14, 1179–1194, doi:10.5194/hess-14-1179-2010, 2010.

Gharari, S., Hrachowitz, M., Fenicia, F., Gao, H., and Savenije, H. H. G.: Using expert knowledge to increase realism in environmental system models can dramatically reduce the need for calibration, Hydrol. Earth Syst. Sci., 18, 4839–4859, doi:10.5194/hess-18-4839-2014, 2014.

Giustolisi, O. and Simeone, V.: Optimal design of artificial neural networks by a multiobjective strategy: groundwater level predictions, Hydrol. Sci. J.-J. Sci. Hydrol., 51, 502–523, 2006.

Gupta, H. V., Clark, M. P., Vrugt, J. A., Abramowitz, G., and Ye, M.: Towards a comprehensive assessment of model structural adequacy, Water Resour. Res., 48, W08301, doi:10.1029/2011WR011044, 2012.

Gupta, H. V., Wagener, T., and Liu, Y. Q.: Reconciling theory with observations: elements of a diagnostic approach to model evaluation, Hydrol. Process., 22, 3802–3813, 2008.

Gupta, H. V., Sorooshian, S., and Yapo, P. O.: Toward improved calibration of hydrologic models: Multiple and noncommensurable measures of information, Water Resour. Res., 34, 751–763, 1998.

Hartmann, A., Wagener, T., Rimmer, A. , Lange, J., Brielmann, H., and Weiler, M.: Testing the realism of model structures to identify karst system processes using water quality and quantity signatures, Water Resour. Res., 49, 3345–3358, doi:10.1002/wrcr.20229, 2013a.

Hartmann, A., Weiler, M., Wagener, T., Lange, J., Kralik, M., Humer, F., Mizyed, N., Rimmer, A., Barberá, J. A., Andreo, B., Butscher, C., and Huggenberger, P.: Process-based karst modelling to relate hydrodynamic and hydrochemical characteristics to system properties, Hydrol. Earth Syst. Sci., 17, 3305–3321, doi:10.5194/hess-17-3305-2013, 2013.

Haught, D. R. W. and van Meerveld, H. J.: Spatial variation in transient water table responses: differences between an upper and lower hillslope zone, Hydrol. Process., 25, 3866–3877, 2011.

Heidbüchel, I., Troch, P. A., Lyon, S. W., and Weiler, M.: The master transit time distribution of variable flow systems, Water Resour. Res., 48, W06520, doi:10.1029/2011WR011293, 2012.

Hopmans, P. and Bren, L. J.: Long-term changes in water quality and solute exports in headwater streams of intensively managed radiata pine and natural eucalypt forest catchments in southeastern Australia, Forest Ecol. Manage., 253, 244–261, 2007.

Hrachowitz, M., Soulsby, C., Tetzlaff, D., and Malcolm, I. A.: Sensitivity of mean transit time estimates to model conditioning and data availabilityle, Hydrol. Process., 25, 980–990, 2011.

Hrachowitz, M., Savenije, H., Bogaard, T. A., Tetzlaff, D., and Soulsby, C.: What can flux tracking teach us about water age distribution patterns and their temporal dynamics?, Hydrol. Earth Syst. Sci., 17, 533–564, doi:10.5194/hess-17-533-2013, 2013a.

Hrachowitz, M., Savenije, H. H. G., Blöschl, G., McDonnell, J. J., Sivapalan, M., Pomeroy, J. W., Arheimer, B., Blume, T., Clark, M. P., Ehret, U., Fenicia, F., Freer, J. E., Gelfan, A., Gupta, H. V., Hughes, D. A., Hut, R. W., Montanari, A., Pande, S., Tetzlaff, D., Troch, P. A., Uhlenbrook, S., Wagener, T., Winsemius, H. C., Woods, R. A., Zehe, E., and Cudennec, C.: A decade of predictions in ungauged basins (PUB) – a review, Hydrol. Sci. J., 58, 1198–1255, 2013b.

Hrachowitz, M., Fovet, O., Ruiz, L., Euser, T., Gharari, S., Nijzink, R., Freer, J, Savenije, H. H. G., and Gascuel-Odoux, C.: Process consistency in models: The importance of system signatures, expert knowledge, and process complexity, Water Resour. Res., 50, 7445–7469, 2014.

Jung, M., Burt, T. P., and Bates, P. D.: Toward a conceptual model of floodplain water table response, Water Resour. Res., 40, W12409, doi:10.1029/2003WR002619, 2004.

Kampf, S. K. and Burges, S. J.: Parameter estimation for a physics-based distributed hydrologic model using measured outflow fluxes and internal moisture states, Water Resour. Res., 43, W12414, doi:10.1029/2006WR005605, 2007.

Kavetski, D. and Fenicia, F.: Elements of a flexible approach for conceptual hydrological modeling: 2. Application and experimental insights, Water Resour. Res., 47, W11511, doi:10.1029/2011WR010748, 2011.

Kirchner, J. W.: Getting the right answers for the right reasons: Linking measurements, analyses, and models to advance the science of hydrology, Water Resour. Res., 42, W03S04, doi:10.1029/2005WR004362, 2006.

Lamb, R., Beven, K., and Myrabo, S.: Use of spatially distributed water table observations to constrain uncertainty in a rainfall-runoff model, Adv. Water Resour., 22, 305–317, 1998.

Lawler, D. M., Petts, G. E., Foster, I. D. L., and Harper, S.: Turbidity dynamics during spring storm events in an urban headwater river system: The Upper Tame, West Midlands, UK, Sci. Total Environ., 360, 109–126, 2006.

Martin, C., Molénat, J., Gascuel-Odoux, C., Vouillamoz, J.-M., Robain, H., Ruiz, L., Faucheux, M., and Aquilina, L.: Modelling the effect of physical and chemical characteristics of shallow aquifers on water and nitrate transport in small agricultural catchments, J. Hydrol., 326, 25–42, 2006.

McNamara, J. P., Tetzlaff, D., Bishop, K., Soulsby, C., Seyfried, M., Peters, N. E., Aulenbach, B. T., and Hooper, R.: Storage as a Metric of Catchment Comparison, Hydrol. Process., 25, 3364–3371, 2011.

Mohamed, Y. A., Savenije, H. H. G., Bastiaanssen, W. G. M., and van den Hurk, B. J. J. M.: New lessons on the Sudd hydrology learned from remote sensing and climate modeling, Hydrol. Earth Syst. Sci., 10, 507–518, doi:10.5194/hess-10-507-2006, 2006.

Molénat, J., Gascuel-Odoux, C., Davy, P., and Durand, P.: How to model shallow water-table depth variations: the case of the Kervidy-Naizin catchment, France, Hydrol. Process., 19, 901–920, 2005.

Molénat, J., Gascuel-Odoux, C., Ruiz, L., and Gruau, G.: Role of water table dynamics on stream nitrate export and concentration. in agricultural headwater catchment (France), J. Hydrol., 348, 363–378, 2008.

Molénat, J., Gascuel-Odoux,C., Aquilina, L., and Ruiz, L.: Use of gaseous tracers (CFCs and SF6) and transit-time distribution spectrum to validate a shallow groundwater transport model, J. Hydrol., 480, 1–9, doi:10.1016/j.jhydrol.2012.11.043, 2013.

Moussa, R., Chahinian, N., and Bocquillon, C.: Distributed hydrological modelling of a Mediterranean mountainous catchment - Model construction and multi-site validation, J. Hydrol., 337, 35–51, 2007.

Nash, J. E. and Sutcliffe, J. V.: River flow forecasting through conceptual models part I -A discussion of principles, J. Hydrol., 10, 282–290, 1970.

Nester, T., Kirnbauer, R., Parajka, J., and Bloschl, G.: Evaluating the snow component of a flood forecasting model, Hydrol. Res., 43, 762–779, 2012.

Parajka, J., Naeimi, V., Blöschl, G., Wagner, W., Merz, R., and Scipal, K.: Assimilating scatterometer soil moisture data into conceptual hydrologic models at the regional scale, Hydrol. Earth Syst. Sci., 10, 353–368, doi:10.5194/hess-10-353-2006, 2006.

Rouxel, M., Molénat, J., Ruiz, L., Legout, C., Faucheux, M., and Gascuel-Odoux, C.: Seasonal and spatial variation in groundwater quality along the hillslope of an agricultural research catchment (Western France), Hydrol. Process., 25, 831–841, 2011.

Ruiz, L., Abiven, S., Durand, P., Martin, C., Vertès, F., and Beaujouan, V.: Effect on nitrate concentration in stream water of agricultural practices in small catchments in Brittany: I. Annual nitrogen budgets, Hydrol. Earth Syst. Sci., 6, 497–506, doi:10.5194/hess-6-497-2002, 2002.

Salant, N. L., Hassan, M. A., and Alonso, C. V.: Suspended sediment dynamics at high and low storm flows in two small watersheds, Hydrol. Process., 22, 1573–1587, 2008.

Savenije, H. H. G.: Equifinality, a blessing in disguise?, Hydrol. Process., 15, 2835–2838, 2001.

Schwientek, M., Osenbruck, K., and Fleischer, M.: Investigating hydrological drivers of nitrate export dynamics in two agricultural catchments in Germany using high-frequency data series, Environ. Earth Sci., 69, 381–393, 2013.

Schwientek, M., Osenbruck, K., and Fleischer, M.: Investigating hydrological drivers of nitrate export dynamics in two agricultural catchments in Germany using high-frequency data series, Environ. Earth Sci., 69, 381–393, doi:10.1007/s12665-013-2322-2, 2013.

Seibert, J.: Multi-criteria calibration of a conceptual runoff model using a genetic algorithm, Hydrol. Earth Syst. Sci., 4, 215–224, doi:10.5194/hess-4-215-2000, 2000.

Seibert, J. and McDonnell, J. J.: On the dialog between experimentalist and modeler in catchment hydrology: Use of soft data for multicriteria model calibration, Water Resour. Res., 38, 1241, doi:10.1029/2001WR000978, 2002.

Seibert, J., Rodhe, A., and Bishop, K.: Simulating interactions between saturated and unsaturated storage in a conceptual runoff model, Hydrol. Process., 17, 379–390, 2003.

Sivapalan, M., Bloschl, G., Merz, R., and Gutknecht, D.: Linking flood frequency to long-term water balance: Incorporating effects of seasonality. Water Resour. Res., 41, W06012, doi:10.1029/2004WR003439, 2005.

Spence, C.: A Paradigm Shift in Hydrology: Storage Thresholds Across Scales Influence Catchment Runoff Generation, Geogr. Compass, 4, 819–833, 2010.

Spence, C., Guan, X. J., Phillips, R., Hedstrom, N., Granger, R., and Reid, B.: Storage dynamics and streamflow in a catchment with a variable contributing area, Hydrol. Process., 24, 2209–2221, 2010.

Sreelash, K., Sekhar, M., Ruiz, L., Buis, S., and Bandyopadhyay, S.: Improved Modeling of Groundwater Recharge in Agricultural Watersheds Using a Combination of Crop Model and Remote Sensing, J. Indian Inst. Sci., 93, 189–207, 2013.

Taghavi, L., Merlina, G., and Probst, J. L.: The role of storm flows in concentration of pesticides associated with particulate and dissolved fractions as a threat to aquatic ecosystems Case study: the agricultural watershed of Save river (Southwest of France), Knowledge and Management of Aquatic Ecosystems(400), 2011.

Tetzlaff, D., McNamara, J. P., and Carey, S. K.: Measurements and modelling of storage dynamics across scales Preface, Hydrol. Process., 25, 3831–3835, 2011.

Velleux, M. L., England, J. F., and Julien, P. Y.: TREX: Spatially distributed model to assess watershed contaminant transport and fate, Sci. Total Environ., 404, 113–128, 2008.

Vereecken, H., Huisman, J. A., Bogena, H., Vanderborght, J., Vrugt, J. A., and Hopmans, J. W.: On the value of soil moisture measurements in vadose zone hydrology: A review, Water Resour. Res., 44, W00D06, doi:10.1029/2008WR006829, 2008.

Wagener, T.: Can we model the hydrological impacts of environmental change?, Hydrol. Process., 21, 3233–3236, 2007.

Wagener, T. and Montanari, A.: Convergence of approaches toward reducing uncertainty in predictions in ungauged basins, Water Resour. Res., 47, W06301, doi:10.1029/2010WR009469, 2011.

Winsemius, H. C., Savenije, H. H. G., and Bastiaanssen, W. G. M.: Constraining model parameters on remotely sensed evaporation: justification for distribution in ungauged basins?, Hydrol. Earth Syst. Sci., 12, 1403–1413, doi:10.5194/hess-12-1403-2008, 2008.

Yadav, M., Wagener, T., and Gupta, H.: Regionalization of constraints on expected watershed response behavior for improved predictions in ungauged basins, Adv. Water Resour., 30, 1756–1774, 2007.

Yilmaz, K. K., Gupta, H. V., and Wagener, T.: A process-based diagnostic approach to model evaluation: Application to the NWS distributed hydrologic model, Water Resour. Res., 44, W09417, doi:10.1029/2007WR006716, 2008.

Climate and hydrological variability: the catchment filtering role

I. Andrés-Doménech[1]**, R. García-Bartual**[1]**, A. Montanari**[2]**, and J. B. Marco**[1]

[1]Instituto Universitario de Investigación de Ingeniería del Agua y Medio Ambiente, Universitat Politècnica de València, Camino de Vera s/n, 46022 Valencia, Spain
[2]Facoltà di Ingegneria, Università di Bologna, Via del Risorgimento 2, 40136 Bologna, Italy

Correspondence to: I. Andrés-Doménech (igando@hma.upv.es)

Abstract. Measuring the impact of climate change on flood frequency is a complex and controversial task. Identifying hydrological changes is difficult given the factors, other than climate variability, which lead to significant variations in runoff series. The catchment filtering role is often overlooked and thus may hinder the correct identification of climate variability signatures on hydrological processes. Does climate variability necessarily imply hydrological variability? This research aims to analytically derive the flood frequency distribution based on realistic hypotheses about the rainfall process and the rainfall–runoff transformation. The annual maximum peak flow probability distribution is analytically derived to quantify the filtering effect of the rainfall–runoff process on climate change. A sensitivity analysis is performed according to typical semi-arid Mediterranean climatic and hydrological conditions, assuming a simple but common scheme for the rainfall–runoff transformation in small-size ungauged catchments, i.e. the CN-SCS model. Variability in annual maximum peak flows and its statistical significance are analysed when changes in the climatic input are introduced. Results show that depending on changes in the annual number of rainfall events, the catchment filtering role is particularly significant, especially when the event rainfall volume distribution is not strongly skewed. Results largely depend on the return period: for large return periods, peak flow variability is significantly affected by the climatic input, while for lower return periods, infiltration processes smooth out the impact of climate change.

1 Introduction

Many of the concerns about climate change are related to its effects on the hydrological cycle (Kundzewicz et al., 2007, 2008; Koutsoyiannis et al., 2009; Bloeschl and Montanari, 2010), and more specifically, its impact on freshwater availability and flood frequency (Milly et al., 2002; Kay et al., 2006; Allamano et al., 2009). However, results from recent studies about climate change impacts on flood frequency have not been conclusive (Kay et al., 2006). Indeed, detecting changes in flood frequency is not easy, because there are factors other than climate variability that may lead to significant changes – for instance, spatial variability of watershed properties or changes in the channel network geometry and land-use change (Milly et al., 2002). In particular, river bed geometry alterations, even if localized, can significantly affect flood magnitude. Therefore, to better identify climate impacts, one should focus on catchments that are in close to pristine conditions (Di Baldassarre et al., 2010).

This research addresses an issue that is often overlooked and which may hinder the proper identification of climate variability effects on hydrological processes – namely, the filtering role played by catchment. In fact, runoff can be interpreted as a smoothed convolution of past and current rainfall, where smoothing is operated over the catchment contributing area and along the concentration time. Depending on the catchment's physical characteristics and meteorological conditions, smoothing may average out changes in rainfall distribution in space and time and hence cancel out climate variability. This is a key reason why climate variability effects might not be clearly visible in the hydrology response. In other words, climate variability does not necessarily imply hydrological variability. This issue has been also investigated for an urban hydrology context. For example, Andrés-Doménech et al. (2012) analysed storm tank resilience to changes in rainfall statistics, proving that the effect of climate variability on storm tank efficiency is likely to be smoothed out by the filtering effect caused by the urban catchment.

In the present study, modelling efforts are basically centred on the role of climatic variability and its effects on catchment hydrological response, with rainfall statistical properties and their future trends representing the major factors controlling flood frequency distribution. It should be noted that other factors, such as land use change, might have a more significant impact than climate change itself under certain hydrological conditions. The present research focuses on climatic impacts alone: interactions at the catchment scale between landscape characteristics (soils, vegetation and geology, for instance) and climatic properties (Troch et al., 2013), or possible climate-vegetation-soil feedbacks are not considered as they may hinder the assessment of climatic effects.

The modelling framework and simulations performed in this study focus on rainfall patterns' variability, using a suitable modelling framework to investigate the extent to which such rainfall variations can actually be buffered by a given standard hydrological catchment, with typical response parameters of a small catchment in a semi-arid Mediterranean region. Thus, heterogeneity in catchment physical properties, which has provided contrasting and sometimes contradictory results (Sangati et al., 2009), is not considered in the presented approach. Runoff statistics sensitivity to spatial heterogeneity is in principle less significant as the catchment area is smaller and therefore more homogeneous. In our case, we assume that the concentration time is short, therefore implying that the catchment area is small. Thus, the lumped modelling assumption can be considered reasonable for the purpose of the study.

To assess climatic impacts, the frequency of occurrence of peak flows is estimated by means of a derived distribution approach, which is particularly useful to obtain probability distributions of peak flows in ungauged or poorly observed basins. In such cases design floods are calculated from a hydrological model, which is driven by historical or synthetic rainfall data (Haberlandt and Radtke, 2014). The derived flood frequency analysis was also used by Gaume (2006) to investigate the asymptotic behaviour of flood peak distributions from rainfall statistical properties, highlighting the strong dependence of peak flow distribution on rainfall statistical properties, and considering a limited and reasonable hypothesis on the rainfall–runoff transformation.

Accordingly, a stochastic process is used here to model rainfall and a simple deterministic lumped model is proposed to simulate the rainfall–runoff transformation. Such an analytical approach, which has a long history of application in hydrology (see, for instance, Eagleson, 1972 and Papa and Adams, 1997), presents several advantages. The most relevant is the opportunity to analytically assess the cause–effect relationships that take place in the rainfall–runoff transformation.

However, the analytical approach requires the use of models that lend themselves to analytical developments, which are obtained by using simplified representations. Therefore our analysis, being based on the use of an analytical model, cannot account for the overall complexity of catchment processes. Consequently, a simplified representation of hydrological processes is considered herein, without including detailed effects.

Under such assumptions, the aim of this research is to quantify the actual extent to which the rainfall–runoff process actually filters the impact of rainfall variability on runoff annual maximum peak flow series. The flood frequency distribution is analytically derived for a hypothetical catchment based on plausible assumptions about the rainfall process and the rainfall–runoff transformation. Having derived the peak flow probability distribution, one may quantify the smoothing brought on by the rainfall–runoff process. A hypothetical case study is developed according to climatic and hydrological conditions typical of the Valencia region (Spain), described in Sect. 2.2. As also described later, the rainfall–runoff model proposed assumes a simple but common scheme for small, fast-responding, ungauged catchments, subjected to erratic hydrological regimes (Ferrer Polo, 1993; Soulis and Valiantzas, 2012).

2 Analytical model

We set up an analytical model to describe the river flow regime for a hypothetical catchment, based on analytical descriptions of rainfall and rainfall–runoff transformation. Under suitable assumptions which are described below, this model allows us to derive the annual maximum flood frequency distribution, depending on climate and catchment behaviour.

The analysis presented herein is an event-based approach, where each rainfall–runoff event is treated as an independent event. In the Valencia region, as in other many semi-arid locations around the Mediterranean, ephemeral rivers are closely related to small and fast-responding catchments. Such regimes, also named as "erratic regimes" according to the classification provided by Botter et al. (2013), occur when rainfall inter-arrival times are somewhat longer than the typical duration of the resulting flow pulses, as the case presented in this study. As pointed out by Andrés-Doménech et al. (2010), antecedent dry periods for the considered climate can be assumed to be exponentially distributed with a 22 h low bound and an 8-day expected mean value. With such a sporadic rainfall regime, antecedent moisture conditions are mainly related to the event itself and rainfall intensities during the initial stages of the storm, so that the assumption of independence for subsequent events is plausible. Moreover, for this type of hydrological event, direct runoff is the dominant component of the hydrograph.

To carry out this analysis, we assume that the rainfall forcing in the present climate can be modelled by a stationary model. Thus, non-stationarity can be accounted for by changing the parameters of the rainfall model at a given time when climate variability is supposed to occur. Such a change in the rainfall model parameters implies a corresponding deterministic change of rainfall statistics and therefore non-stationarity (Koutsoyiannis and Montanari, 2014; Montanari and Koutsoyiannis, 2014). Non-stationarity in the river flow is assumed to occur for the presence of the above non-stationarity in rainfall and thus is quantified through the proposed approach.

2.1 Rainfall description

A rainfall analytical model is used to describe the occurrence of the rainfall process over time. We adopt a stochastic rectangular-pulse model that simulates rainfall dynamics by assuming that rainfall events occur as independent rectangular pulses over time. Events are assumed to occur according to a Poisson process (Madsen and Rosbjerg, 1997; Madsen et al., 1997) and thus the probability of experiencing n rainfall events in the time span $[0, t]$ is given by

$$P[n] = \frac{(\beta t)^n}{n!} e^{-\beta t}, \tag{1}$$

where β is the mean number of rainfall events per unit time. Event rainfall depth (v) is assumed to be independent and the result of a generalized Pareto distribution (Andrés-Doménech et al., 2010). This model provided a good fit for the rainfall series of Valencia (Spain), recorded with 5 min resolution by the Júcar River basin hydrological service (SAIH) during the period 1990–2006. Andrés-Doménech et al. (2010) also found the model to be accurate for other locations in Spain. Other authors have also reported good results in other Mediterranean locations (Tzavelas et al., 2010).

The distribution function of the generalized Pareto distribution is given by

$$F_V(v) = 1 - \left(1 - \kappa \frac{v}{\alpha}\right)^{1/\kappa} \quad v \geq 0, \tag{2}$$

where $\kappa < 0$ and $\alpha > 0$ are the shape and scale parameters, respectively.

For the region that is considered in the study, convective storms usually occur during autumn, particularly in September and October, while frontal events mostly occur during winter and spring. Thus, maximum rainfall peaks occur systematically during autumn. The rainfall model that we use can potentially reproduce both frontal and convective events (see, for instance, Andrés-Doménech et al., 2010). Consequently, seasonality is not specifically accounted for. We assume that climatic variability may occur through an intensification of rainfall events, and we investigate the conditions

under which it may imply or not an amplification of annual maximum floods – that is, to what extent the rainfall–runoff transformation may filter out or amplify the effects of climate variability.

2.2 Rainfall–runoff description

To conceptualize rainfall–runoff transformation, the SCS-CN event-based model was adopted. This model has been widely used in Spain (Ferrer Polo, 1993) and other Mediterranean countries (Soulis and Valiantzas, 2012). In this model, runoff volume, $r(v)$, is related to event rainfall volume v by the following relationship:

$$\begin{cases} r(v) = 0 & \text{if } v \leq I_a \\ r(v) = \frac{(v - I_a)^2}{v - I_a + S} & \text{if } v > I_a, \end{cases} \tag{3}$$

where $I_a = k\,S$ is the initial rainfall abstraction, S is the catchment storage capacity and k is the initial abstraction coefficient. By assuming the dimensionless SCS unit hydrograph (SCS, 1971), each rainfall event produces a single-peak triangular hydrograph. The specific peak river flow can be expressed as

$$q_P(v) = \lambda_P \frac{r(v)}{t_C}, \tag{4}$$

where $r(v)$ is the runoff event volume computed by Eq. (3), t_C is the concentration time of the catchment and λ_P is a dimensionless peak factor.

The original SCS model recommends a standard value $\lambda_P = 9/8$, implying that $3/8$ of the total runoff volume occurs before the peak, being the time to peak equal to $2\,t_C/3$ from the beginning of net rainfall. For the particular case of semiarid regions in Spain, a value $\lambda_P = 5/3$ is recommended (Ferrer Polo, 1993) to take into account the faster hydrological response.

2.3 Deriving the peak flow probability distribution

The rainfall and rainfall–runoff analytical descriptions allow for the analytical derivation of the probability distribution function (PDF) of all events peak flow. Assuming that no runoff occurs if $v < I_a$,

$$F_{Q_P}(0) = F_V(I_a) = 1 - (1 - \kappa I_a/\alpha)^{1/\kappa}, \tag{5}$$

where Q_P indicates the stochastic process whose outcome is the event peak flow $q_P(t)$. On the other hand, when initial abstraction I_a is exceeded then $Q_P > 0$, and the related cumulative probability distribution is

$$F_{Q_P}(q_P) = \int_0^{q_P} f_{Q_P}(q_P)\,dq_P = F_{Q_P}(0)$$

$$+ \int_{I_a}^{v} f_V(v)dv = 1 - (1 - \kappa vt/\alpha)^{1/\kappa}. \tag{6}$$

Combining these expressions with Eqs. (3) and (4) leads to

$$F_{Q_P}(q_P) = \begin{cases} 1 - (1 - \kappa I_a/\alpha)^{1/\kappa} & q_P = 0 \\ 1 - \left\{1 - \frac{\kappa}{\alpha}\left[I_a + \frac{t_C q_P}{2\lambda_P}\left(1 + \sqrt{1 + \frac{4\lambda_P S}{t_C q_P}}\right)\right]\right\}^{1/\kappa} & q_P > 0. \end{cases} \tag{7}$$

As previously explained, it should be noted that these rainfall and rainfall–runoff models assume statistical independence of peak river flow over time. Therefore, the distribution function of maximum annual floods Q_{Pm} can be expressed as (see, for instance, Viglione and Blöschl, 2009)

$$F_{Q_{Pm}}(q_{Pm}) = e^{-\beta(1 - F_Q(q_P))}, \tag{8}$$

where β is the annual number of rainfall events. In terms of return period, the T-year maximum peak flow can be expressed as:

$$q_{Pm,T} = F_{Q_P}^{-1}\left[\frac{1}{\beta}\ln\left(1 - \frac{1}{T}\right) + 1\right]. \tag{9}$$

This analysis is equivalent to an Annual Maximum Series analysis of flood flows, as the flood events are assumed to be independent (Andrés-Doménech et al., 2010).

2.4 Confidence intervals of peak flow PDF

Asymptotic properties of the maximum likelihood estimators (MLEs) of the generalized Pareto distribution (Eq. 2) such as consistency, normality and efficiency were obtained by Smith (1984). The MLEs (κ, α) are asymptotically normal (De Zea Bermudez and Kotz, 2010) with a variance–covariance matrix given by

$$\begin{bmatrix} \sigma_\kappa^2 & \sigma_{\kappa\alpha} \\ \sigma_{\kappa\alpha} & \sigma_\alpha^2 \end{bmatrix} = \frac{1}{n}\begin{bmatrix} (1-\kappa)^2 & \alpha(1-\kappa) \\ \alpha(1-\kappa) & 2\alpha^2(1-\kappa) \end{bmatrix}, \tag{10}$$

where n is the sampling size. Consequently, the correlation coefficient is

$$\rho_{\kappa\alpha} = \frac{1}{\sqrt{2(1-\kappa)}}. \tag{11}$$

Monte Carlo simulations are performed to generate 1000 pairs (κ, α) normally distributed according to Eq. (10) and also to the MLEs of Eq. (2). Thus, 1000 discrete probability functions are obtained according to Eqs. (7) and (8). For a specific value q_{Pmi}, 1000 normally distributed values F_{Qpmi} are calculated so that for each q_{Pmi}, percentiles $F_{Qpmi}(\xi)$ and $F_{Qpmi}(1-\xi)$ corresponding to ξ and $1-\xi$ probabilities are derived. These values are then transformed with Eq. (9) into their corresponding return periods, T_ξ and $T_{1-\xi}$, which represent the confidence interval limits for a ξ significance level.

3 Qualitative sensitivity analysis for peak flows to climate change

Based on the previously established assumptions, the analysis shows that the following parameters affect the magnitude of the annual maximum peak river flow $q_{Pm,T}$:

1. expected number of rainfall events per year, β [yr^{-1}];

2. shape and scale parameters, κ [–] and α [mm], respectively, of the generalized Pareto distribution for event rainfall depth;

3. storage capacity of the catchment, S [mm];

4. initial abstraction of the catchment, I_a [mm];

5. concentration time of the catchment t_C [h];

6. SCS peak factor λ_P [–];

7. return period, T [year].

Parameters 1 and 2 are directly related to climate input; parameters 3 and 4 are related to the runoff production process in the catchment; parameters 5 and 6 affect the temporal catchment response; finally, parameter 7 is conditioned by the scope of the analysis.

The dependence of $q_{Pm,T}$ on these eight parameters is dictated by Eqs. (7)–(9). In particular, Eq. (9) dictates the dependence of $q_{Pm,T}$ on the return period and β. An increase in the annual number of rainfall events implies an increase in the mean annual rainfall if all other climatic behaviours remain unchanged. Consequently, an increase in β does not affect the distribution of flood peaks as long as the events remain distant enough in time and therefore independent, but only affects the number of flood peaks sampled per unit of time. This implies a relevant effect on the flood return period. According to Eq. (9), a 20 % increase in β implies a decrease in the flood return period ranging from 0 % (for low T values) to 16.7 % (for high T values). This result is counterintuitive, but one should note that a relevant change in the return period does not necessarily imply a significant change in the flood quantile. As a matter of fact, changes in $q_{Pm,T}$ can be negligible after a change in β, especially if the Pareto distribution for event rainfall depth is not strongly skewed. The hypothetical case study presented herein will prove this first conclusion, as shown later. Therefore, it can be concluded that the filtering role of the catchment with regard to changes in β is particularly significant when the distribution of event rainfall volume is not strongly skewed.

The sensitivity to the other climatic and catchment parameters is to be analysed through Eq. (7). Specifically, an increase in the flood quantile is induced by an increase in parameters α and t_C. The latter is raised to a power less than 1 and therefore is less effective than α. Conversely, an increase in k, S, I_a and λ_P leads to a decrease in the flood quantile value. These considerations are somewhat intuitive, but it is interesting to quantitatively analyse the sensitivity of the flood quantile to production parameters 3 and 4 to quantify the actual filtering role of the catchment on climate variability. The case study is developed with data from Valencia (Spain) presented as a quantitative sensitivity analysis.

4 Quantitative sensitivity analysis for peak flows to climate variability: a hypothetical case study

Rainfall model parameters are estimated by maximum likelihood for the 1990–2006 data series in Valencia. Resulting values are $\beta = 27.29 \, \mathrm{yr}^{-1}$, $\alpha = 8.46 \, \mathrm{mm}$ and $\kappa = -0.411$. Consequently, the average event depth per event is $\mu_V = 14.36 \, \mathrm{mm}$ and the coefficient of variation is $CV_V = 2.37$. Further details regarding the rainfall model can be found in Andrés-Doménech et al. (2010). This climate scenario constitutes the reference situation (scenario 0) to perform the sensitivity analysis.

Parameters defining the catchment are adopted in a dimensionless form. This analysis focuses on how the production parameters influence the peak flow statistics. Thus, the storage capacity is considered through the ratio S/μ_V, with an initial abstraction coefficient $k = 0.2$ (as in the original version of the SCS-CN model and also mentioned by Ferrer Polo, 1993).

Peak flows are expressed per unit area ($\mathrm{mm\,h^{-1}}$), so no particular catchment area is assumed.

4.1 Sensitivity to β and to the skewness of the rainfall depth distribution

The first quantitative analysis performed corresponds to flood quantile sensitivity to β and to the skewness of the Pareto distribution governing event rainfall depth. Catchment parameters are set to $S/\mu_V = 3.5$ and $t_C = 1 \, \mathrm{h}$, corresponding to typical values for small catchments in the Valencia region. Concentration time has been set to a representative value, based on a wide hydrological experience in many small catchments of rapid response in the eastern Mediterranean and southeast coast of Spain (Olivares Guillem, 2004; Camarasa Belmonte, 1990). It can be considered a realistic and representative value for a typical ephemeral river in fast-responding small catchments in semi-arid Mediterranean regions.

Relative changes in 10- and 100-year flood quantiles compared to scenario 0 are evaluated for different situations, combining variations in β and CV_V. It should be noted that changes in β mean that μ_V should be scaled accordingly. Lowering CV_V brings the Pareto event rainfall depth distribution close to the exponential distribution (Koutsoyiannis, 2005), while increasing CV_V progressively increases skewness. Given CV_V variations, the κ parameter of the Pareto distribution, as well as its skewness, vary (Singh and Guo, 1995). Pareto parameters (κ, α) for the modified scenarios

can be analytically derived from their relationships with CV_V (Andrés-Doménech et al., 2012).

Figure 1 summarizes the results obtained and shows that changes in β do not lead to significant flood quantile variations, unless the distribution of rainfall event depth is highly skewed (higher CV_V values). As stated in the previous section, the less skewed the rainfall regime is, the less significant the filtering role of the catchment. Conversely, changes in CV_V are not filtered at all.

4.2 Sensitivity to the runoff production process

Catchment production is highly influenced by the balance between rainfall depth and the catchment storage capacity. Thus, sensitivity to the production process should be analysed by introducing variability in rainfall event depth for different S/μ_V situations.

Arbitrary variations in $v(t)$ statistics from the reference situation (scenario 0) are considered as plausible climate variability scenarios for rainfall event depth. Instead of evaluating the effects of changes on the distribution parameters, changes in the rainfall statistic μ_V of rainfall event depth are considered. The analysis is now performed by changing μ_V in the range $\pm 30\%$ of its reference value (scenarios 1.a, $+30\%$ and 1.b, -30%). This is in accordance with the maximum expected variability in annual amounts of rainfall for the predicted climate change scenarios in Spain (Brunet et al., 2009). In this scenario CV_V remains unchanged. It follows that both the κ parameter of the Pareto distribution and its skewness also remain unchanged (Singh and Guo, 1995). The modified α values for the modified scenarios can be derived from α dependence on μ_V (Andrés-Doménech et al., 2012). As stated before, physical parameters defining the catchment are adopted in a dimensionless form. To analyse the filtering role of the catchment depending on production parameters, three realistic storage capacity scenarios are considered, namely, $S/\mu_V = 3.5, 5$ and 10.

For each S/μ_V scenario, Fig. 2 depicts flood quantile variations for scenarios 1.a ($+30\% \; \mu_V$) and 1.b ($-30\% \; \mu_V$). Unchanged climatic conditions (scenario 0) yield a flow quantile decrease as S/μ_V increases. Hence, considering scenario 1.a and 1.b leads to quantile increments associated to S/μ_V increments. In fact, flood quantile reductions caused by higher S/μ_V values (scenario 0) are more relevant than the variation resulting from μ_V changes (scenarios 1.a and 1.b).

Another point to be noted is the magnitude of relative variations depending on the return period T. For higher return periods, relative changes in flood quantiles tend to be very close to those imposed by the climatic input (mean rainfall event depth μ_V). This result reinforces the thesis supported by Gaume (2006) who demonstrated that, for large return periods, the rainfall PDF behaviour is decisive on the catchment response and determines the asymptotic behaviour of the flood peak distribution. On the other hand, for low return

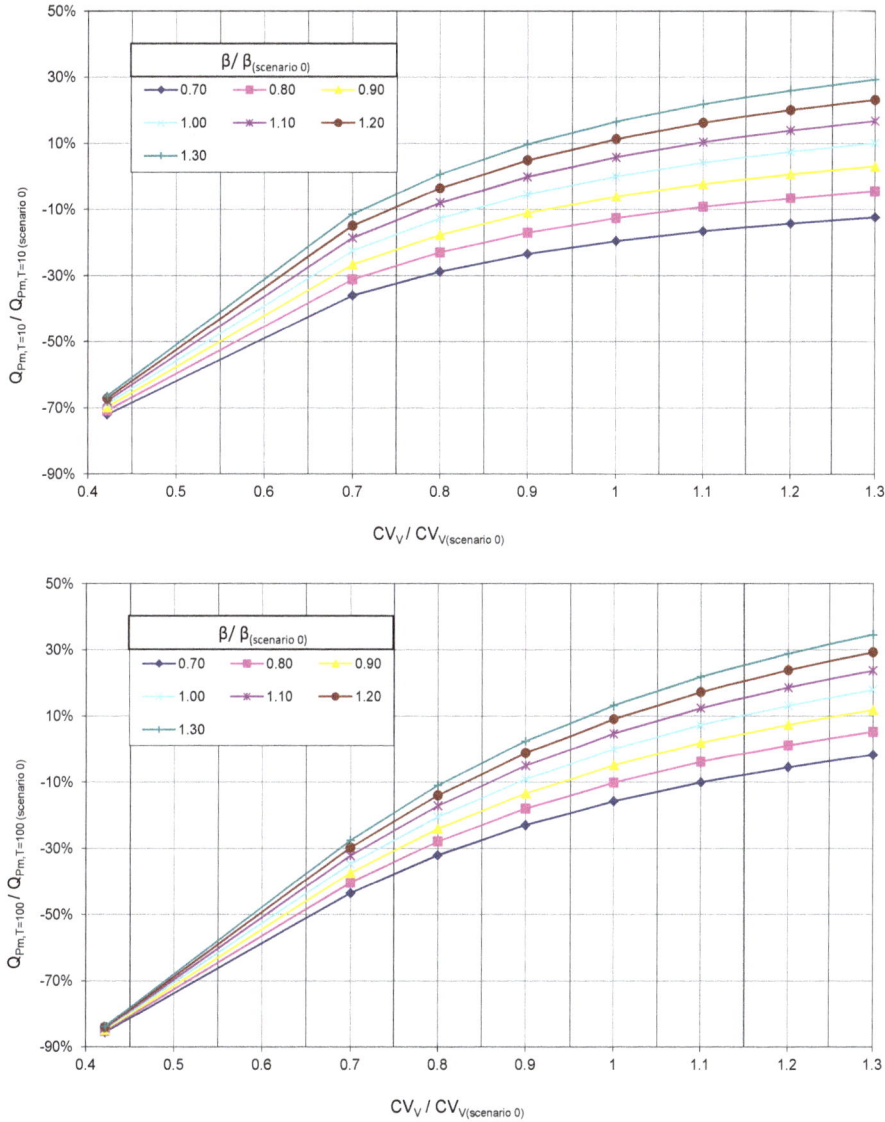

Figure 1. Annual maximum flood quantile variations for changes in β and CV_V. Catchment parameters are set to $S/\mu_V = 3.5$ and $t_C = 1$ h. Cases $T = 10$ years (top panel) and $T = 100$ years (bottom panel).

periods, catchment infiltration parameters strongly influence the derived peak flows for each scenario considered. This result is in accordance with typical Mediterranean catchment behaviour (Gioia et al., 2008; Preti et al., 2011).

4.3 Peak flow confidence intervals

Confidence interval limits for a $\xi = 0.05$ significance level are obtained for annual maximum peak flow quantiles corresponding to climatic scenario 0. In order to quantify the statistical significance of peak flow variations after considering various scenarios, eight different climatic scenarios are selected from amongst those previously analysed. These account for climatic variations induced by changes in μ_V, β and CV_V (Table 1). Annual maximum peak flow quantiles

are evaluated for each scenario and variations with regard to scenario 0 are calculated. Figure 3 summarizes the results obtained for each scenario and for the confidence interval limits for scenario 0. As observed, all results corresponding to β and/or CV_V variations (scenarios 2.a to 4.b) lie within the 90 % confidence interval for scenario 0. Therefore, results show that there is no concluding evidence from the statistical point of view concerning the significance of peak flow variability induced by these parameters. Nevertheless, when considering peak flow variations due to changes in μ_V (scenarios 1.a and 1.b), our results confirm the conclusions already drawn in Sect. 3. For low return periods, changes are significant because they are strongly influenced by the runoff production process in the catchment. For larger T, the significance of peak flow variations drastically decreases.

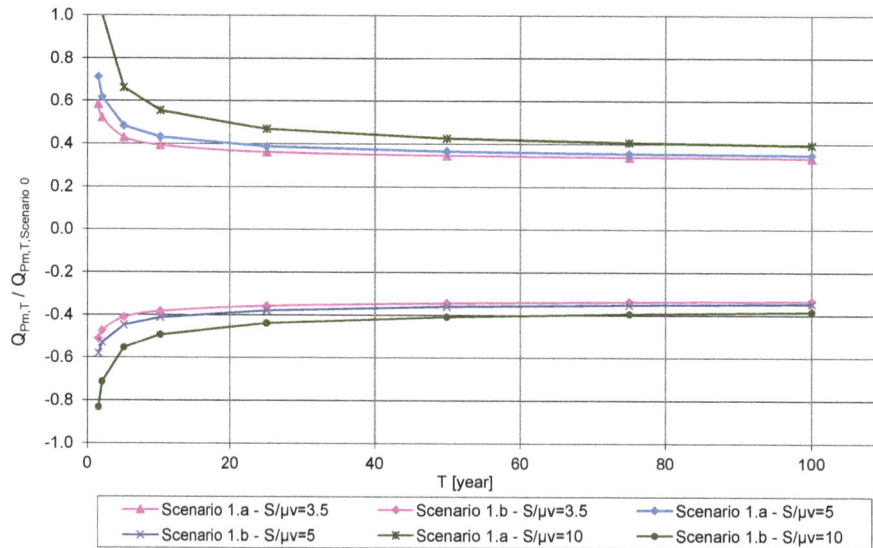

Figure 2. Annual maximum flood quantile variations for scenarios 1.a ($+30\%\ \mu_V$) and 1.b ($-30\%\ \mu_V$) and for $S/\mu_V = 3.5, 5$ and 10.

Table 1. Climate scenarios considered for significance analysis.

Climatic scenario	μ_V hypothesis	CV_V hypothesis	β hypothesis	μ_V [mm]	CV_V	α [mm]	κ	β
0	Reference scenario	Reference scenario	Reference scenario	14.36	2.37	8.46	0.411	27.29
1a	30 % increase in μ_V	Reference scenario	Reference scenario	18.67	2.37	11.00	0.411	27.29
1b	30 % decrease in μ_V	Reference scenario	Reference scenario	10.05	2.37	5.92	0.411	27.29
2a	Reference scenario	30 % increase in CV_V	Reference scenario	14.36	3.08	7.94	0.447	27.29
2b	Reference scenario	30 % decrease in CV_V	Reference scenario	14.36	1.66	9.79	0.318	27.29
3a	Reference scenario	30 % increase in CV_V	30 % increase in β	14.36	3.08	7.94	0.447	35.48
3b	Reference scenario	30 % decrease in CV_V	30 % increase in β	14.36	1.66	9.79	0.318	35.48
4a	Reference scenario	Reference scenario	30 % increase in β	14.36	2.37	8.46	0.411	35.48
4b	Reference scenario	Reference scenario	30 % decrease in β	14.36	2.37	8.46	0.411	19.11

5 Conclusions

The research presented herein highlights the filtering role brought on by catchment processes through a simple rainfall–runoff transfer function. The peak flow distribution is analytically derived from a rainfall model using the CN-SCS hydrological conceptualization. Variability of annual maximum peak flows is quantitatively analysed when changes in climatic input are introduced.

Such a modelling approach involves certain limitations, and yet it benefits from the analytical simplicity and practical applicability. Consequently, numerical results obtained after simulations cannot be transferred to hydrological regimes that differ from the type of Mediterranean catchments specified here. Nevertheless, the proposed methodology represents a useful modelling framework for further studies, and may constitute a first step forward towards a more complex analysis after relaxing some of the initial assumptions.

Although certain dominant drivers of the hydrological response, like variability of watershed properties or land use changes, have not been explicitly considered in this study, the proposed modelling framework has the potential to incorporate those drivers to a certain extent, and thus, allow for the effect of such variability to be assessed and compared in future studies.

The results obtained from the sensitivity analysis can be summarized as follows:

1. The filtering role of the catchment with regard to changes in the annual number of rainfall events is particularly significant when the rainfall event volume distribution is not strongly skewed.

2. Sensitivity to the runoff production parameters in the catchment is highly influenced by the balance between rainfall depth and catchment storage capacity. For higher return periods, relative changes in annual maxi-

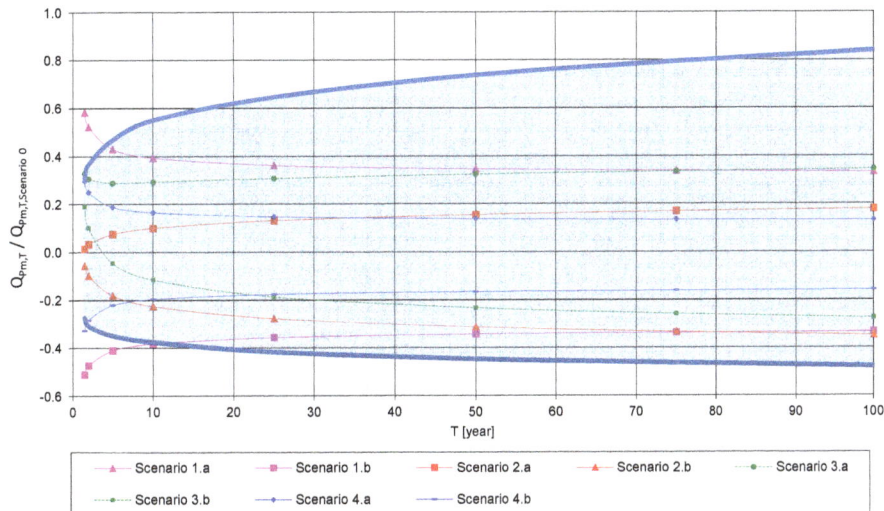

Figure 3. Annual maximum flood quantile variations for scenarios defined in Table 1 and $\xi = 0.05$ confidence interval for scenario 0 peak flow distribution (shaded area). Catchment parameters are set to $S/\mu_V = 3.5$ and $t_C = 1$ h.

mum flood quantiles tend to be asymptotically similar to those imposed by the climatic input. For low return periods, the infiltration process strongly influences the derived peak flow distribution, which is in accordance with typical Mediterranean catchment hydrological behaviour.

3. In the range of low return periods (1 to 10 years), the only parameter of the rainfall model which actually affects significantly peak flows is the mean rainfall event depth. The other parameters involved in the rainfall modelling approach play a negligible role in this case, mainly due to the threshold-based conceptualization used in the CN-SCS model.

Although these conclusions were derived under simplified assumptions, results correspond to a rigorous sensitivity analysis performed for realistic hydrological conditions of typical ephemeral, fast-responding rivers, and thus provide indications of general validity for small Mediterranean catchments responding under these simple rainfall–runoff models. Further research should focus on the limitations of such a simple model for high and very high return periods and on the dependence of peak flow variability on time-dependent parameters of the rainfall–runoff transformation. On the other hand, the research could be extended by including in the rainfall–runoff deterministic model additional climatic perturbations and land use changes, as well as by exploring possible parameter interaction effects.

Acknowledgements. The authors wish to thank Debra Westall for revising the paper. The present work was (partially) developed within the framework of the Panta Rhei Research Initiative of the International Association of Hydrological Sciences (IAHS).

Edited by: S. Attinger

References

Allamano, P., Claps, P., and Laio, F.: Global warming increases flood risk in mountainous areas, Geophys. Res. Lett., 36, L24404, doi:10.1029/2009GL041395, 2009.

Andrés-Doménech, I., Montanari, A., and Marco, J. B.: Stochastic rainfall analysis for storm tank performance evaluation, Hydrol. Earth Syst. Sci., 14, 1221–1232, doi:10.5194/hess-14-1221-2010, 2010.

Andrés-Doménech, I., Montanari, A., and Marco, J. B.: Efficiency of Storm Detention Tanks for Urban Drainage Systems under Climate Variability, J. Water Resour. Pl. Manage., 138, 36–46, doi:10.1061/(ASCE)WR.1943-5452.0000144, 2012.

Bloeschl, G. and Montanari, A.: Climate change impacts – throwing the dice?, Hydrol. Process., 24, 374–381, doi:10.1002/hyp.7574, 2010.

Botter, G., Basso, S., Rodriguez-Iturbe, I., and Rinaldo, A.: Resilience of river flow regimes, P. Natl. Acad. Sci. USA, 110, 12925–12930, doi:10.1073/pnas.1311920110, 2013.

Brunet, M., Casado, M. J., de Castro, M., Galán, P., López, J. A., Martín, J. M., Pastor, A., Petisco, E., Ramos, P., Ribalaygua, J., Rodríguez, E., Sanz, I., and Torres, L.: Generación de escenarios regionalizados de cambio climático para España, Agencia Estatal de Meteorología (AEMET), Ministerio de Medio Ambiente y Medio Rural y Marino, Madrid, 2009.

Camarasa Belmonte, A. M.: Génesis de avenidas en pequeñas cuencas semiáridas: la Rambla de Poyo (Valencia), Cuad. De Geogr., 48, 81–104, 1990.

De Zea Bermudez, P. and Kotz, S.: Parameter estimation of the generalized Pareto distribution – Part I, J. Stat. Plan. Infer., 140-6, 1353–1373, doi:10.1016/j.jspi.2008.11.019, 2010.

Di Baldassarre, G., Montanari, A., Lins, H., Koutsoyiannis, D., Brandimarte, L., and Blöschl, G.: Flood fatalities in Africa:

From diagnosis to mitigation, Geophys. Res. Lett., 37, L22402, doi:10.1029/2010GL045467, 2010.

Eagleson, P. S.: Dynamics of flood frequency, Water Resour. Res., 8, 878–898, doi:10.1029/WR008i004p00878, 1972.

Ferrer Polo, J.: Recomendaciones para el cálculo hidrometeorológico de avenidas, Centro de Estudios y Experimentación de Obras Públicas, Madrid, 1993.

Gaume, E.: On the asymptotic behavior of flood peak distributions, Hydrol. Earth Syst. Sci., 10, 233–243, doi:10.5194/hess-10-233-2006, 2006.

Gioia, A., Iacobellis, V., Manfreda, S., and Fiorentino, M.: Runoff thresholds in derived flood frequency distributions, Hydrol. Earth Syst. Sci., 12, 1295–1307, doi:10.5194/hess-12-1295-2008, 2008.

Haberlandt, U. and Radtke, I.: Hydrological model calibration for derived flood frequency analysis using stochastic rainfall and probability distributions of peak flows, Hydrol. Earth Syst. Sci., 18, 353–365, doi:10.5194/hess-18-353-2014, 2014.

Kay, A. L., Jones, R. G., and Reynard, N. S.: RCM rainfall for UK flood frequency estimation, II. Climate change results, J. Hydrol., 318, 163–172, doi:10.1016/j.jhydrol.2005.06.013, 2006.

Koutsoyiannis, D.: Uncertainty, entropy, scaling and hydrological stochastics, 1. Marginal distributional properties of hydrological processes and state scaling, Hydrolog. Sci. J., 50, 381–404, doi:10.1623/hysj.50.3.381.65031, 2005.

Koutsoyiannis, D. and Montanari, A.: Negligent killing of scientific concepts: the stationarity case, Hydrolog. Sci. J., doi:10.1080/02626667.2014.959959, in press, 2014.

Koutsoyiannis, D., Montanari, A., Lins, H. F., and Cohn, T. A.: Discussion of "The implications of projected climate change for freshwater resources and their management". Climate, hydrology and freshwater: towards an interactive incorporation of hydrological experience into climate research, Hydrolog. Sci. J., 54, 394–405, doi:10.1623/hysj.54.2.394, 2009.

Kundzewicz, Z. W., Mata, L. J., Arnell, N. W., Döll, P., Kabat, P., Jiménez, B., Miller, K. A., Oki, T., Sen, Z., and Shiklomanov, I. A.: Freshwater resources and their management, Climate Change 2007: Impacts, Adaptation and Vulnerability, in: Contribution of Working Group II to the Fourth Assessment Report of the Intergovernmental Panel on Climate Change, edited by: Parry, M. L., Canziani, O. F., Palutikof, J. P., van der Linden, P. J., and Hanson, C. E., Cambridge University Press, Cambridge, UK, 173–210, 2007.

Kundzewicz, Z. W., Mata, L. J., Arnell, N. W., Döll, P., Jimenez, B., Miller, K., Oki, T., Sen, Z., and Shiklomanov, I.: The implications of projected climate change for freshwater resources and their management, Hydrolog. Sci. J., 53, 3–10, doi:10.1623/hysj.53.1.3, 2008.

Madsen, H. and Rosbjerg, D.: The partial duration series method in regional index-flood modeling, Water Resour. Res., 33, 737–746, doi:10.1029/96WR03847, 1997.

Madsen, H., Rasmussen, P. F., and Rosbjerg, D.: Comparison of annual maximum series and partial duration series for modeling extreme hydrologic events, 1. At-site modeling, Water Resour. Res., 33, 747–757, doi:10.1029/96WR03848, 1997.

Milly, P. C. D., Wetherald, R. T., Dunne1, K. A., and Delworth, T. L.: Increasing risk of great floods in a changing climate, Nature, 415, 514–517, doi:10.1038/415514a, 2002.

Montanari, A. and Koutsoyiannis, D.: Modeling and mitigating natural hazards: Stationarity is immortal!, Water Resour. Res., 50, 9748–9756, doi:10.1002/2014WR016092, 2014.

Olivares Guillem, A.: Modelación hidrológica pseudo-distribuida del barranco del Carraixet: aplicación al episodio de octubre de 2000, Cuad. De Geogr., 76, 155–182, 2004.

Papa, F. and Adams, B. J.: Application of derived probability and dynamic programming techniques to planning regional stormwater management systems, Water Sci. Technol., 36, 227–234, 1997.

Preti, F., Forzieri, G., and Chirico, G. B.: Forest cover influence on regional flood frequency assessment in Mediterranean catchments, Hydrol. Earth Syst. Sci., 15, 3077–3090, doi:10.5194/hess-15-3077-2011, 2011.

Sangati, M., Borga, M., Rabuffetti, D., and Bechini, R.: Influence of rainfall and soil properties spatial aggregation on extreme flash flood response modelling: An evaluation based on the Sesia river basin, North Western Italy, Adv. Water Resour., 32, 1090–1106, 2009.

SCS: National Engineering Handbook, Section 4: Hydrology, Soil Conservation Service, USDA, Washington, D.C., 1971.

Singh, V. P. and Guo, H.: Parameter estimation for 3-parameter generalized Pareto distribution by the principle of maximum entropy (POME), Hydrolog. Sci. J., 40, 165–181, doi:10.1080/02626669509491402, 1995.

Smith, R. L.: Threshold methods for sample extremes, in: Statistical Extremes and Applications, edited by: de Oliveira, J. T., Reidel, Dordrecht, 621–638, 1984.

Soulis, K. X. and Valiantzas, J. D.: SCS-CN parameter determination using rainfall–runoff data in heterogeneous watersheds – the two-CN system approach, Hydrol. Earth Syst. Sci., 16, 1001–1015, doi:10.5194/hess-16-1001-2012, 2012.

Troch, P. A., Carrillo, G., Sivapalan, M., Wagener, T., and Sawicz, K.: Climate–vegetation–soil interactions and long-term hydrologic partitioning: signatures of catchment co-evolution, Hydrol. Earth Syst. Sci., 17, 2209–2217, doi:10.5194/hess-17-2209-2013, 2013.

Tzavelas, G., Paliatsos, A. G., and Nastos, P. T.: Brief communication "Models for the exceedances of high thresholds over the precipitation daily totals in Athens, Greece", Nat. Hazards Earth Syst. Sci., 10, 105–108, doi:10.5194/nhess-10-105-2010, 2010.

Viglione, A. and Blöschl, G.: On the role of storm duration in the mapping of rainfall to flood return periods, Hydrol. Earth Syst. Sci., 13, 205–216, doi:10.5194/hess-13-205-2009, 2009.

3

Using high-resolution phosphorus data to investigate mitigation measures in headwater river catchments

J. M. Campbell, P. Jordan, and J. Arnscheidt

School of Environmental Sciences, University of Ulster, Coleraine, Northern Ireland, BT52 1SA, UK

Correspondence to: J. M. Campbell (jm.campbell@ulster.ac.uk)

Abstract. This study reports the use of high-resolution water quality monitoring to assess the influence of changes in land use management on total phosphorus (TP) transfers in two $5\,km^2$ agricultural sub-catchments. Specifically, the work investigates the issue of agricultural soil P management and subsequent diffuse transfers at high river flows over a 5-year timescale. The work also investigates the phenomenon of low flow P pollution from septic tank systems (STSs) and mitigation efforts – a key concern for catchment management. Results showed an inconsistent response to soil P management over 5 years with one catchment showing a convergence to optimum P concentrations and the other an overall increase. Both catchments indicated an overall increase in P concentration in defined high flow ranges. Low flow P concentration showed little change or higher P concentrations in defined low flow ranges despite replacement of defective systems and this is possibly due to a number of confounding reasons including increased housing densities due to new-builds. The work indicates fractured responses to catchment management advice and mitigation and that the short to medium term may be an insufficient time to expect the full implementation of policies (here defined as convergence to optimum soil P concentration and mitigation of STSs) and also to gauge their effectiveness.

1   Introduction

With the introduction of the European Union (EU) Water Framework Directive (OJEC, 2000), and similar international legislation that seeks to improve and maintain the water quality of rivers, lakes and coastlines, assessing the magnitude of nutrient losses from agricultural catchments has become a priority (Cherry et al., 2008). Phosphorus (P) is particularly related to eutrophication problems in freshwaters and is prioritised in certain states for mitigation (Wall et al., 2011; Hirsch, 2012). Despite this requirement, an equally important need is to maintain a healthy agri-economy which depends on P inputs to catchments.

The P transferred from agricultural catchments can be from diffuse and/or point sources and in recent years the role of septic tank systems (STSs), as rural point sources, has been identified in elevating low flow P concentrations (Withers et al., 2012, 2014). The flow paths taken from source to impact point are affected by the complexity of catchments at both temporal and spatial scales (Turcotte, 1997; Kirchner et al., 2004) resulting in river P that has dependencies and signals according to hydrological conditions (Jordan et al., 2007a).

Low-resolution grab sampling for chemical parameters remains the norm for most river monitoring programmes, at least in the European Union (EU) (Bowes et al., 2009). However, it is recognised that in order to completely understand the full range of river chemical signals, which are influenced by catchment processes and entrained in the range of river discharges, higher resolution monitoring is preferred (Palmer-Felgate et al., 2008; Harris and Heathwaite, 2012).

The use of bankside analysers is a relatively recent stepped change in technology for high-resolution P monitoring of both diffuse and point source signals. Jordan et al. (2005a, 2007a) used this technology in the border area of Ireland to provide a typology of river P transfers. Arnscheidt et al. (2007) used data from the same sub-catchments to define the sources of low flow P transfers. Several other studies have also used bankside analysis to define the magnitude of P transfers from point and diffuse sources at diurnal, event, seasonal and annual scales (Palmer-Felgate et al., 2008; Melland et al., 2012; Wade et al., 2012; Donn et al., 2012).

However, despite these innovations and increased understanding of P dynamics in rivers, there has been no reported use of high-resolution data sets to monitor the influence of catchment change on river P patterns. Other workers, using coarser resolution data, have proposed Load Apportionment Models to determine changes in the proportions of point and diffuse P transfers using assumptions based on relationships between discharge and P concentration (Greene et al., 2011). Major limitations of this method include the reliance on sparse concentration data at the high end of discharge records (Bowes et al., 2008).

In this investigation, a method is used to detect catchment change influences on P transfer which integrates all the discharge-P concentration patterns that might occur, in toto, during different stages of river discharges including low and storm flows. The method is applied to a 5-year P and discharge data set in two agricultural sub-catchments which were subject to a series of voluntary (soil P) and mandatory (closed periods for fertiliser/slurry application) management regimes and changes to STSs and monitored between 2006 and 2011. This study sought to assess if there was an improvement in chemical water quality in the sub-catchments as a result of changes in management regimes and the implementation of a series of mitigation measures.

Study area

The study areas for this project were two $5 \, km^2$ sub-catchments in Co. Monaghan and Co. Tyrone in the Irish border area (Fig. 1). These are situated on two major tributaries, the Mountain Water and Oona Water, (respectively) of the Blackwater River, which has a cross border catchment ($1480 \, km^2$). The Blackwater is the largest catchment of the six rivers which flow into the hypertrophic Lough Neagh (Griffiths, 2007).

Land use is 90 % agricultural with mixed livestock (beef, dairy and sheep enterprises in order of importance) (Land cover 2000 classification) and a stocking density of $1.5 \, LU \, ha^{-1}$ (livestock unit per hectare). The sub-catchments had a rural population density in 2005 of 13.8 and 3.4 households km^{-2} in Co. Monaghan and Co. Tyrone, respectively, with an average household size of 3 persons (Arnscheidt et al., 2007).

The underlying geology is mostly Carboniferous (Dinantian) sandstones, limestone, shale and mudstones (Cruickshank, 1997) overlaid by a till of pro-glacial boulder clay forming drumlins and Rogën moraines (Knight et al., 1999). These boulder clays effectively inhibit surface and ground water interactions resulting in isolated aquifers (O'Dochartaigh, 2003). Soils are, therefore, poorly drained with a seasonally perched water table resulting in winter saturation excess overland flow following heavy rains (Kane, 2009). The rainfall in both catchments is 800–1200 mm yr^{-1},

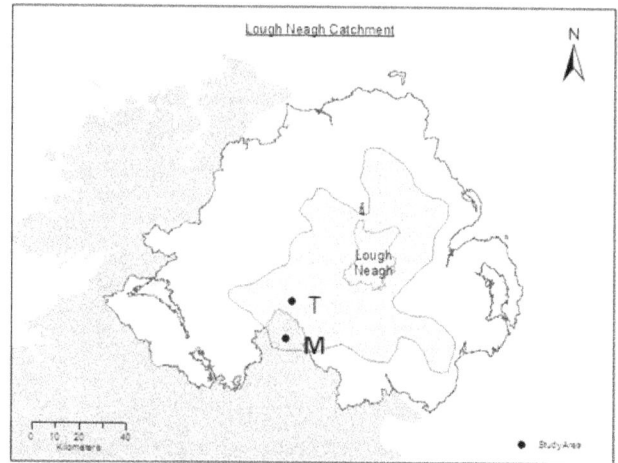

Figure 1. Map of Northern Ireland and part of the Republic of Ireland showing the Lough Neagh catchment and location of Co. Tyrone (T) and Co. Monaghan (M) study areas.

the upper end of which is above average for this part of Ireland, based on the current 30-year average (Met Eirann, Clones, Co. Cavan).

The hydrology of the soil is driven by high run-off rates, low water storage capacity, perched water tables and low permeability (Jordan et al., 2007b). These hydrological factors combine to create fast run-off pathways during storm events leading to increased diffuse pollution risk. In addition to this, the lack of water storage capacity suppresses summer base flows and increases point source pollution risk.

2 Methods

2.1 Surveys

Three surveys were conducted in this study; soil P status and changes, which were subsequently mapped; septic tank system changes; and P concentration delivered to stream systems using a semi-continuous high-resolution approach.

2.2 Soil phosphorus surveys

Between December 2004 and February 2005 composite soil samples were taken from 170 fields in the Co. Monaghan sub-catchment and 276 fields in the Co. Tyrone sub-catchment, following a reasonable period since the application of organic/inorganic fertiliser to ensure soil P equilibrium (6 weeks: Agbenin and Tiessen, 1995); this represented 62 and 69 % by land area, respectively. The composites consisted of thirty sub-samples that were randomly extracted to 7.5 cm depth from each field using a soil corer and across a "W" pattern in order to allow for heterogeneity of soil nutrient status at intra-field scale, but avoiding areas where animals would have sheltered or congregated close to feeding troughs. The composite samples were prepared for analysis by air drying and sieving through a 2 mm mesh. Plant available Olsen-P concentrations were determined on 5 cm^3 samples in both catchments and, in accordance with standard soil P testing methods for the Republic of Ireland in Co. Monaghan, plant available Morgan-P concentration (Olsen et al., 1954; Morgan, 1941). Nutrient management plans were provided to farmers on a field basis and based on respective requirements for optimum grassland agriculture in each catchment and jurisdiction (equivalent to SI 31, 2014 and SRNI 488, 2006). Of the farmers participating, formal nutrient management plans were either a new concept or had been previously provided at the very coarse scale with several field blocks integrated as single land use units.

Between December 2009 and February 2010 a percentage of the same fields was resampled and analysed to assess any change in soil nutrient concentration subsequent to nutrient management recommendations. Sampling fewer fields was necessary due to budget and time constraints. The resample percentage size was determined using an a priori Monte Carlo power analysis, using a 5 % possible change as a realistic value based on the time between samples and a zero P application rate (Watson et al., 2007), to compute the minimum sample size required to be confident in detecting a change in soil concentration. Therefore, 91/170 and 143/276 fields, approximately 53 % of the original number, were re-sampled in Co. Monaghan and Co. Tyrone, respectively.

The soil nutrient data were mapped using GIS (ArcGIS v10) and calculations made of the length of stream in each catchment which was potentially impacted by bankside soils with P concentrations which were well above the Olsen agronomic optimum i.e. soils at index 4 > 45 mg P L^{-1} (according to RB209; MAFF, 2000). Such sites, recognised as having a high P transport potential, are known as critical source areas (CSAs) and, although they may have a low proportion of the catchment area, can contribute a majority of annual P loss (Sharpley and Reikolainen, 1997).

2.3 Septic tank system surveys

In a companion study, STSs were surveyed, ranked and a subset chosen for mitigation. The results of this study are largely reported by Macintosh et al. (2011) although the Co. Tyrone catchment and a more detailed analysis are included (with the Co. Monaghan catchment) in the present study.

2.4 Water quality

In 2005 Hach-Lange Phosphax and Sigmatax bankside analyser stations were installed in kiosks at the outlet of the sub-catchments (Jordan et al., 2005a). The bankside analyser takes a 100 mL sample and determines total phosphorus (TP) on a default 20 minute cycle. The sample is homogenised with ultrasonic pulses for approximately 120 s before a 10 mL sub-sample is extracted and superheated with

sulfuric acid and sodium peroxydisulfate. The TP concentration is then determined photometrically using the method of Eisenreich et al. (1975). Regular calibration and analytical checks were carried out in conjunction with the automated analysis and weekly maintenance was carried out to ensure optimum suite performance (Cassidy and Jordan, 2011). Details of a quality control study were also included in the present investigation with TP samples analysed in the laboratory from stream samples immediately adjacent to the Sigmatax intake and also samples from the Sigmatax homogenisation jar (prior to Phosphax sub-sampling). These tests were designed to test the robustness of both the Sigmatax sample collection stage and also the Phosphax digestion and colorimetric stages. Total P was deemed an appropriate parameter for analysis in these catchments as previous studies had highlighted a process of fast entrainment of soluble P to particulate P owing to high iron and aluminium sources and transfers (Jordan et al., 2005b; Douglas et al., 2007).

River discharges were recorded using an OTT Thalimedes, measuring stage height at 1 min intervals and recording the average over 15 mins. Stage heights were converted to discharge (m^3 s^{-1}) using a rating curve that was constructed for each stream by measuring discharge over a range of stage heights above controls using the velocity area method. Velocity was measured using an OTT Nautilus electromagnetic flowmeter and a Streampro ADCP. Rainfall was monitored at both sites using an ARG-100 gauge but not over the full record in Co. Monaghan.

2.5 Data analysis

Having determined normality of the soil test P data, all changes in potential soil P test concentration were assessed using a two-tailed matched pair t test. Olsen-P results were compared based on whole catchment, farm block, soil P index and individual field status. The assumption here is that, as no P fertiliser was being applied to fields where there was found to be an excess in the 2005 testing, both catchments would see a reduction in soil P concentrations and a convergence on an optimum index equivalent to Olsen 16–25 mg P kg^{-1} (Morgan 5–8 mg P L^{-1}). The indices considered were 0 (0–9 mg P kg^{-1}), 1 (10–15 mg P kg^{-1}), 2 (16–25 mg P kg^{-1}), 3 (26–45 mg P kg^{-1}) and > 4 (> 46 mg P kg^{-1}).

The data from the bankside analysers in both catchments for the five hydrological years (determined as two water-half years), April 2006–March 2011, were extracted and, prior to analysis, were screened. Erroneous values were removed and the completeness of the data was determined. Once the data for each year had been cleaned hourly averages for each parameter were calculated and load (kg P ha^{-1}) was reported as a product of the hourly discharge and its associated TP concentration.

The discharge data from the continuous time series were ranked and the discharge ranges for each catchment were extracted from Q_5–Q_{95} in a flow-duration curve. The method of comparison over the 5-year period was to identify similar discharge ranges in each year corresponding to the discharge range, and extract the concurrent P concentration data. Following tests of normality (Lloyd et al., 2014), the mean and variance of TP concentrations, for each discharge range, were compared in a year on year basis by one-way ANOVA in order to determine the magnitude and significance of any change.

Low and extreme-low flow P concentration changes were investigated in Q_{90}–Q_{80} and Q_{95}–Q_{90} discharges ranges, respectively. These ranges, below which for example, 20 % of the record fell, were deemed most likely to show a TP influence from rural point sources (Withers et al., 2012). Additionally, previous studies on the catchments (Douglas et al., 2007) had shown that > 70 % of the annual P load can occur at Q_{10} or greater; the infrequent high flows above 90 % of the record. For this reason part of the study focuses on the Q_{50} and above, the higher flows above 50 % of the record, during which the majority of the annual P load were deemed to be delivered from the catchments. These were extracted as Q_{40}–Q_{50}, Q_{20}–Q_{30} and Q_5–Q_{10} discharges.

As total annual P load is highly influenced by annual discharge (Edwards and Withers, 2007), which varies with climate (rainfall) variability, it was assumed that this metric (annual P load) would be an unreliable predictor of change if catchment mitigation strategies had influenced P in run-off in the short term. Nevertheless, the metric was calculated so the overall magnitude, and subsequent downstream risk, of these sub-catchment P exports could be compared with others.

3 Results

3.1 Soil phosphorus changes

When soil P concentrations were compared for each sub-catchment (Table 1a and b) the fields in Co. Tyrone showed a small and non-significant increase in Olsen-P concentration across all soil indices (expressed using the Olsen data) between 2005 and 2010, whereas an increase in Co. Monaghan was larger and significant; 1.6 mg P L^{-1} ($P = 0.1$, $n = 143$) and 5.7 mg P L^{-1} ($P = 0.9 \times 10^{-4}$, $n = 91$) in Co. Tyrone and Co. Monaghan, respectively.

On an index basis in Co. Tyrone over the 5-year period, the index 1 and 2 fields showed a significant increase in mean P concentration from 16.28 to 22.24 mg P L^{-1} ($P = 0.161 \times 10^{-10}$, $n = 78$). However, when all fields with 2005 concentration > 26 mg P L^{-1} (index 3 and above) were analysed for change there was a significant decrease from 41.8 to 37.7 mg P L^{-1} ($P = 0.008$, $n = 65$). When only fields with a 2005 concentration > 46 mg P L^{-1} (index 4 and above) were considered there was also a significant mean decrease from 56.0 to 48.9 mg P L^{-1} ($P = 0.045$, $n = 30$).

Table 1. Summary of change in Olsen-P concentration for all soil indices in the Co. Tyrone (**a**) and Co. Monaghan (**b**) sub-catchments.

Catchment	Olsen-P index	Mean change in Olsen-P conc. mg P/L	Matched pairs two tailed t test P value	No. of fields
		(a)		
Co.	1, 2, 3, 4	1.6	0.1	143
Tyrone	1, 2	5.96	1.61×10^{-9}	78
	3, 4	−4.1	0.008	65
	4	−7.2	0.045	30
		(b)		
Co.	1, 2, 3, 4	5.7	0.9×10^{-4}	91
Monaghan	1, 2	7.1	0.8×10^{-3}	32
	3, 4	4.5	0.01	59
	4	4.9	0.033	27

The Co. Monaghan soil P changes reflected the same pattern as Co. Tyrone in the lower index fields 1 and 2; there was a significant increase from a mean value of 19.8 to 26.9 mg P L^{-1} ($P = 0.8 \times 10^{-3}$, $n = 32$). Unlike Co. Tyrone, however, the higher index fields in Co. Monaghan showed an increase in mean concentration. For example, between 2005 and 2010 the mean concentration rose significantly from 46.2 to 50.7 mg P L^{-1} for index 3 and above ($P = 0.01$, $n = 59$). When only index 4 and above fields were considered there was a mean increase of 56.4 to 61.3 mg P L^{-1} ($P = 0.033$, $n = 27$).

3.2 Critical source areas

Despite both catchments having the same area (5 km^2) and similar agricultural management, the soil survey carried out in 2010 determined that the percentage of fields in the Co. Monaghan sub-catchment with index 4 and above was higher than in Co. Tyrone. Furthermore, the adjacent location of these high index fields in relation to the stream network (Fig. 2a and b) was higher in Co. Monaghan (2552 m of the 36852 m total stream network – 6.9 %) than Co. Tyrone (724 m of the 19375 m total stream network – 3.7 %) (Table 2).

3.3 Phosphax quality control

The results of the quality control investigation of the Sigmatax and Phosphax stages are shown in Fig. 3. The sub-samples extracted from the homogenisation jar prior to the Phosphax stage, and analysed using standard laboratory techniques, are in very close agreement ($b = 0.97$; $R^2 = 0.98$) with the Phosphax measurements. This is essentially the same parcel of sampled water analysed by two techniques but using the same method (acid digestion and colorimetry) and indicating 97 % recovery. The same Phosphax measure-

Table 2. Length of stream network potentially impacted in each catchment by Olsen-P index 4 fields (critical source areas).

Sub-catchment	Percentage of catchment fields Olsen-P index ≥ 4	Length of stream network (m)	Length of stream network impacted by high index fields (m)	Percentage stream network impacted by high index fields
Co. Tyrone	21	19 375	724	3.7
Co. Monaghan	29	36 852	2552	6.9

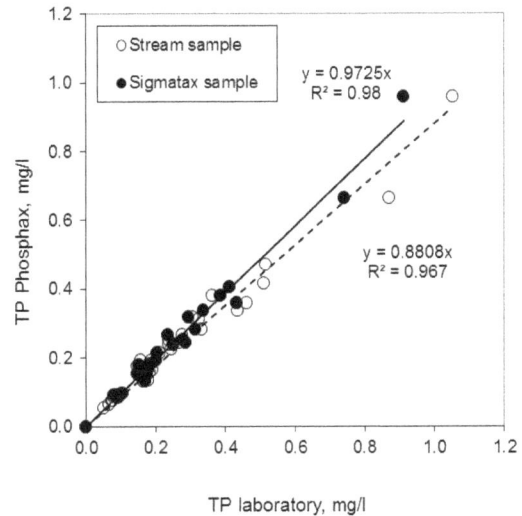

Figure 3. Quality control check on the Phosphax colorimetry process (closed symbols) and the Sigmatax sampling followed by the Phosphax colorimetry process.

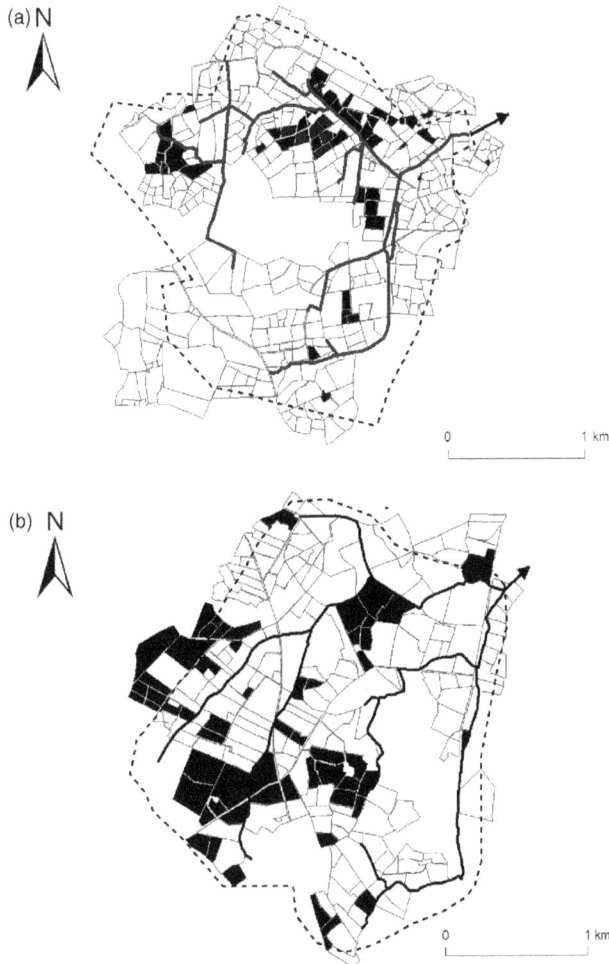

Figure 2. Location of index 4 fields in the Co. Tyrone (**a**) and Co. Monaghan (**b**) sub-catchments (dashed lines) showing their proximity to the stream network (black lines).

ments compared with samples taken adjacent to the Sigmatax intake show a small deviation ($b = 88$; $R^2 = 0.97$) from unity. This deviation may be due to either each method not directly sampling the same parcel of moving water and/or due to loss of material in the transfer of water from the Sigmatax intake to the Sigmatax jar – a horizontal length of 10 m and a vertical height of 2 m – as noted in other automatic sampling methods (Clarke et al., 2009).

3.4 High-resolution data analysis

There were some periods of data loss during the course of this study due to extreme weather conditions and instrument failure as a result of interruptions to power supply or breakdown (Table 3). However, up to 99 % of TP records were recorded. Nevertheless, in Tyrone the final year's data ends on 21 January 2011 due to extreme freezing conditions and so this full year is incomplete for this catchment.

Despite this, the nature of the time series data of this study is of a resolution that all states of hydrological discharge, including storm flow, had been captured by the data set. This means that, within every discharge range, rising, falling and peak discharge states are integrated, with each discharge range containing several hundred data points.

The time-series discharge, TP concentration and cumulative load for both catchments (Figs. 4 and 5) clearly show the lower base flows in Co. Monaghan sub-catchment compared with Co. Tyrone and also the larger increases in cumulative TP load (in Co. Monaghan) with seasonal storms, indicative of a flashier catchment.

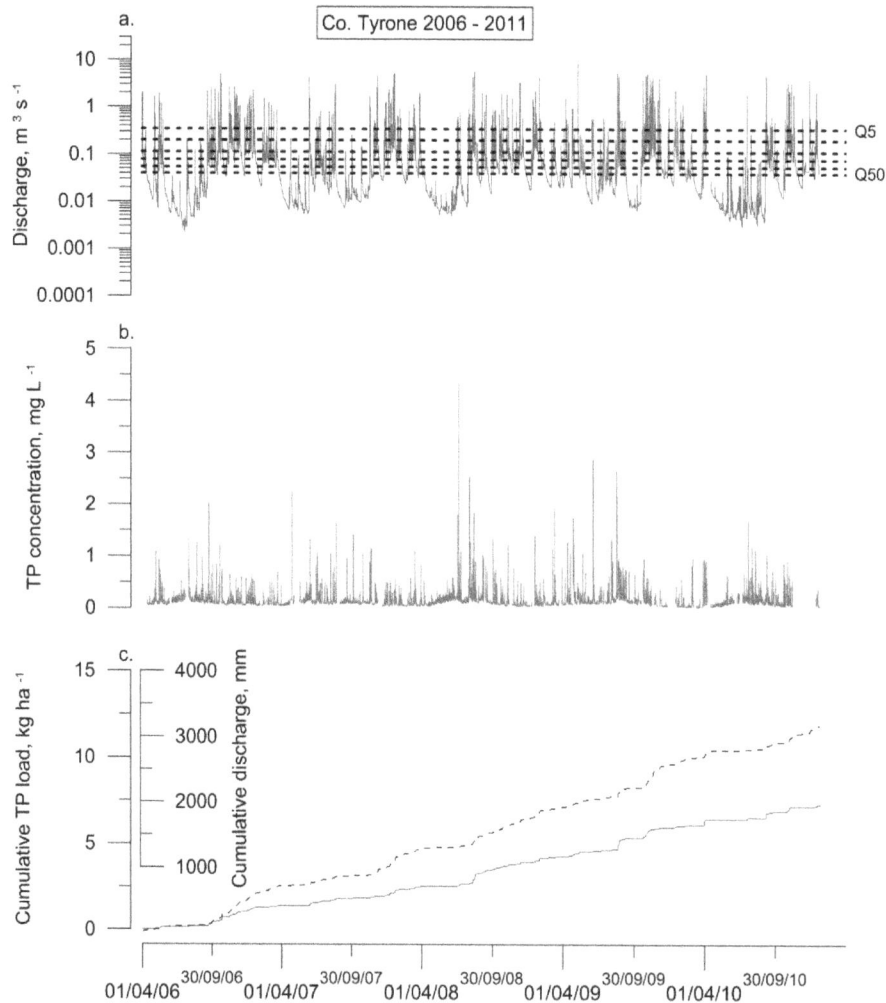

Figure 4. Time-series charts for discharge (**a**), TP concentration (**b**) and cumulative TP load and cumulative discharge (solid line) (**c**) from 2006–2011 in the Co. Tyrone sub-catchment. Discharge ranges (Q_{50}, Q_{40}, Q_{30}, Q_{20}, Q_{10} and Q_5) are shown on the discharge time series as dashed lines.

Table 3. Completeness of discharge and TP data for both catchments from 2006–2011.

Year	Tyrone % of data recorded for year		Monaghan % of data recorded for year	
	Discharge	TP	Discharge	TP
2006–2007	100	87	100	94
2007–2008	100	86	100	99
2008–2009	100	91	91	91
2009–2010	97	80	100	90
2010–2011	78	54	100	77

Flow duration curves (Fig. 6) show that the Co. Tyrone sub-catchments had consistently higher discharges than Co. Monaghan up to approximately Q_{20} ($0.113\,\mathrm{m^3\,s^{-1}}$). Although the total annual discharges from the catchments were not significantly different ($P < 0.05$) the P loads in the Co. Monaghan sub-catchment were more than double those

from Co. Tyrone highlighting TP concentrations differences in these similar discharges (Table 4).

For both catchments the highest TP load was in 2009/2010; in the Co. Tyrone sub-catchment this was also the year with the highest discharge but in Co. Monaghan the highest discharge was in 2006/2007, the year with the highest recorded rainfall (1299 mm).

Total P concentration, load and discharge distributions for the Q_{40}–Q_{50}, Q_{20}–Q_{30} and Q_5–Q_{10} discharge ranges over the 5 years showed that both catchments had a significant increase in TP concentration at the higher flows ($> Q_{10}$) between 2006 and 2011, although there were some small decreases in concentrations at Q_{40}–Q_{50} and Q_{20}–Q_{30} in both catchments in the interim years (Table 5a and b).

The high load from the catchments in 2009 does not appear to be linked to the frequency of high flow events, as events of Q_{20} and above are no more frequent in this year than any other from 2006–2011. Additionally, the rainfall in 2009

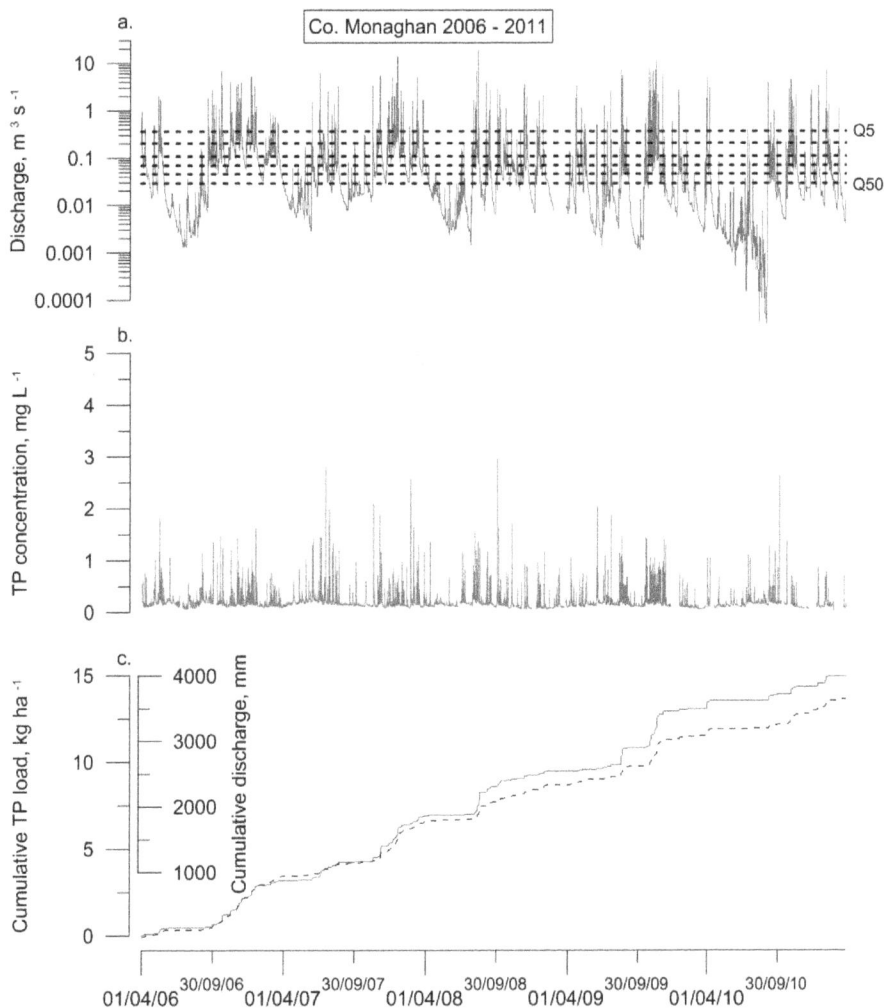

Figure 5. Time-series charts for discharge (**a**), TP concentration (**b**) and cumulative TP load and cumulative discharge (solid line) (**c**) from 2006–2011 in the Co. Monaghan sub-catchment Discharge ranges (Q_{50}, Q_{40}, Q_{30}, Q_{20}, Q_{10} and Q_5 are shown on the discharge time series as dashed lines.

Table 4. Annual discharge and TP load from catchments 2006–2011.

Year	Tyrone		Monaghan	
	Discharge, mm	TP load, kg P ha^{-1}	Discharge, mm	TP load, kg P ha^{-1}
2006–2007	711	1.35	944	3.18
2007–2008	574	1.17	828	3.74
2008–2009	637	1.73	564	2.55
2009–2010	809	2.12	814	3.98
2010–2011	439	0.89	494	1.50
Total	3170	7.26	3644	14.94

was only the second highest recorded throughout the study (1123 mm) but when the timing of the rainfall for this year was analysed it was noted that the summer of 2009 (323 mm) was considerably wetter than 2006 (209 mm). This persistent rainfall would have constantly wetted the ground and may have resulted in continued saturation of fields particularly at times of year when fields are actively managed.

Results for the low flow analysis were confounded as, although there had been a strategic replacement of four and eleven, older and potentially polluting septic tank systems in Co. Tyrone and Co. Monaghan, respectively, there had also been an increase in household density in both catchments since 2005. Co. Tyrone households had increased from 3.4 to 4.2 km^{-2} in 2010 and Co. Monaghan from 13.8 to 17.2 km^{-2} in 2010 (Arnscheidt et al., 2007; Macintosh et al., 2011).

The pair-wise comparison analysis of yearly concentrations of the low and extreme low discharges (Q_{80-90} and Q_{90-95}), in Monaghan showed there was a significant increase in TP concentration between 2006 and 2010 for Q_{90}–Q_{95}, from 0.120 to 0.148 mg P L^{-1} (Table 6). For Q_{80}–Q_{90} there was a significant decrease in TP concentration between 2006 and 2010, from 0.183 to 0.155 mg P L^{-1}. The same

Table 5. Mean TP concentrations at stated discharges (Q ranges based on Flow Duration Curve analysis) in Tyrone and Monaghan sub-catchments from 2006 to 2011 showing significance of change in concentration from the initial 2006 concentration (significant at ≤ 0.05 value).

Year	Discharge range Tyrone					
	Q_{40-50}		Q_{20-30}		Q_{5-10}	
	Mean TP concentration, $\mathrm{mg\,TP\,L^{-1}}$	Sig. of change	Mean TP concentration, $\mathrm{mg\,TP\,L^{-1}}$	Sig. of change	Mean TP concentration, $\mathrm{mg\,TP\,L^{-1}}$	Sig. of change
(a)						
2006–07	0.094		0.118		0.152	
2007–2008	0.109	0.503	0.115	1.000	0.204	0.001
2008–2009	0.089	1.000	0.103	0.058	0.227	< 0.001
2009–2010	0.152	< 0.001	0.162	< 0.001	0.280	< 0.001
2010–2011	0.097	1.000	0.103	0.226	0.214	0.001
(b)						
2006–2007	0.190		0.189		0.228	
2007–2008	0.171	0.026	0.180	0.575	0.276	< 0.001
2008–2009	0.151	< 0.001	0.179	0.382	0.280	< 0.001
2009–2010	0.157	< 0.001	0.208	0.012	0.391	< 0.001
2010–2011	0.171	0.050	0.171	0.026	0.330	< 0.001

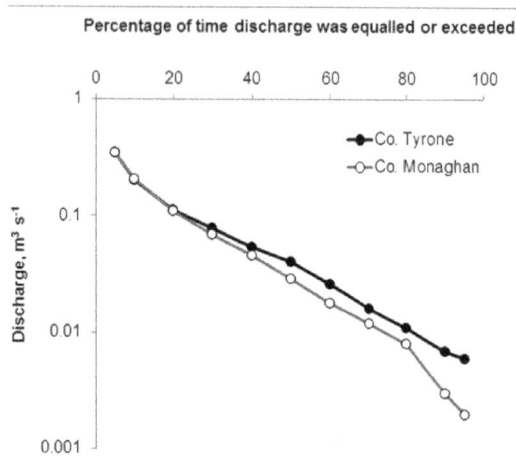

Figure 6. Flow duration curves for streams in Co. Tyrone (2006–2011) and Co. Monaghan (2006–2011).

Table 6. Pairwise comparisons between years for TP concentrations. The mean difference is significant at the 0.05 level (*), with TP as the dependent variable.

(I) Year	(J) Year	Mean 95 % confidence difference (I–J)	(I) Year	(J) Year	Mean 95 % confidence difference (I–J)
Mon. 2006 Q_{90-95}	2007	0.099*	Tyrone 2006 Q_{90-95}	2007	0.002
	2008	0.024*		2008	0.049*
	2009	0.016*		2009	0.017*
	2010	0.028*		2010	0.000
Mon. 2006 Q_{80-90}	2007	0.002	Tyrone 2006 Q_{80-90}	2007	−0.018*
	2008	−0.037*		2008	0.027*
	2009	−0.058*		2009	0.014*
	2010	−0.028*		2010	0.010*

analysis in Tyrone showed there was an initial significant increase between 2006 and 2008 for Q_{90}–Q_{95} but no overall change between 2006 and 2010. For Q_{80}–Q_{90} there was a small, significant increase from 2006 to 2010 from 0.122 to $0.133\,\mathrm{mg\,P\,L^{-1}}$. The changes are small and there is perhaps a risk of including instrument and sample "noise" in the comparisons; however, the comparisons are also made on filtered data based on an average of three data-points per hour which should give confidence in the statistical tests used.

4 Discussion

Soil nutrient status data is an important inventory of management within catchments (Maguire and Sims, 2002) and was one of the metrics used to assess the effectiveness of the best management practice measures for diffuse P mitigation throughout this study. The results from the Co. Tyrone catchment indicate that significant soil P source status reductions in the index 4 high risk category were achievable within the 5-year time frame of this study and are indicative of a convergence to optimum soil P status (index 2: 16–25 $\mathrm{mg\,P\,L^{-1}}$) in this catchment. In Co. Monaghan the picture is less clear with an overall increase in Olsen soil P across all indices, including the high risk category. Reasons for this increase in Co. Monaghan are unclear, and, in similar catchments, further assessment may be needed on issues such as resistance to advice (Falconer, 2000), insufficient spreadlands on smaller farm holdings (Carton and Magette, 1999; Buckley and Fealy, 2012), intensification and/or land use change (Edwards and Withers, 1998; Jarvie et al., 2010).

Within both sub-catchments there appeared to be a fractured or inconsistent relationship between any changes in soil P nutrient status and those in run-off concentration at this time scale. Previous studies have sought to determine the relationships between input and impact; the Phosphorus Transfer Continuum concept proposed by Haygarth et al. (2005) suggests that the amount of P lost from soils is actually less than 5 % of that applied annually and that high soil test P concentrations alone do not necessarily predict potential for loss of P. While it is possible to improve chemical water quality almost immediately at low flows following removal of a point source input in river catchments (Jarvie et al., 2006; Withers et al., 2014), due to historic reserves of P in sediments and soil, the time it will take for improvements in water quality from diffuse sources to occur may be difficult to estimate (Meals et al., 2010; Schulte et al., 2010). This fractured response is shown by assessment of the influence of changes in the sub-catchments on diffuse nutrient transfer which on the one hand did not reflect the reductions made in soil P status of the Co. Tyrone sub-catchment but, on the other hand, were more indicative of the soil P increases seen in Co. Monaghan. It may, however, be that any reductions in P transfer from diffuse sources have been offset by P losses as the result of, for example, bank erosion. A study by Kronvang et al. (2005) showed that this process could contribute 17–25 % of the P load from a catchment and so any reductions as a result of decreases in soil test P may not be seen.

Improvement in water quality by reducing point source inputs, although straightforward in principle, was found in this case to be more complicated in practice. Despite the mitigation measures introduced to the catchments, the Q_{80}–Q_{90} and Q_{90}–Q_{95} results in Counties Tyrone and Monaghan suggest that few improvements to point source inputs to the streams have taken place. Whether this is due to the improvements not including the STSs which were actually those impacting on the streams, or additional systems offsetting any improvements which would otherwise have taken place, is hard to determine.

Arnscheidt et al. (2007) stated that there was a significant correlation in the two catchments between STS condition and density, and low flow TP concentration. The increase in STS density in Tyrone from 3.4 to 4.6 km^{-2} may, therefore, have had some impact on the P concentrations in the catchment during the post mitigation period. Despite the Monaghan tank density also increasing, from 13.8 to 17.2 km^{-2}, there was no significant increase in TP concentration.

Changes in the overall annual sub-catchment TP loads as a result of any improvements to STSs would not be expected to be significant in this study as the changes which could potentially be achieved would be too small.

Although it was proposed that annual TP load would not be the optimum metric for assessing change in catchment water quality, it is, however, a useful measure when comparing catchment with catchment. For example, TP loads from the grassland catchments of this study were 0.89 to 2.12 kg P ha^{-1} and 1.50 to 3.98 kg P ha^{-1}, in Co. Tyrone and Co. Monaghan, respectively. These are high in relation to two arable catchments in Ireland reported by Melland et al. (2012) which had TP loads ranging from 0.12 to 0.83 kg P ha^{-1} but are similar to TP loads observed in Danish arable catchments by Kronvang et al. (2005) which ranged from 0.5 to 5.8 kg P ha^{-1}. Both sub-catchments of this study are the same area (5 km^2) yet it is clear from the results that the loads from Co. Monaghan are consistently almost double those from Co. Tyrone – even accounting for missing data in the last year of the study.

High soil test P concentration in near stream fields is known to impact on P transfers to the stream during rainfall events which cause overland and sub-surface flow (Gburek, 2000). GIS analysis showed that the Co. Monaghan sub-catchment had a larger percentage of high index fields (≥ 4) and also a greater length stream network adjacent to these fields (Fig. 2a and b). These fields, which can be classified as CSAs using the broadest definition (high soil P and low flow-path length) (McDowell and Sharpley, 2003), may have been the source of the higher load delivery in the Co. Monaghan catchment. The Co. Monaghan catchment was also determined to be flashier than Co. Tyrone, having lower base flows and periodically higher storm flows. The longer stream length in the Co. Monaghan sub-catchment would also increase risk of P transfer from bank erosion (Kronvang et al., 2005).

Although CSAs are now being recognised as major sources of nutrient and other pollutant transfers from land to water they are not specifically targeted within any of the directives pertaining to the Water Framework Directive or daughter directives at least at farm and field scale (Wall et al., 2012). This omission is also critically reviewed in more recent studies (Doody et al., 2012; Thompson et al., 2012).

The increased source and transport pressures from the Co. Monaghan sub-catchment can, therefore, be determined as a main factor in the increased load from this catchment. In order to reduce these pressures, and therefore the potential for loss, a policy of reducing inputs should be continued. Indeed, the P regulations in both Irish jurisdictions do promote, inter alia, convergence to optimum soil P ranges. In support of the policy, a long term catchment study in Norway showed that a reduction in P inputs as a result of change in soil management could produce a decrease in run-off P concentration with time; although there could be a considerable time lag between implementation and improvements from P saturated soils (Bechmann et al., 2008) and this is noted by Schulte et al. (2010) for Irish soils. However, even with a reduction in source pressure, the flashier hydrology of the Co. Monaghan catchment will likely continue to drive nutrient transfers more so than Co. Tyrone. This was also noted in similar catchments by Jordan et al. (2012), and will most likely mean that lag effects will be longer.

The two wettest years were 2006 (1299 mm) and 2009 (1123 mm) with the latter associated with the highest TP loads but with the summer of 2009 being more than 50 % wetter than 2006. When the timing of the rainfall events in both years, 2006 with the most storms and 2009 with the highest loads, was analysed it was determined that the majority of major storm events ($> Q_5$) in 2006 occurred between 30 October and the 30 January, with no storm discharges at all between May and September of that year; only 4 h of Q_5 discharges were recorded in the Co. Monaghan sub-catchment outside of this period for this year. In 2009, however, there were frequent Q_5 flows throughout the summer with the maximum flow for Tyrone for that year recorded on 20 August (5.2 m^3 s^{-1}). The small percentage of Q_{95} flows for this year, especially in the Co. Tyrone sub-catchment, indicated that run-off from the land was sufficient to maintain the stream above extreme (5-year) baseflow for over 99 % of the year. This is indicative of there being sufficient precipitation to maintain fields in a constantly wetted state throughout the year.

The Co. Tyrone and Co. Monaghan sub-catchment response in 2009 to this persistent rainfall throughout the entire year resulted in the highest TP load and also, importantly, the highest TP concentration for the Q_{5-10} discharges over the 5 years. From this the data indicate that antecedent weather conditions can add to the potency of any rainfall event's impact on catchment water quality at subsequently higher discharges (as proposed by Kirchner et al., 2004; Neal et al., 2012). This finding adds to previous studies (Turner and Haygarth, 2001; Styles and Coxon, 2006) where, at laboratory scale, dry soils when wetted were found to release more TP than persistently wet soils. However, this study showed that at larger scales, and in these particular catchments, already wet soil has greater potency for TP transfer during storm events. This is due to sustained wetting (prolonged events maintaining saturation excess flows and an increase in effective rainfall) and, in particular, during periods which may coincide with nutrient management and incidental losses during the summer.

5 Conclusions

This study used high-resolution TP monitoring in two catchment streams to assess changes in two principle source pressures over a 5-year period (2006–2011); diffuse source pressure changes allied to high soil P and rural point source pressures relating to STS changes. The findings were that:

- There were inconsistent changes in high soil P status between catchments with one showing an increase in high

soil P fields and the other a convergence towards optimum, despite similar soil nutrient management advice.

- Despite changes in STS pressure, companion work indicated an overall increase in STS pressure due to new-builds during the study period.

- There were inconsistent, or fractured, responses between changes in soil P pressure, point source pressure and P delivered during both high and low flow periods as analysed in the high-resolution TP concentration data set. None of the analysis for TP delivered indicated that decreased high soil P pressures (in one catchment) had made an impact on water quality; more so, the apparent increase in soil P pressure in the other catchment, and offset or increased STS pressures in both catchments, appeared to be more reflected in the high-resolution TP data set.

- The diffuse source pressures in one catchment were found to be higher than the other catchment due to a higher magnitude of high soil P fields adjacent to water courses coupled with a flashier run-off regime. This lead to consistently double the annual TP load, despite similar sized catchments, soil type and land use.

- The potency of TP loss in any similar discharge range between years was found to be reflective of antecedent conditions with already wet soils delivering more P per unit discharge than similar discharges following drier conditions. At least some of this loss may have been due to incidental P pressures during the summer but highlights wet summers as being particularly vulnerable to P loss and perhaps offsetting the response of other catchment mitigation measures.

This study shows the utility of high-resolution water quality analysis in the short to medium term for studying changes in both catchment pressures and response. It also firmly suggests that the short to medium term is an inappropriate time scale to gauge both the capacity of policy to influence a catchment change and also for that change to be detected in metrics of water quality.

Acknowledgements. We acknowledge the contribution of Rachel Cassidy (formerly University of Ulster) Queen's University, Belfast, for undertaking Monte-Carlo analysis for soil re-sampling. We also acknowledge Dr Suzanne Higgins (formerly University of Ulster), Agri-Food Biosciences Institute (AFBI), Belfast, for access to the 2005 soil P data set. We thank Hugo McGrogan and Pete Devlin for technical help with the water and soil sampling work and catchment farmers for access to land. This research was funded by the Special European Programmes Board under INTERREG IIIA (020204 – Blackwater TRACE), Northern Ireland Environment Agency, Lisburn, Irish Environmental Protection Agency, Dublin, and a Department of Employment and Learning (Northern Ireland) PhD scholarship.

Edited by: B. Kronvang

References

Agbenin, J. O. and Tiessen, H.: Phosphorus sorption at field capacity and soil ionic strength: Kenetics and transformation, Soil Sci. Soc. Am. J., 59, 998–1005, 1995.

Arnscheidt, J., Jordan, P., Li, S., McCormick, S., McFaul, R., McGrogan, H. J., Neal, M., and Sims, J. T.: Defining the sources of low-flow phosphorus transfers in complex catchments, Sci. Total Environ., 382, 1–13, 2007.

Bechmann, M., Deelstra, J., Stalnacke, P., Eggestad, H. O., Øygarden, L., and Pengerud, A.: Monitoring catchment scale agricultural pollution in Norway: policy instruments, implementation of mitigation methods and trends in nutrient and sediment losses, Environ Sci. Policy, 11, 102–114, 2008.

Bowes, M. J., Smith, J. T., Jarvie, H. P., and Neal, C.: Modelling of phosphorus inputs to rivers from diffuse and point sources, Sci. Total Environ., 395, 125–138, 2008.

Bowes, M. J., Smith, J. T., and Neal, C.: The value of high-resolution nutrient monitoring: A case study of the River Frome, Dorset, UK, J. Hydrol., 378, 82–96, 2009.

Buckley, C. and Fealy, R.: Intra-national importation of pig and poultry manure, Int. J. Agr. Manage., 1, 41–47, 2012.

Carton, O. T. and Magette, W. L.: The implications of Irish legislationand regulations for the land spreading of manures from intensive agricultural enterprises, Teagasc, Dublin, Ireland, 1999.

Cassidy, R. and Jordan, P.: Limitations of instantaneous water quality sampling in surface-water catchments: Comparison with near-continuous phosphorus time-series data, J Hydrol., 405, 182–193, 2011.

Cherry, K. A., Shepard, M., Withers, P. J. A., and Mooney, S. J.: Assessing the effectiveness of actions to mitigate nutrient loss from agriculture: A review of methods, Sci. Total Environ., 406, 1–23, 2008.

Clarke, S. E., Siu, C. Y. S., Pitt, R., Roenning, C. D., and Treese, D. P.: Peristaltic pump autosamplers for solids measurement in stormwater runoff, Water Environ Res., 81, 192–200, 2009.

Criuckshank, J. G.: Soil and Environment: Northern Ireland, DANI/QUB., Belfast, 1997.

Donn, M. J., Barron, O. V., and Barr, A. D.: Identification of phosphorus export from low-runoff yielding areas using combined application of high frequency water quality data and MODHMS modelling, Sci. Total Environ., 426, 264–272, 2012.

Doody, D. G., Archbold, M., Foy, R. H., and Flynn, R.: Approaches to the implementation of the Water Framework Directive: Targeting mitigation measures at critical source areas of diffuse phosphorus in Irish catchments, J. Environ. Manage., 93, 225–234, 2012.

Douglas, R. W., Menary, W., and Jordan, P.: P and sediment transfers in a grassland river catchment, Nutr. Cy. Agroecosyst., 77, 199–212, 2007.

Edwards, A. C. and Withers, P. J. A.: Soil phosphorus management and water quality: a UK perspective, Soil Use Manage., 14, 124–130, 1998.

Edwards, A. C. and Withers, P. J. A.: Linking phosphorus sources to impacts in different types of waterbody, Soil Use Manage., 23, 133–143, 2007.

Eisenreich, S. J., Bannerman, R. T., and Armstrong, D. E.: A simplified P analysis technique, Environ. Lett., 9, 43–53, 1975

Falconer, K.: Farm-level constraints on agri-environmental scheme participation: a transactional perspective, J. Rural Stud., 16, 379–394, 2000.

Gburek, W. J., Sharpley, A. N., Heathwaite, L., and Folmar, G. J.: Phosphorus management at the watershed scale: A modification of the phosphorus index, J. Environ. Qual., 29, 130–144, 2000.

Greene, S., Taylor, D., McElarney, Y. R., Foy, R. H., and Jordan, P.: An evaluation of catchment-scale phosphorus mitigation using load apportionment modelling, Sci. Total Environ., 409, 2211–2222, 2011.

Griffiths, D.: Effects of climatic change and eutrophication on the glacial relict, Mysis relicta, in Lough Neagh, Freshwater Biol., 52, 1957–1967, 2007.

Harris, G. P. and Heathwaite, A. L.: Why is achieving good ecological outcomes in rivers so difficult?, Freshwater Biol., 57, 91–107, 2012.

Haygarth, P. M., Condron, L. M., Heathwaite, A. L., Turner, B. L., and Harris, G .P.: The phosphorus transfer continuum: Linking source to impact with an interdisciplinary and multi-scaled approach, Sci. Total Environ., 344, 5–14, 2005.

Hirsch, R. M.: Flux of nitrogen, phosphorus, and suspended sediment from the Susquehanna River Basin to the Chesapeake Bay during Tropical Storm Lee, September 2011, as an indicator of the effects of reservoir sedimentation on water quality, USGS Scientific Investigations Report 5185, USGS, USA, p. 17, 2012.

Jarvie, H. P., Neal, C., and Withers, P. J. A.: Sewage-effluent phosphorus: A greater risk to river eutrophication than agricultural phosphorus?, Sci. Total Environ., 360, 246–253, 2006.

Jarvie, H. P., Withers, P. J. A., Bowes, M. J., Palmer-Felgate, E. J., Harper, D. M., Wasiak, K., Wasiak, P., Hodgkinson, R. A., Bates, A., Stoate, C., Neal, M., Wickham, H. D., Harman, S. A., and Armstrong, L. K.: Streamwater phosphorus and nitrogen across a gradient in rural–agricultural land use intensity, Agr. Ecosyst. Environ., 135, 238–252, 2010.

Jordan, P., Arnscheidt, J., McGrogan, H., and McCormick, S.: High-resolution phosphorus transfers at the catchment scale: the hidden importance of non-storm transfers, Hydrol. Earth Syst. Sci., 9, 685–691, doi:10.5194/hess-9-685-2005, 2005a.

Jordan, P., Menary, W., Daly, K., Kiely, G., Morgan, G., Byrne, P., and Moles, R.: Patterns and processes of phosphorus transfer from Irish grassland soils to rivers – integration of laboratory and catchment studies, J. Hydrol., 304, 20–34, 2005b.

Jordan, P., Arnscheidt, A., McGrogan, H., and McCormick, S.: Characterising phosphorus transfers in rural catchments using a continuous bank-side analyser, Hydrol. Earth Syst. Sci., 11, 372–381, doi:10.5194/hess-11-372-2007, 2007a.

Jordan, C., Higgins, A., and Wright, P.: Slurry acceptance mapping of Northern Ireland for run-off risk assessment, Soil Use Manage., 23, 245–253, 2007b.

Jordan, P., Melland, A. R., Mellander, P., Shortle, G., and Wall, D.: The seasonality of phosphorus transfers from land to water: Implications for trophic impacts and policy evaluation, Sci. Total Environ., 434, 101–109, 2012.

Kane, D.: Hydrograph seperation using end member mixing models in the Oona water river catchment, Co. Tyrone, PhD Edn., University of Ulster, Ulster, 2009.

Kirchner, J. W., Feng, X., Neal, C., and Robson, A. J.: The fine structure of water-quality dynamics: the (high-frequency) wave of the future, Hydrol. Process., 18, 1353–1359, 2004.

Knight, J., McCarron, S. G., and McCabe, A. M.: Landform modifications by paleao ice streams in east central Ireand, Ann. Glaciol., 28, 161–167, 1999.

Kronvang, B., Bechmann, M., Lundekvam, H., Behrendt, H., Rubaek, G. H., Schoumans, O. F., Syversen, N., Andersen, H. E., and Hoffmann, C. C.: Phosphorus Losses from Agricultural Areas in River Basins: Effects and Uncertainties of Targeted Mitigation Measures, J. Environ. Qual., 34, 2129–2144, 2005.

Kronvang, B., Audet, J., Baattrup-Pedersen, A., Jensen, H., and Larsen, S.: Phosphorus Load to Surface Water from bank erosion in a Danish Lowland River Basin, J. Environ. Qual., 41, 304–313, 2012.

Lloyd, C. E. M., Freer, J. E., Collins, A. L., Johnes, P. J., and Jones, J. I.: Methods for detecting change in hydrochemical time series in response to targeted pollutant mitigation in river catchments, J. Hydrol., 514, 297–312, 2014.

Macintosh, K. A., Jordan, P., Cassidy, R., Arnscheidt, J., and Ward, C.: Low flow water quality in rivers; septic tank systems and high-resolution phosphorus signals, Sci. Total Environ., 412–413, 58–65, 2011.

MAFF: Fertiliser Recommendations for Agricultural and Horticultural Crops (RB209), RB209 Edn., The Stationary Office, London, UK, 2000.

Maguire, R. O. and Sims, J. T.: Measuring agronomic and environmental soil phosphorus saturation and predicting phosphorus leaching with Mehlich 3, Soil Sci. Am. J., 66, 2033–2039, 2002.

Meals, D. W., Dressing, S. A., and Davenport, T. E.: Lag time in water quality response to best management practices: A review, J. Environ Qual., 39, 85–96, 2010.

McDowell, R. W. and Sharpley, A. N.: Phosphorus solubility and release kinetics as a function of soil test P concentration, Geoderma, 112, 143–154, 2003.

Melland, A. R., Mellander, P., Murphy, P. N. C., Wall, D. P., Mechan, S., Shine, O., Shortle, G., and Jordan, P.: Stream water quality in intensive cereal cropping catchments with regulated nutrient management, Environ. Sci. Policy, 24, 58–70, 2012.

Morgan, M. F.: Chemical Soil Diagnosis by the Universal Soil Testing System, Connecticut Agricultural Experimental Station Bulletin, 450, New Haven, CT, 1941.

Neal, C., Reynolds, B., Rowland, P., Norris, D., Kirchener, J. W., Neal, M., Sleep, D., Lawlor, A., Woods, C., Thacker, S., Guyatt, H., Vincent, C., Hockenhull, K., Wickham, H., Harman, S., and Armstrong, L.: High-frequency water quality time series in precipitation and streamflow: From fragmentary signals to scientific challenge, Sci. Total Environ., 434, 3–12, 2012.

O'Dochairtaigh, B. E.: The Oona water catchment: Trial Groundwater Body Characterisation for Water Framework Directive Implementation, BGS, 45 pp., 2003.

OJEC – Official Journal of the European Communities: Establishing a Framework for Community Action in the Field of Water Policy, Water Framework Directive, 2000/60/EC, Brussels, 2000.

Olsen, S. R., Cole, C., Watanabe, F. S., and Dean, L.: Estimation of available phosphorus in soils by extraction with sodium bicarbonate, US Department of Agriculture, Washington, D.C., 1954.

Palmer-Felgate, E. J., Jarvie, H. P., Williams, R. J., Mortimer, R. J., Loewenthal, M., and Neal, C.: Phosphorus dynamics and productivity in a sewage-impacted lowland chalk stream, J. Hydrol., 351, 87–97, 2008.

Schulte, R. P. O., Melland, A. R., Fenton, O., Herlihy, M., Richards, K., and Jordan, P.: Modelling soil phosphorus decline: Expectations of Water Framework Directive policies., Environ. Sci. Policy, 13, 472–484, 2010.

Sharpley, A. N. and Reikolainen, S.: Phosphorus in agriculture, in: Phosphorus loss from soil to water, 1st Edn., edited by: Tunney, H., CAB International, New York, 1997.

SRNI 488 – Statutory Rules Northern Ireland 488, Phosphorus (Use in Agriculture) Regulations (Northern Ireland): Department of the Environment Northern Ireland, Belfast, p. 8, 2006.

Styles, D. and Coxon, C.: Laboratory drying of organic-matter rich soils: Phosphorus solubility effects, influence of soil characteristics, and consequences for environmental interpretation, Geoderma, 136, 120–135, 2006.

Thompson, J. J. D., Doody, D. G., Flynn, R., and Watson, C. J.: Dynamics of critical source areas: Does connectivity explain chemistry?, Sci. Total Environ., 435–436, 499–508, 2012.

Turcotte, D. L.: Fractals and chaos in geology and geophysics, 2nd Edn., Cambridge University Press, Cambridge, 1997.

Turner, B. L. and Haygarth, P. M.: Phosphorus mobilisation in rewetted soils, Nature, 411, 258, 2001.

Wade, A. J., Palmer-Felgate, E. J., Halliday, S. J., Skeffington, R. A., Loewenthal, M., Jarvie, H. P., Bowes, M. J., Greenway, G. M., Haswell, S. J., Bell, I. M., Joly, E., Fallatah, A., Neal, C., Williams, R. J., Gozzard, E., and Newman, J. R.: Hydrochemical processes in lowland rivers: insights from in situ, high-resolution monitoring, Hydrol. Earth Syst. Sci., 16, 4323–4342, doi:10.5194/hess-16-4323-2012, 2012.

Wall, D., Jordan, P., Melland, A. R., Mellander, P., Buckley, C., Reaney, S. M., and Shortle, G.: Using the nutrient transfer continuum concept to evaluate the European Union Nitrates Directive National Action Programme, Environ. Sci. Policy, 14, 664–674, 2011.

Wall, D. P., Murphy, P. N. C., Melland, A. R., Mechan, S., Shine, O., Buckley, C., Mellander, P.-E., Shortle, G., and Jordan, P.: Evaluating nutrient source regulations at different scales in five agricultural catchments, Environ. Sci. Policy, 24, 34–43, 2012.

Watson, C. J., Smith, R. V., and Matthews, D. I.: Increase in phosphorus losses from grassland in response to olsen-P accumulation, J. Environ Qual., 36, 1452–1460, 2007.

Withers, P. J. A., May, L., Jarvie, H. P., Jordan, P., Doody, D., Foy, R. H., Bechmann, M., Cooksley, S., Dils, R., and Deal, N.: Nutrient emissions to water from septic tank systems in rural catchments: Uncertainties and implications for policy, Environ. Sci. Policy, 24, 71–82, 2012.

Withers, P. J. A., Jordan, P., May, L., Jarvie, H. P., and Deal, N. E.: Do septic tank systems pose a hidden threat to water quality?, Front. Ecol. Environ., 12, 123–130, 2014.

4

Monitoring hillslope moisture dynamics with surface ERT for enhancing spatial significance of hydrometric point measurements

R. Hübner[1], K. Heller[1], T. Günther[2], and A. Kleber[1]

[1]Institute of Geography, Dresden University of Technology, Helmholtzstr. 10, 01069 Dresden, Germany
[2]Leibniz Institute for Applied Geophysics (LIAG), Stilleweg 2, 30655 Hanover, Germany

Correspondence to: R. Hübner (rico.huebner@tu-dresden.de)

Abstract. Besides floodplains, hillslopes are basic units that mainly control water movement and flow pathways within catchments of subdued mountain ranges. The structure of their shallow subsurface affects water balance, e.g. infiltration, retention, and runoff. Nevertheless, there is still a gap in the knowledge of the hydrological dynamics on hillslopes, notably due to the lack of generalization and transferability. This study presents a robust multi-method framework of electrical resistivity tomography (ERT) in addition to hydrometric point measurements, transferring hydrometric data into higher spatial scales to obtain additional patterns of distribution and dynamics of soil moisture on a hillslope. A geoelectrical monitoring in a small catchment in the eastern Ore Mountains was carried out at weekly intervals from May to December 2008 to image seasonal moisture dynamics on the hillslope scale. To link water content and electrical resistivity, the parameters of Archie's law were determined using different core samples. To optimize inversion parameters and methods, the derived spatial and temporal water content distribution was compared to tensiometer data. The results from ERT measurements show a strong correlation with the hydrometric data. The response is congruent to the soil tension data. Water content calculated from the ERT profile shows similar variations as that of water content from soil moisture sensors. Consequently, soil moisture dynamics on the hillslope scale may be determined not only by expensive invasive punctual hydrometric measurements, but also by minimally invasive time-lapse ERT, provided that pedo-/petrophysical relationships are known. Since ERT integrates larger spatial scales, a combination with hydrometric point measurements improves the understanding of the ongoing hydrological processes and better suits identification of heterogeneities.

1 Introduction

The knowledge of system-internal water flow pathways and the response to precipitation on different spatial and temporal scales is essential for the prediction of hydrological and hydrochemical dynamics within catchments (Uhlenbrook et al., 2008; Wenninger et al., 2004). Understanding the processes involved is of particular importance for improving precipitation-runoff and pollutant-transport models (Di Baldassarre and Uhlenbrook, 2012).

Hillslopes are important links between the atmosphere and the water input into catchments. They mainly control different runoff components and residence times (Uhlenbrook et al., 2008). Several studies have addressed hillslope hydrology (Anderson and Burt, 1990; Kirkby, 1980; Kleber and Schellenberger, 1998; McDonnell et al., 2001; Tromp-van Meerveld, 2004; Uchida et al., 2006). A major problem is that the spatial and temporal variability of the hydrological response due to different natural settings – e.g. geomorphological, pedological, lithological characteristics and the spatial heterogeneity – make it difficult to generalize and to transfer results to ungauged basins (McDonnell et al., 2007).

In catchments of Central European subdued mountain ranges, the shallow subsurface of hillslopes is mostly covered by Pleistocene periglacial slope deposits (Kleber and Terhorst, 2013). These slope deposits have developed in different layers. In the literature normally three layers are classified (upper layer – LH, intermediate layer – LM, basal layer – LB: classification according to AD-hoc AG-Boden, 2005; Kleber and Terhorst, 2013). Sometimes locally a fourth layer (*Oberlage* AD-hoc AG-Boden, 2005) could be found. The occurrence of these layers can vary spatially and has different regional and local characteristics. Due to the sedimentolog-ical and substrate-specific properties, e.g. grain-size distri-bution, clast content and texture, they remarkably influence near-surface water balance (e.g. infiltration, percolation) and are of particular importance for near-surface runoff, e.g. in-terflow (Chifflard et al., 2008; Kleber, 2004; Kleber and Schellenberger, 1998; Sauer et al., 2001; Scholten, 1999; Völkel et al., 2002a, b; Heller, 2012; Moldenhauer et al., 2013).

Most of the prior studies were based on invasive and ex-tensive hydrometric point measurements or on tracer inves-tigations. Punctual hydrometric measurements may modify flow pathways and are not sufficient in the case of signif-icant spatial heterogeneities in the subsurface. Tracer ex-periments, e.g. using isotopes, integrate much larger scales up to entire catchments but provide less direct insights into ongoing processes. Internal hydrological processes may be complex and due to the spatio-temporal interlinking of near-surface processes and groundwater dynamics, there is still a lack of knowledge regarding runoff generation in watersheds (McDonnell, 2003; Tilch et al., 2006; Uhlenbrook, 2005). For an efficient and accurate modelling of the hydrological behaviour at the crucial hillslope scale, additional methods are needed especially to improve the understanding of these complex processes in order to enhance the model hypotheses. Hydrogeophysical methods are capable of closing the gap be-tween large-scale depth-limited remote-sensing methods and invasive punctual hydrometric arrays (Robinson et al., 2008a, b; Lesmes and Friedman, 2006; Uhlenbrook et al., 2008).

Many studies show the potential of electrical resistivity tomography (ERT) for hydrological investigation by means of synthetic case studies for aquifer transport characteriza-tion (Kemna et al., 2004; Vanderborght et al., 2005), imag-ing water flow on soil cores (Bechtold et al., 2012; Binley et al., 1996a, b; Garré et al., 2010, 2011; Koestel et al., 2008, 2009a, b), cross-borehole imaging of tracers (Daily et al., 1992; Oldenborger et al., 2007; Ramirez et al., 1993; Singha and Gorelick, 2005; Slater et al., 2000), or imaging of tracer injection or irrigation with surface ERT (Cassiani et al., 2006; De Morais et al., 2008; Descloitres et al., 2008a; Michot et al., 2003; Perri et al., 2012). However, some re-search has been conducted under natural conditions to char-acterize water content change, infiltration or discharge by use of cross-borehole ERT (French and Binley, 2004), sur-face ERT (Brunet et al., 2010; Benderitter and Schott, 1999; Descloitres et al., 2008b; Massuel et al., 2006; Miller et al., 2008) or a combined surface cross-borehole ERT array (Beff et al., 2013; Zhou et al., 2001).

Besides hydrogeophysical methods such as electromag-netics (EM) (Popp et al., 2013; Robinson et al., 2012; Tromp-van Meerveld and McDonnell, 2009), time-lapse ERT have been frequently applied to hillslope investigation in the runoff and interflow (Uhlenbrook et al., 2008; Cassiani et al., 2009) or preferential flow context (Leslie and Heinse, 2013).

However, the use of ERT for monitoring hydrological dy-namics on hillslopes with layered structure is still rare.

The objective of this paper is to show the potential of min-imally invasive surface time-lapse ERT as a robust method-ological framework for monitoring long-term changes in soil moisture and to improve the spatial resolution of punctual hydrometric measurements (e.g. tensiometer and soil mois-ture sensors) on a hillslope with periglacial cover beds. Fur-thermore, we want to show the ability of ERT for mapping spatially heterogeneous structures and water content distri-butions of the shallow subsurface. With a multi-method ap-proach, we attempt to demonstrate the possibility to ade-quately transfer hydrometric data to higher spatial scales and to obtain additional patterns of soil water dynamics on a hill-slope. These scales are fundamental for achieving a better un-derstanding of the influence of the layered subsurface on wa-ter fluxes (e.g. infiltration, percolation or interflow) and the response to different amounts of precipitation on hillslopes.

2   Material and methods

2.1   Study site

The study area covers 6 ha of a forested spring catchment in the Eastern Ore Mountains, eastern Germany, which is lo-cated in the Freiberger Mulde catchment (Fig. 1).

Annual precipitation averages 930 mm, mean annual temperature is 6.6 °C. The altitude ranges from 521 to 575 m a.s.l. with a predominant land cover of spruce for-est (*Picea abies*, approx. 30 years). The slope angle of the catchment ranges from 0.05 to 22.5° with an average of 7°. Bedrock is gneiss overlain by periglacial cover beds with up to three layers (LH, LM, LB, with no occurrence of the *Ober-lage*, see Heller, 2012). The upper layer (LH) with a thick-ness of 0.3–0.65 m consists of silty–loamy material with a bulk density of 1.2 g m^{-3} and many roots (see Table 1). In the central part of the catchment, a silty–loamy intermedi-ate layer (LM) follows with higher bulk density and a thick-ness of up to 0.55 m. The ubiquitous sandy–loamy basal layer (LB) is characterized by even higher bulk density and longi-tudinal axes of coarse clasts oriented parallel to the slope. Downslope it may reach a thickness of at least 3 m (see Fig. 1).

2.2   Laboratory work

Quality, amount and distribution of pore water exert a huge influence on resistivity[1] and form the link between electri-cal and hydrological properties. The empirical relationship of Archie's law (Archie, 1942) describes the connection be-tween electrical resistivity and saturation in porous media.

---

[1]In this context the term "resistivity" always refers to " specific electrical resistivity".

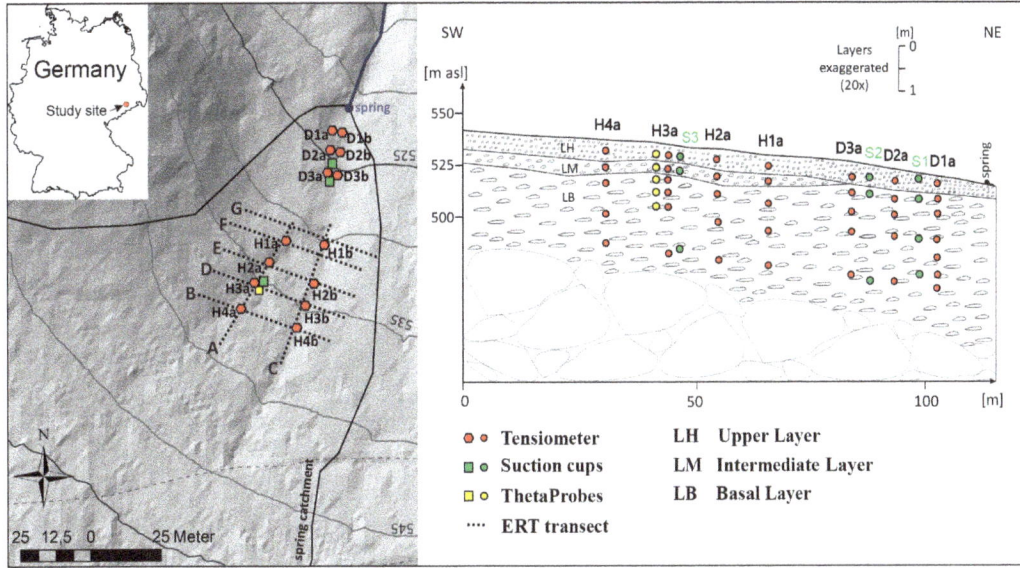

Figure 1. Study site with locations of ERT profiles and hydrometric stations (left panel; data source: ATKIS®-DGM2, Landesvermessungsamt Sachsen, 2008) and profile section with installation depths of tensiometers, soil moisture sensors and suction cups (right panel).

Table 1. Properties of cover beds from the study site ($n \geq 15$ per layer) – adapted from Moldenhauer et al. (2013).

| Layer | Soil horizon | Colour (moist) | Soil texture | | | | Bulk density ($\mathrm{g\,m^{-3}}$) | Porosity | Hydraulic conductivity ($\mathrm{cm\,d^{-1}}$)* |
			Clay (%)	Silt (%)	Sand (%)	Clasts (%)			
LH	A/Bw	10YR/5/8	14	52	34	36	1.2	0.55	27
LM	2Bg	10YR/5/4	12	53	35	43	1.5	0.43	9
LB	3CBg	10YR/5/3	7	22	71	56	1.7	0.36	52

* Field-saturated hydraulic conductivity measured using the compact constant head permeameter (CCHP) method (Amoozegar, 1989).

Instead of saturation we use the volumetric water content θ with

$$\rho_{\mathrm{eff}} = F_\theta \rho_{\mathrm{w}} \theta^{-n_\theta}, \tag{1}$$

where ρ_{eff} is the bulk resistivity of the soil probe and ρ_{w} is the resistivity of the pore fluid. The formation factor F_θ describes the increase of resistivity due to an insulating solid matrix and constitutes an intrinsic measure of material microgeometry (Schön, 2004; Lesmes and Friedman, 2006). The exponent n_θ is an empirical constant, which depends on the distribution of water within the pore space (Schön, 2004).

This model disregards the surface conductivity, which may occur due to interactions between pore water and soil matrix, especially with a high percentage of small grain sizes. In our study the curve fitting could be carried out very well without accounting for surface conductivity.

To investigate the pedo-/petrophysical relationship between resistivity and water content, 14 undisturbed soil core specimens (diameter = 36 mm, length = 40 mm) taken at different depths (0.3–1.4 m) were analysed. After dehydration in a drying chamber, the samples were saturated. The satura-

tion was done successively by stepwise injection in the middle of the soil core to achieve a better moisture distribution within the sample. Using a four-point array, electrical resistivity was measured for different saturation conditions during the saturation process. A calibrating solution with known resistivity was used to determine the geometric factor. Particle sizes were determined by sieving and the pipette method, using $Na_4P_2O_7$ as a dispersant (Klute, 1986, p. 393, 399–404, but with the sand–silt boundary at 0.063 mm).

Brunet et al. (2010) described remarkable conductivity increases of low mineralized water due to contact with the soil matrix. This may cause variation of resistivity with time. To minimize this effect we used spring water with high conductivity (approx. $\sigma_{\mathrm{w}25} = 150\,\mu\mathrm{S\,cm^{-1}} / \rho_{\mathrm{w}25} = 66\,\Omega\,\mathrm{m}$ for $T = 25\,°\mathrm{C}$). This corresponds to the mean conductivity of soil water in the study area, which is influenced by long-term contact with the subsoil.

Aside from the invariant parameters F_θ and n_θ, the resistivity of the pore water must be known to calculate the water content from resistivity values. Because it was not possible to extract pore water under dry conditions in summer, only

Table 2. Median pore water conductivity $\tilde{\sigma}_w$, resistivity $\tilde{\rho}_w$ and mean resistivity $\bar{\rho}_w$ with standard deviation (SD$_w$) per depth ($n \geq$ 11 per sampling depth).

Depth [m]	0.3	0.6	0.85	1.05	1.65	2.3
$\tilde{\sigma}_w$ [μS cm^{-1}]	72.4	107.8	111.6	114.7	135	156.7
$\tilde{\rho}_w$ [Ω m]	138.1	92.8	89.6	87.2	74.1	63.8
$\bar{\rho}_w$ [Ω m]	135.7	92.3	88.9	86.8	75.1	63.7
SD$_w$ [Ω m]	16.9	5.0	4.5	4.8	5.4	7.3

a few measurements of pore water conductivity could be carried out in late spring and early autumn. To calculate water content from resistivity obtained by field surveys, the median value over the entire time period of ρ_w for each depth was used (see Table 2). Interim values between the extraction depths were linearly interpolated.

After reforming Eq. (1) it is possible, with known parameters F_θ and n_θ and measured variables ρ_{eff} and ρ_w, to calculate volumetric water content:

$$\frac{\rho_{eff}}{F_\theta \rho_w}^{\frac{1}{-n_\theta}} = \theta. \tag{2}$$

As water saturation (S) is defined as the ratio between water content and porosity (Φ), it is also possible to calculate the degree of saturation using

$$\frac{\rho_{eff}}{F_\theta \rho_w}^{\frac{1}{-n_\theta}} \frac{1}{\Phi} = S. \tag{3}$$

The porosity (Φ) was calculated with

$$\Phi = 1 - \frac{\rho_{bulk}}{\rho_{particle}}. \tag{4}$$

The bulk density (ρ_{bulk}) was determined using undisturbed core samples. The particle density ($\rho_{particle}$) was measured with a capillary-stoppered pycnometer. The maximum sample depth for undisturbed soil cores was < 2 m. Below, the porosity had to be transferred according to grain size distribution, clasts and compaction from percussion drilling.

2.3 Field work

2.3.1 ERT mapping

In addition to conventional percussion drilling, at the end of October 2008 we measured seven ERT profiles to survey the subsurface resistivity distribution (A–G in Fig. 1). A and C are parallel to the slope inclination of approx. 9°, connecting inflection points of contour lines. B, D, E, F and G are perpendicular to these profiles ($\angle$ A102.5°, $\angle$ C90°). This arrangement allows identifying potential 3-D effects, which may cause inaccurate interpretation of the subsurface resistivity distribution. To improve the mapping results aided by hydrometric data, the profiles were located close to the tensiometer stations (distance < 2 m). For all resistivity measurements, the instrument "4 Point light hp" from "LGM –

Lippmann Geophysical Equipment" with 50 electrodes was used. Because of the expected interferences (e.g. by roots or clasts) and the multiple-layered stratification of periglacial cover beds, a Wenner array was found to be the most suitable configuration for the study area. This is characterized by low geometric factors (K), a high vertical resolution for laterally bedded subsurface structures, and a good signal-to-noise ratio (Dahlin and Zhou, 2004). To improve the spatial resolution, a Wenner-β array was measured additionally. With an electrode spacing of 1 m, this results in a combined data set with 784 data points for each pseudo-section with a maximum depth of investigation of 9.36 m (Wenner-β: depth of investigation for radial dipole in homogeneous ground 0.195L with L the maximum electrode separation in metres, according to Roy and Apparao, 1971; Apparao, 1991; Barker, 1989).

Horizontal resolution of a multi-electrode array is for shallow parts of the subsurface of the order of electrode distances. However, vertical resolution is far better, as the depth-of-investigation curves indicate (Roy and Apparao, 1971; Barker, 1989). This is further improved by measuring two electrode arrays (Wenner and Wenner-β) with different sensitivity curves so that we can expect a vertical resolution of the order of about 0.2 m in the case of excellent data quality.

2.3.2 Joint hydrometric and ERT monitoring

Since November 2007 soil water tension has been measured using 76 recording tensiometers (T8, UMS) arranged in 14 survey points along the slope at 5–7 different depths (see Fig. 1). Additionally, at the survey point H3a five soil moisture sensors (ThetaProbe, ML2x, Delta-T) were installed to measure volumetric water content. A V-notch weir with a pressure meter was used to quantify spring discharge. Rainfall was recorded by four precipitation gauges with tipping bucket (R. M. Young Co., 200 cm^2, resolution: 0.1 mm with max. 7 mm min^{-1}). For determination of pore water conductivity and resistivity, soil water was extracted with suction cups (VS-pro, UMS) at four depths at three locations (S1, S2, S3; Fig. 1) and cumulated as a weekly mixed sample.

Time lapse ERT measurements were performed with the same equipment, electrode array and spacing used for the mapping. The two time lapse profiles are congruent with profiles A and B (see Fig. 1). From May to December 2008, twenty seven time lapse measurements were carried out within almost weekly intervals. Contact resistance was checked before each measurement and was within the range of 0.2 kΩ to max. 1 kΩ over the whole measuring period. This range is very favourable and does not influence the measurements as numerical studies show (Rücker and Günther, 2011).

To compare time lapse measurements and to apply sophisticated inversion routines, the location of electrodes needs to remain constant. For current injection we used stainless steel electrodes (diameter 6 mm, length 150 mm), completely plunged into the ground, thus avoiding shifting of electrodes, except for natural soil creep. In the numerical computations, electrodes are considered points, which is not the case for the present ratio of length to distance. However, numerical computations with real electrode lengths show that the deviations are negligible, particularly if the points are placed at about half the electrode length (Rücker and Günther, 2011).

Subsoil temperature, especially in the upper layers, is characterized by distinct annual and daily variations. Therefore, the temperature dependence of resistivity must be considered when comparing different time steps. The installed tensiometers are able to measure soil temperature simultaneously. These data have been used to correct resistivity measurements to a standard temperature. Comparing several existing models for the correction of soil electrical conductivity measurements, Ma et al. (2011) conclude that the model (Eq. 5) proposed by Keller and Frischknecht (1966) is practicable within the temperature range of environmental monitoring:

$$\rho_{25} = \rho_t \left(1 + \delta \left(T - 25\,^{\circ}\mathrm{C}\right)\right). \tag{5}$$

With this equation the inverted resistivity (ρ_t) at the temperature (T) was corrected to a resistivity at a soil temperature of $25\,^{\circ}\mathrm{C}$ (ρ_{25}). The empirical parameter δ is the temperature slope compensation, with $\delta = 0.025\,^{\circ}\mathrm{C}^{-1}$ being commonly used for geophysical applications (Keller and Frischknecht, 1966; Hayashi, 2004; Ma et al., 2011).

2.3.3 ERT data inversion

For inversion of the ERT data, we used the BERT Code (Günther et al., 2006). In order to account for the present topography, we used an unstructured triangular discretization of the subsurface and applied finite element forward calculations. For static inversion, a smoothness-constraint objective function is minimized that consists of the error-weighted misfit between measured data $\mathbf{d}$ and model response $\mathbf{f(m)}$, and a model roughness:

$$\Phi = \|\mathbf{D}(\mathbf{d} - \mathbf{f(m)})\|_2^2 + \lambda \|\mathbf{Cm}\|_2^2 \to \min. \tag{6}$$

The regularization parameter λ defines the strength of regularization imposed by the smoothness matrix $\mathbf{C}$ and needs to be chosen such that the data are fitted within expected accuracy, which is incorporated in the data weighting matrix $\mathbf{D}$. In our case, values of $\lambda = 30$ provided sufficient data fit. See Günther et al. (2006) for details of the minimization procedure, and Beff et al. (2013) or Bechtold et al. (2012) for specific modifications in hydrological applications.

For time lapse inversion, i.e. calculating the temporal changes in resistivity, there are three different methodical approaches: (i) inverting the models for each point in time separately, (ii) using the initial model as reference model for the time step, (iii) or inverting the differences of the two data sets (Miller et al., 2008). With our data, each method generates insufficient results with unsubstantiated artifacts. An

increase on the surface was always followed by a decrease below and vice versa. These systematic changes cannot be explained or related to any natural process. Descloitres et al. (2003, 2008b) showed with synthetic data that time lapse inversion may produce artifacts due to the smoothness constraints especially with changes caused by shallow infiltration (decrease of resistivity), as mostly expected in our case.

As smoothness constraints are the main reason of these problems, we avoid the smoothness operator in the time lapse inversion and minimize a different objective function for the subsequent time steps:

$$\Phi = \|\mathbf{D}(\mathbf{d}^n - \mathbf{f(m}^n))\|_2^2 + \lambda \|\mathbf{m}^n - \mathbf{m}^{n-1}\|_2^2 \to \min. \tag{7}$$

Beginning from the static inversion, the subsequent models are found by reference model inversion. Only the total difference between the models of subsequent time steps $n-1$ and n is used for regularization (minimum-length constraints). A higher regularization parameter of $\lambda = 100$ proved optimal for time lapse inversion concerning both data fit and in comparison to the hydrometric results.

In order to find representative resistivity values as a function of depth, which are independent on small-scale heterogeneities, we subdivide the model down to a depth of 3 m into seven layers according to the boundaries of the described layering (see Table 1) and installation depth of hydrometric devices (see Fig. 1) (0–0.2, 0.2–0.4, 0.4–0.9, 0.9–1.2, 1.2–1.5, 1.5–2.0 and 2.0–3.0 m). The representative values are median resistivities in the layers from the stations H1a–H4a and H4b–H4a for profiles A and B, respectively.

3 Results

3.1 Laboratory

Within the separately analysed samples, non-linear curve fitting was carried out. Using the method of least squares, the data could be fitted using a power function in the form of Archie's law (Eq. 1, $0.973 < r < 0.999$).

The exponent n_θ shows a positive correlation to small grain sizes, primarily medium silt (6.3–$20\,\mu\mathrm{m}$, $r = 0.909$), but in the same case a negative correlation to grain sizes $> 630\,\mu\mathrm{m}$ including clast content ($r = -0.852$) (see Fig. 2).

The amount of silt as well as the clast content are important distinctive attributes to differentiate the basal layer from the overlying intermediate or upper layer (Table 1). Two different "electrical" layers may be identified. This is due to the fact that the exponent is strongly influenced by grain size, which shows a remarkable change at the upper boundary of LB. On the other hand, grain size distribution and clast content are very similar between LH and LM, so that these may not be differentiated using ERT. Figure 3 shows the aggregation of the 14 single samples into two regions with different depth ranges.

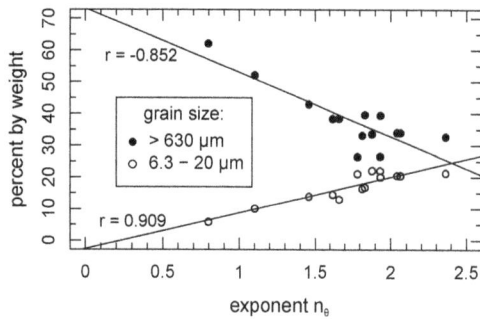

Figure 2. Exponent n_θ in dependence of different grain sizes.

Table 3. Fitted water content formation factor (F_θ), water content exponent (n_θ) and mean squared error for ρ_{eff}/ρ_w (MSE) for Eqs. (1)–(3) of the two different depth ranges.

Depth range	F_θ	n_θ	r^*	MSE
< 0.9 m	0.577	1.83	0.895	2.8
$\geq$ 0.9 m	0.587	1.34	0.888	1.4

* $p < 0.01$.

The first depth range comprises the upper and the intermediate layer. These two periglacial layers are characterized by a high amount of silt (mostly medium silt) and comparatively low clast content. The exponent n_θ ranges from 1.8 to 2.3. The second depth range is represented by the basal layer. This is characterized by a higher amount of coarse material at the expense of fine grain sizes. In this depth range n_θ ranges from 0.7 to 1.8. Within each of these two depth ranges, we assume, analogously to the properties of the substrate, similar electrical properties with a threshold at 0.9 m. The threshold depth of 0.9 m is not developed as an exact, continuous boundary. Rather it is a short transition zone, because the samples right from this depth may have properties of the shallow or the deeper region, similar to the geomorphological differentiation between the basal and intermediate layer, whose boundary varies between depths of 0.8 to 1 m. By combining samples from different depths into two regions, it was possible to derive the parameter for Eqs. (1)–(3) for each region (Table 3).

This relationship between water content and resistivity, shown in Fig. 3 and Table 3, is only a mean value for each depth range. In the first depth range (0–0.9 m), especially close to the surface, the differences in soil or electrical properties between the samples even at the same depth may vary. This higher variation may be explained by intense biotic activity near the surface, enhancing small-scale heterogeneity compared to deeper parts of the soil.

The fitted curves of both regions are quite similar, except for n_θ. The adapted values for F_θ are almost identical (0.577 vs. 0.587, Table 3). With high saturation, the difference of resistivity between the depth ranges is small and

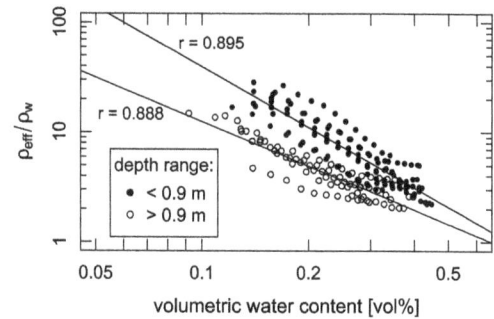

Figure 3. Volumetric water content in dependence of resistivity ratio (ρ_{eff}/ρ_w) for two different depth ranges.

primarily influenced by the conductivity of the pore fluid, but increases with decreasing water content. As a result of the higher exponent, LH and LM react more sensitively to water content changes than LB, especially at low presaturations. Related to this, small water content changes cause larger changes in resistivity than in the deeper region.

3.2 ERT mapping

At our study site the resistivity of the subsoil ranges from nearly $100\,\Omega$ m up to more than $4000\,\Omega$ m. The distribution may be divided in two main areas, the "inner" area between the depression lines and the "outer" area at the hillsides which differ in their depth profiles. (see Fig. 4)

At the intersection between the longitudinal and diagonal profiles, a good match of the calculated resistivity models can be found at shallow depth. With increasing depth, the differences become more notable – e.g. A × B: depth < 1 m, average deviation 8% ($\sigma = 5.4\%$); depth 1–7 m, average deviation 20% ($\sigma = 10\%$); and depth > 7 m, average deviation 43% ($\sigma = 6.6\%$). To exclude potential errors (e.g. electrode positioning errors), the data quality may be evaluated by comparing normal and reciprocal measurements, i.e. interchanging potential and current electrodes (LaBrecque et al., 1996; Zhou and Dahlin, 2003). For profiles A and B repeated measurements with reciprocal electrode configuration were conducted. Thereby, no large errors (max ±1.2 %) could be found between normal and reciprocal measurements. Because of the absence of large potential errors, the increasing deviation with depth may be only explained by the inversion process, decreasing sensitivity, less spatial resolution or potential 3-D effects.

The resistivity distribution of the subsurface is characterized by large-scale and small-scale heterogeneities, but also distinct patterns may be identified. At shallow depth up to 0.9 m, the study area is characterized by high resistivity. This comprises the upper and the intermediate layer.

Since the laboratory results indicate similar electrical properties, remarkable differences between upper and intermediate layers only occur if water content deviates. There

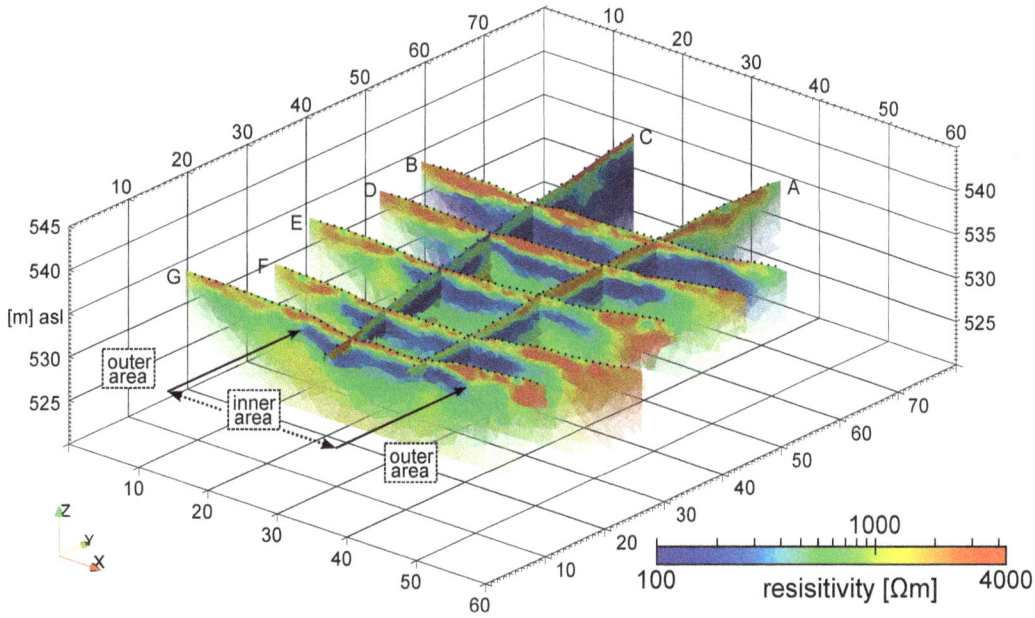

Figure 4. Resistivity results from ERT mapping (October 2008) of the study area: pseudo 3-D view of the profiles A to G.

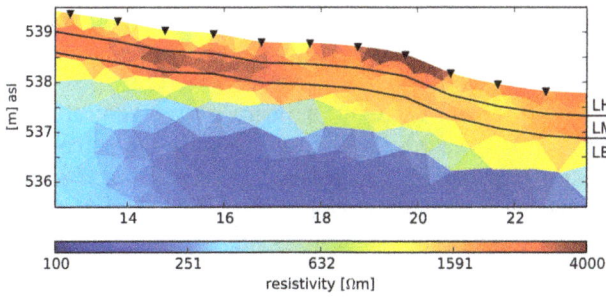

Figure 5. ERT section of profile A with plotted layer boundaries (date: 21 October 2008).

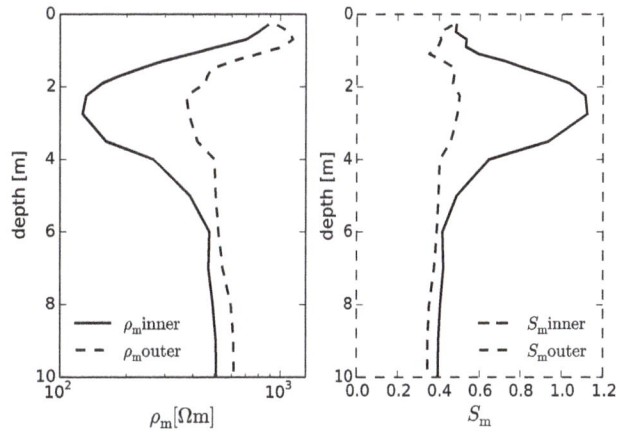

Figure 6. Median resistivity (left panel) and median water saturation (right panel) per depth for the inner region (between the depression lines) and outer region (hillslopes) (date: 21 October 2008).

are areas where the intermediate layer has higher resistivity, suggesting lower water content (see Fig. 5).

The hydrometric data show the driest conditions in 0.55–0.65 m (see Fig. 8) which is consistent with the high median resistivity of the intermediate layer at the time of data acquisition (see Fig. 6).

Resistivity decreases in greater depths (starting at 1 m). Thus, the basal layer is characterized by lower resistivity compared to the overlying layers. However, this is not constant in the lateral direction. Two different patterns are found. In the "inner" area between the two depression lines (approx. between profiles A and C), the resistivity of the basal layer is lower than in the "outer" area (the hillsides) (see Fig. 6). Between the depression lines LB is characterized as a connected zone of low resistivity. A calculation of saturation using Eq. (3) and the porosity from Table 1 indicates that this may be interpreted as a connected saturated zone (Fig. 6).

Due to the slope gradient, water from the hillsides and upper parts of the catchment flows into the direction of the depression lines, where it concentrates and forms a local slope groundwater reservoir. This results in a maximum decrease of resistivity in this zone as observed in all measured profiles at depths of 1.5–4.5 m (see Figs. 4 and 6). Percussion drilling confirmed that the thickness of LB exceeds 3.5 m downslope. Therefore, we assume that the entire saturated zone is located within the basal layer and since it is connected to the spring, it is also the source of the base flow. According to this, the shape of the surface may be partially transferred to the subsurface to identify regions of different hydrogeological conditions. Convex areas indicate dryer conditions in

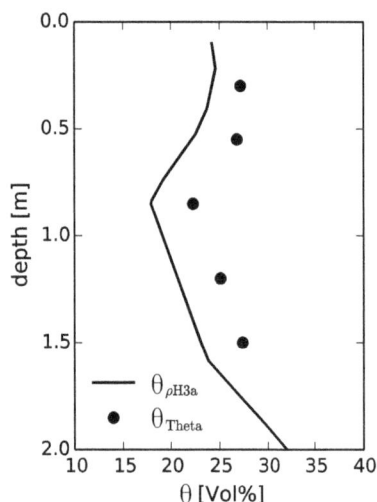

Figure 7. Volumetric water content calculated from resistivity data close to H3a in comparison with soil moisture sensors (date: 21 October 2008).

the basal layer in comparison to concave or elongated parts of the hillslope, which may act as local aquifers.

It is not feasible to relate a resistivity to the underlying gneiss or its regolith. Percussion drilling was only realized down to 4 m depth where bedrock could not be reached. If the maximum thickness of the basal layer is equal to the saturated zone, as obtained by resistivity data, the change from basal layer to underlying gneiss may be set at a depth around 4.5 m.

ERT mapping of the spatial distribution of periglacial cover beds is associated with several restrictions. In our study area, stratification is concealed by the influence of pore water, the main factor driving resistivity. On the other hand this fact may be used to improve the understanding of the moisture conditions of the subsurface.

To check the equations obtained in the lab and also to compare directly with hydrometric data, we used the water contents from the soil moisture sensors at H3a. Figure 7 compares water content, calculated with temperature-corrected resistivity ($\theta_{\rho H3a}$ profile A close to H3a), with water content from the ThetaProbes (θ_{Theta}) at time of mapping.

The values of $\theta_{\rho H3a}$ and θ_{Theta} show depth profiles of similar shape, but the values differ slightly. The resistivity depth profile shows a shift of -4.5 Vol% in comparison to the ThetaProbes. The different positions of the two probe locations could be one reason for this mismatch. Other reasons could be the inversion process of the resistivity data or differing pore water resistivity from the used median value (Table 2).

Because the data of the resistivity measurements and also the ThetaProbes may contain biased errors (e.g. caused by clast content or by the installation procedure), it is difficult to draw reliable conclusion as to which depth profile is more accurate.

3.3 Joint hydrometric and ERT monitoring

During the period May to December 2008 the spring discharge varied between 0.07 and $1.67 \, \mathrm{L\,s^{-1}}$. Median soil water tension of the study area, related to depth and time (see Fig. 8), indicates the impact of soil moisture on spring discharge. During summer increasing evapotranspiration causes the drying-out of soil. The spring showed only a slight reaction to precipitation events. Rainfall could only balance the soil water deficit and caused no runoff. Therefore, there is almost no runoff generation in the summer season. Primarily base flow dominates and decreasing discharge is mainly caused by saturation excess overland flow from the area surrounding the spring.

In contrast, during winter season (starting in November) at all depths lower tensions ($< 90 \, \mathrm{hPa}$) were measured. Less evapotranspiration results in a replenishment of the storage water reservoirs in the subsurface. Due to the moist conditions, high presaturations predominate and cause a rapid runoff response with rain and the high discharges within the winter season.

Furthermore, there is an influence of the layered subsurface on soil moisture and runoff response. Until the beginning of May and again from December the low tensions of the upper parts of LB indicate saturated conditions, in contrast to the deeper LB with higher tensions (see Fig. 8). Due to the anisotropic hydraulic properties (low vertical compared to horizontal hydraulic conductivity) the percolation into deeper parts of LB decreases. The seepage water is concentrated as backwater in the LM and the upper parts of LB. Because of the high lateral hydraulic conductivity this saturated depth range is mainly involved in runoff and causes strong interflow.

As the hydrometric data show, the first period from May to October was mainly characterized by drying of the subsurface. After that, humid conditions began to dominate (see Fig. 8). Major changes occur at shallow depth and proceed to depth, though remarkably attenuated. Each depth has its own characteristics, its own variation in time and shows different hydrological and electrical response. To better distinguish the results and to deal with the subsurface layered structure, a depth- or layer-based analysis is appropriate.

Figure 9 shows the trend of median resistivity for each depth range for the entire time series of profile A between H1a and H4a and profile B between H4b and H4a, in comparison with daily accumulated precipitation.

The resistivity of profile A clearly correlates with profile B (Table 4). This correlation is more pronounced at shallow depths. The absolute values are similar, except for the near-surface part of LH (0–0.2 m) and parts of LB (1.5–2.0 m). These two depth ranges have higher resistivity values at profile B than A at all points in time, due to the different positions. Profile A is completely situated in one of the depression lines, in which higher soil water contents can be expected in general.

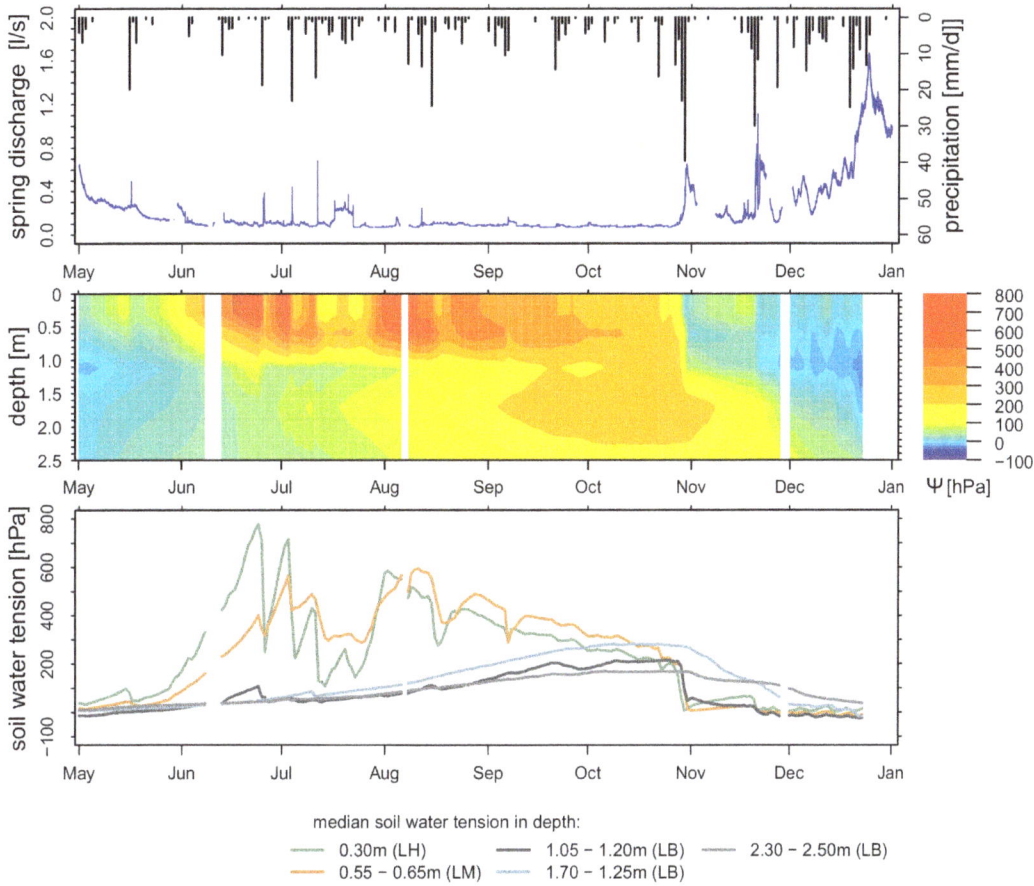

Figure 8. Spring discharge in comparison with daily precipitation (top panel), image of median soil water tension of the shallow subsurface (middle panel) and median soil water tension for different depths (bottom panel) – adapted from Heller (2012).

Table 4. Correlation between median resistivity of profiles A ($\rho_{profile\ A}$) and B ($\rho_{profile\ B}$) and between subsequent resistivity ratio of profile A ($\rho_{timestep}/\rho_{initial}$) and cumulative precipitation during the time step (ppt).

Depth [m]	$r_{(\rho_{profile\ A},\rho_{profile\ B})}{}^a$	$r\left(\frac{\rho_{timestep}}{\rho_{initial}},ppt\right)$
0–0.2	0.977	−0.773[a]
0.2–0.4	0.988	−0.770[a]
0.4–0.9	0.987	−0.804[a]
0.9–1.2	0.987	−0.586[a]
1.2–1.5	0.852	−0.378[b]
1.5–2.0	0.831	−0.078[b]
2.0–3.0	0.878	0.173[b]

[a] $p < 0.01$; [b] $p > 0.01$.

During the measuring period, the upper layer (0–0.2 and 0.2–0.4 m) reacts with similar resistivity variations as the intermediate layer (0.4–0.9 m). Resistivity of the intermediate layer may temporarily exceed the upper layer (e.g. profile A, October–December).

The temporal changes in resistivity decrease with depth. Short time variations are limited down to 2 m. Below, the differences are marginal with only a continuous slight increase during the investigated period.

The variation of resistivity is significantly influenced by rainfall. As shown in Table 4, the upper and intermediate layers (< 0.9 m) show a strong negative correlation with the cumulated amount of precipitation (ppt). This correlation decreases with depth. Upper parts of the basal layer (0.9–1.5 m) respond slightly and with a delay to intense rain events or enduring dry periods. Depths > 1.5 m show no direct correlation with rainfall. Water cannot infiltrate straight to greater depths because of decreasing hydraulic conductivity, evaporation, storage, or consumption of water by roots.

One problem is the temporal resolution. Because of the time intervals (usually ≥ 1 week), we are not able to resolve the entire temporal heterogeneity of the subsurface, which may lead to misinterpretation. For example, during the period from 3 to 16 September, the amount of 33 mm rain seems not to affect the resistivity of profile A. However, 32 of these 33 mm had already been fallen by 7 September. At profile B with an additional measurement on 9 September, resistivity

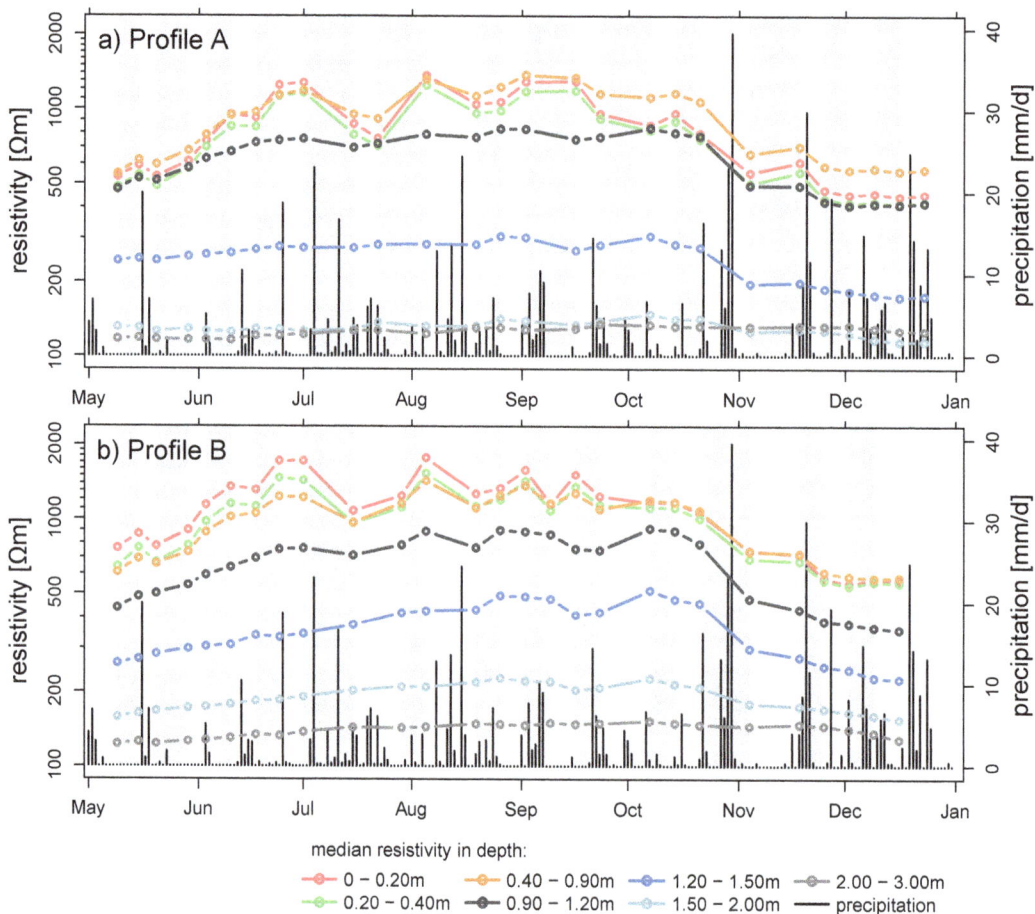

Figure 9. Trend of median resistivity for different depth ranges for **(a)** profile A and **(b)** profile B in comparison with daily precipitation (grey and white shaded regions for visualization of ERT time intervals).

at shallow depth decreases first and after that increases back to the initial level of 3 September (see Fig. 9b). Due to the missing time step, this alteration is not traced in profile A (see Fig. 9a).

This issue is also evident when comparing the resistivity with the soil suction data. With the higher temporal resolution of the tensiometer it is possible to resolve short time events, e.g. single rain events (Fig. 8), which cannot be rendered with the resistivity survey (see Fig. 9).

During the investigation period, different trends could be identified. The initial conditions in April and early May are characterized by a highly saturated subsurface. This is indicated by low soil water tension, high spring discharge and high water content. Due to the humid conditions at the beginning of the measurements, the conductivity of the shallow subsurface is high and the observed resistivity is low relative to the seasonal variations.

The first period between May and October is mainly characterized by increasing resistivity. The accumulated precipitation from 9 May to 21 October is only 337 mm. In combination with increasing evapotranspiration, this causes a drying

of the subsurface (see Fig. 8). As a result of drying, at shallow depths (< 0.9 m) resistivity quickly increases until July. Below, the increase proceeds slightly, but continuously until October.

As mentioned above, resistivity, especially of LH and LM (up to 0.9 m), shows a high short time variability and is strongly associated with the amount of precipitation (ppt) (Table 4). During the investigated period three different response types could be identified that are exemplarily illustrated in Fig. 10 and compared to soil water tension.

1. A small amount of precipitation (see 23 September–7 October, ppt = 23 mm) causes a short deferment of increasing resistivity of LH and LM during the summer period. The values of initial state and time step are of the same order of magnitude. Within the temporal resolution, only a slight decrease could be recorded. Deeper parts are not affected and dry continuously. Constant discharge indicates that there is no runoff generation during this period. This amount of rain is only able to balance the deficit caused by evaporation at shallow

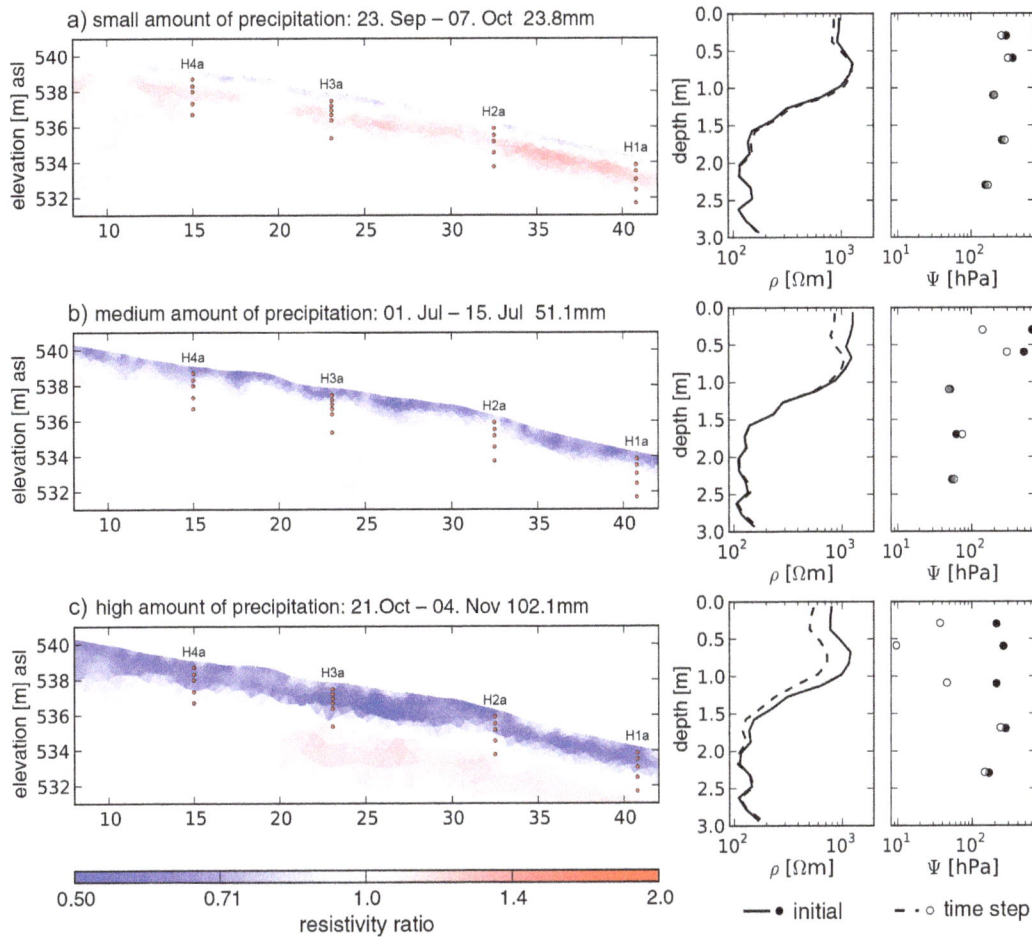

Figure 10. Ratio between subsequent resistivity (first column), median resistivity as a function of depth (second column) and median soil water tension (third column) for three exemplary precipitation responses: (**a**) small amount, (**b**) medium amount and (**c**) high amount.

depths, at least within the temporal resolution of measurements.

2. A medium amount of precipitation (see 1 July–15 July, ppt = 51.1 mm) causes a distinctive reaction at shallow depth. Resistivity at these depths shows a sharp decrease by comparatively the same ratio (~ 0.7). However, the signal is not traced into the deeper ground (> 1.2 m), which remains completely unaffected. So vertical seepage dominates in LH and LM, which leads to recharge of soil water. The water is predominantly fixed by capillary force; hence it does not percolate into deeper layers. The short rise of discharge is caused by saturation overland flow in the spring bog.

3. A high amount of precipitation (see 22 October–4 November, ppt = 102.1 mm) results in a strong response down to 2.0 m and affects LH, LM as well as parts of LB. Such a heavy rain period does not induce larger resistivity changes in LH and LM than the medium rain period, but influences deeper regions at the same order of magnitude as above. The water infiltrates to the

upper, but does not reach the deeper parts of the basal layer (2–3 m). The vertical seepage is limited and therefore the increasing spring discharge may only be caused by lateral subsurface flow, such as interflow in the unsaturated subsoil.

After the major rain event at the end of October, resistivity values remain constant until the next time step. Due to precipitation of 102.1 mm during the period from 19 November to 16 December, resistivity drops below the initial state and shows highly saturated conditions.

A comparison of water content obtained by soil moisture sensors (θ_{Theta}) and water content calculated from resistivity data for different depths over time at profile A close to the hydrometric station H3a ($\theta_{\rho H3a}$) using Eq. (2) is shown in Fig. 11

At shallow depth (≤ 0.85 m), $\theta_{\rho H3a}$ correlates closely with θ_{Theta} (Table 5). However, there is a shift of the curves during the whole period. The volumetric water content from resistivity data is consequently smaller than from the soil moisture

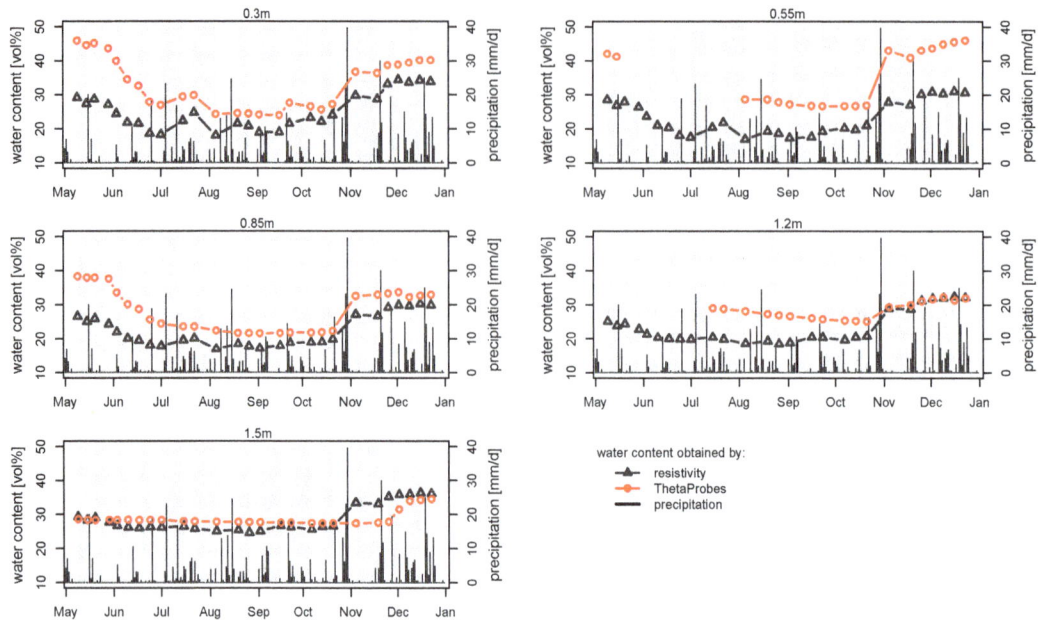

Figure 11. Trend of volumetric water content, obtained by resistivity data and soil moisture sensors (ThetaProbes) with daily precipitation for different depths.

Table 5. Correlation of volumetric water content calculated from resistivity values ($\theta_{\rho H3a}$) and water content from soil moisture sensors (θ_{Theta}) and correlation of resistivity at H3a (ρ_{H3a}) and soil suction at H3a (Ψ_{H3a}).

Depth [m]	$r_{(\theta_{\rho H3a}, \theta_{Theta})}$ [a]	$r_{(\rho_{H3a}, \Psi_{H3a})}$
0.30	0.863	0.993[a]
0.55	0.957	0.904[a]
0.85	0.885	0.905[a]
1.20	0.136	0.120[b]
1.50	0.619	0.566[a]

[a] $p < 0.01$; [b] $p > 0.01$.

sensors. In dry periods (e.g. July–October), the difference is less than under humid conditions (e.g. May).

In deeper parts the variations are attenuated. At a depth of 1.2 m there is almost no response over the year, until the heavy rain period at the end of October.

At 1.5 m depth the response of the soil moisture sensor is marginal until December, but thereafter shows an increase. In contrast, $\theta_{\rho H3a}$ shows already in late October a reaction to the heavy rain event, which is not reproducible with the ThetaProbe.

The same holds true for the correlation between resistivity (ρ_{H3a}) and soil suction at H3a (Ψ_{H3a}) (see Table 5). The resistivity of LH and LM fits well to the tensiometer data at the same depth, but in deeper parts it deviates.

These deviations between resistivity data and hydrometric measurements may have different causes. Both methods

contain measuring errors, just as the laboratory and other hydrometric (e.g. soil–water resistivity) measurements. Furthermore, the soil moisture sensors and tensiometers measure punctual values. Heller (2012) demonstrated with dye infiltration experiments that preferential flow is an important process in our study area. Hence, hydrometric point measurements may over- or underestimate soil moisture, depending on whether they are inside or outside a preferential pathway. Therefore the data are very limited, with restricted validity for the entire depth range or layer. In contrast, ERT has the advantage of integrating over a larger measuring volume, which makes it more suitable for extensive depth-related interpretations.

4   Conclusions

In drainage basins, hillslopes link precipitation to river runoff. Runoff components, different flow pathways and residence times are mainly influenced by the properties of the hillslope, especially the shallow subsurface. The knowledge of these properties is one of the keys to characterize the runoff dynamics in catchments. According to this, we used ERT for mapping the spatial heterogeneity of the subsurface structure on a hillslope with particular focus on mid-latitude slope deposits (cover beds).

ERT makes it possible to differentiate between LH and LM as one unit and LB as another. Like the intrinsic properties (e.g. sedimentological), LH and LM have very similar electrical characteristics. Therefore, they may only be distinguished by ERT if water contents are different or change differently with time.

In contrast, the sediments within LB have their own electrical characteristics. The pedo-/petrophysical relationship, with neglecting surface conductivity, shows equal formation factors to LH and LM, but different exponents. With the lower exponent, LB is characterized by lower resistivity at the same water content. Therefore, the resistivity of LB is lower in the entire study area, which is further reinforced by the increasing mineralization of pore water with depth.

From the results of field measurements and pedo-/petrophysical parameter determination in the laboratory we have been able to monitor seasonal changes in subsurface resistivity and its relationship to precipitation and soil moisture on the hillslope scale with a minimally invasive method directly. In combination with commonly used hydrometric approaches, we improved our understanding of the allocation, distribution, and movement of water in the subsurface. Different amounts of precipitation affect the subsurface moisture conditions differently and accordingly different depths take part in runoff generation.

Because pore water (saturation and conductivity) is the main driver for resistivity, we have arrived at some comprehensive interpretations of the subsurface moisture conditions. The high resistivities of LH and LM indicate low water contents, whereas LB is divided into two different moisture zones. On the hillsides water saturation of LB is less than between the depression lines, where low resistivity shows high water saturation and implies a local slope groundwater reservoir.

During the investigation period, temperature-corrected resistivity showed distinct seasonal variations due to changes in moisture conditions, primarily influenced by precipitation and evapotranspiration. Close to the surface, these variations are very evident and decline with increasing depth, mainly limited to a depth of 2 m. This primarily affects LH, LM, and the upper parts of LB, since it may be assumed that deeper parts are already saturated and changes are only possible due to changes in water conductivity.

In summer the subsurface continuously dries, starting at the surface and proceeding to depth. This drying is temporarily interrupted by precipitation. Penetration depth and intensity of the response strongly depend on the amount of precipitation. During periods with a small amount of precipitation, infiltration is limited to LH. There is no runoff generation, and greater depths remain unaffected which leads after repeated occurrence to drier conditions within LM compared to LH. In contrast to this, a response caused by a medium amount of precipitation includes LM and a small increase in spring discharge. The main source of this runoff is saturated overland flow from the surface surrounding the spring. With a high amount of precipitation, changes in resistivity point to vertical seepage down to 2 m. Due to lateral subsurface flow within LH, LM and the upper parts of LB, the discharge of the spring strongly increases.

The results from ERT measurements show a strong correlation to the hydrometric data. The average resistivity response is congruent to the average soil tension data. Water content obtained with soil moisture sensors shows similar variations as calculated from the closest ERT profile. Consequently, soil moisture on the hillslope scale may be determined not only by punctual hydrometric measurements, but also by minimally invasive ERT monitoring, provided pedo-/petrophysical relationships are known. By the use of ERT, expansive invasive hydrometric measurements may be reduced or partially substituted without losing information – but rather enhancing the spatial significance of these conventional point measurements. A combination improves the spatial understanding of the ongoing hydrological processes and is more suitable for the identification of heterogeneities.

Cassiani et al. (2009) pointed out that a combination of geophysical and hydrometric data may be used for quantitative estimation of hillslope moisture conditions. Our study has shown that this may also be applied to mid-latitude hillslopes covered by periglacial slope deposits. Nevertheless, there are some restrictions requiring further improvements.

One shortcoming is the temporal resolution. Some hydrological responses especially at hillslopes may proceed very quickly. The major goal for further research should be to increase the temporal resolution of ERT measurements to at least trace single rain events. This could be realized with automated data acquisitions as described in Kuras et al. (2009).

Another aim should be to improve the spatial resolution. A high-resolution ERT in combination with additional cross-borehole measurements would be more suitable to deal with small-scale heterogeneities and to overcome the problem of decreasing sensitivity with depth.

Acknowledgements. We acknowledge support by the German Research Foundation and the Open Access Publication Funds of the TU Dresden. We would also like to thank M. Leopold, J. Boaga and two anonymous reviewer for their helpful comments and suggestions that improved the quality of the paper.

Edited by: N. Romano

References

AD-hoc AG-Boden: Bodenkundliche Kartieranleitung, 5th Edn., Bundesanst. für Geowiss. und Rohstoffe in Zusammenarb. mit den Staatl. Geol. Diensten, Hannover, 2005.

Amoozegar, A.: A compact constant-head permeameter for measuring saturated hydraulic conductivity of the Vadose Zone, Soil Sci. Soc. Am. J., 53, 1356–1361, 1989.

Anderson, M. G. and Burt, T. P. (Eds.): Process studies in hillslope hydrology, Wiley, Chichester, West Sussex, England, New York, 1990.

Apparao, A.: Geoelectric profiling, Geoexploration, 27, 351–389, 1991.

Archie, G.: The electrical resistivity log as an aid in determining some reservoir characteristics, T. Am. I. Min. Met. Eng., 146, 54–61, 1942.

ATKIS®-DGM2, Landesvermessungsamt Sachsen – Amtliches Topographisch-Kartographische Informationssystem – Digital elevation model, resolution: 2m, 2008

Barker, R.: Depth of investigation of collinear symmetrical four-electrode arrays, Geophysics, 54, 1031–1037, 1989.

Bechtold, M., Vanderborght, J., Weihermueller, L., Herbst, M., Günther, T., Ippisch, O., Kasteel, R., and Vereecken, H.: Upward transport in a three-dimensional heterogeneous laboratory soil under evaporation conditions, Vadose Zone J., 11, doi:10.2136/vzj2011.0066, 2012.

Beff, L., Günther, T., Vandoorne, B., Couvreur, V., and Javaux, M.: Three-dimensional monitoring of soil water content in a maize field using Electrical Resistivity Tomography, Hydrol. Earth Syst. Sci., 17, 595–609, doi:10.5194/hess-17-595-2013, 2013.

Benderitter, Y. and Schott, J. J.: Short time variation of the resistivity in an unsaturated soil: the relationship with rainfall, Eur. J. Environ. Eng. Geophys., 4, 37–49, 1999.

Binley, A., Henry-Poulter, S., and Shaw, B.: Examination of solute transport in an undisturbed soil column using electrical resistance tomography, Water Resour. Res., 32, 763–769, 1996a.

Binley, A., Shaw, B., and Henry-Poulter, S.: Flow pathways in porous media: electrical resistance tomography and dye staining image verification, Meas. Sci. Technol., 7, 384–390, 1996b.

Brunet, P., Clément, R., and Bouvier, C.: Monitoring soil water content and deficit using Electrical Resistivity Tomography (ERT) a case study in the Cevennes area, France, J. Hydrol., 48, 146–153, 2010.

Cassiani, G., Bruno, V., Villa, A., Fusi, N., and Binley, A.: A saline trace test monitored via time-lapse surface electrical resistivity tomography, J. Appl. Geophys., 59, 244–259, 2006.

Cassiani, G., Godio, A., Stocco, S., Villa, A., Deiana, R., Frattini, P., and Rossi, M.: Monitoring the hydrologic behaviour of a mountain slope via time-lapse electrical resistivity tomography, Near Surf. Geophys., 7, 475–486, 2009.

Chifflard, P., Didszun, J., and Zepp, H.: Skalenübergreifende Prozess-Studien zur Abflussbildung in Gebieten mit periglazialen Deckschichten (Sauerland, Deutschland), Grundwasser, 13, 27–41, 2008.

Dahlin, T. and Zhou, B.: A numerical comparison of 2-D resistivity imaging with 10 electrode arrays, Geophys. Prospect., 52, 379–398, 2004.

Daily, W., Ramirez, A., LaBrecque, D. J., and Nitao, J.: Electrical resistivity tomography of vadose water movement, Water Resour. Res., 28, 1429–1442, 1992.

De Morais, F., De Almeida Prado Bacellar, L., and Aranha, P. R. A.: Study of flow in vadose zone from electrical resistivity surveys, Rev. Bras. Geofis., 26, 115–122, 2008.

Descloitres, M., Ribolzi, O., and Le Troquer, Y.: Study of infiltration in a Sahelian gully erosion area using time-lapse resistivity mapping, Catena, 53, 229–253, 2003.

Descloitres, M., Ribolzi, O., Troquer, Y. L., and Thiébaux, J. P.: Study of water tension differences in heterogeneous sandy soils using surface ERT, J. Appl. Geophys., 64, 83–98, 2008a.

Descloitres, M., Ruiz, L., Sekhar, M., Legchenko, A., Braun, J., Mohan Kumar, M. S., and Subramanian, S.: Characterization of seasonal local recharge using electrical resistivity tomography and magnetic resonance sounding, Hydrol. Process., 22, 384–394, 2008b.

Di Baldassarre, G. and Uhlenbrook, S.: Is the current flood of data enough? A treatise on research needs for the improvement of flood modelling, Hydrol. Process., 26, 153–158, 2012.

French, H. and Binley, A.: Snowmelt infiltration: monitoring temporal and spatial variability using time-lapse electrical resistivity, J. Hydrol., 297, 174–186, 2004.

Garré, S., Koestel, J., Günther, T., Javaux, M., Vanderborght, J., and Vereecken, H.: Comparison of heterogeneous transport processes observed with electrical resistivity tomography in two soils, Vadose Zone J., 9, 336–349, 2010.

Garré, S., Javaux, M., Vanderborght, J., Pages, L., and Vereecken, H.: Three-dimensional electrical resistivity tomography to monitor root zone water dynamics, Vadose Zone J., 10, 412–424, 2011.

Günther, T., Rücker, C., and Spitzer, K.: Three-dimensional modelling and inversion of dc resistivity data incorporating topography – II. Inversion, Geophys. J. Int., 166, 506–517, 2006.

Hayashi, M.: Temperature-electrical conductivity relation of water for environmental monitoring and geophysical data inversion, Environ. Monit. Assess., 96, 119–128, 2004.

Heller, K.: Einfluss periglazialer Deckschichten auf die oberflächennahen Fließwege am Hang – eine Prozessstudie im Osterzgebirge, Sachsen, Ph.D. thesis, Faculty of Environmental Sciences, TU Dresden, available at: http://nbn-resolving.de/urn: nbn:de:bsz:14-qucosa-98437 (last access: 30 April 2014), 2012.

Keller, G. V. and Frischknecht, F. C.: Electrical Methods in Geophysical Prospecting, Pergamon Press, Oxford, 1966.

Kemna, A., Vanderborght, J., Hardelauf, H., and Vereecken, H.: Quantitative imaging of 3-D solute transport using 2-D time-lapse ERT: a synthetic feasibility study, in: 17th EEGS Symposium on the Application of Geophysics to Engineering and Environmental Problems, Colorado, USA, 342–353, 2004.

Kirkby, M. J. (Ed.): Hillslope Hydrology, Wiley, Chichester, 1980.

Kleber, A.: Lateraler Wasserfluss in Hangsedimenten unter Wald, in: Stoff- und Wasserhaushalt in Einzugsgebieten, Beiträge zur EU-Wasserrahmenrichtlinie, edited by: Lorz, C. and Haase, D., Springer Verlag, Heidelberg, 7–22, 2004.

Kleber, A. and Schellenberger, A.: Slope hydrology triggered by cover-beds, With an example from the Frankenwald Mountains, northeastern Bavaria, Z. Geomorphol., 42, 469–482, 1998.

Kleber, A. and Terhorst, B. (Eds.): Mid-Latitude Slope Deposits (Cover Beds), vol. 66 of Developments in sedimentology, 1st Edn., Elsevier, Amsterdam, Boston, Heidelberg, London, New York, Oxford, Paris, San Diego, San Francisco, Singapore, Sydney, Tokyo, 2013.

Klute, A. (Ed.): Methods of Soil Analysis, Part 1: Physical and Mineralogical Methods, 2nd Edn., ASA, SSA, Madison, Wisconsin, 1986.

Koestel, J., Kemna, A., Javaux, M., Binley, A., and Vereecken, H.: Quantitative imaging of solute transport in an unsaturated and undisturbed soil monolith with 3-D ERT and TDR, Water Resour. Res., 44, W12411, doi:10.1029/2007WR006755, 2008.

Koestel, J., Vanderborght, J., Javaux, M., Kemna, A., Binley, A., and Vereecken, H.: Noninvasive 3-D transport characterization in a sandy soil using ERT: 1. Investigating the validity of ERT-derived transport parameters, Vadose Zone J., 8, 711–722, 2009a.

Koestel, J., Vanderborght, J., Javaux, M., Kemna, A., Binley, A., and Vereecken, H.: Noninvasive 3-D transport characterization in a sandy soil using ERT: 2. Transport process inference, Vadose Zone J., 8, 723–734, 2009b.

Kuras, O., Pritchard, J. D., Meldrum, P. I., Chambers, J. E., Wilkinson, P. B., Ogilvy, R. D., and Wealthall, G. P.: Monitoring hydraulic processes with automated time-lapse electrical resistivity tomography (ALERT), CR Geosci., 341, 868–885, 2009.

LaBrecque, D. J., Miletto, M., Daily, W., Ramirez, A., and Owen, E.: The effects of noise on Occam's inversion of resistivity tomography data, Geophysics, 61, 538–548, 1996.

Leslie, I. N. and Heinse, R.: Characterizing soil-pipe networks with pseudo-three-dimensional resistivity tomography on forested hillslopes with restrictive horizons, Vadose Zone J., 12, doi:10.2136/vzj2012.0200, 2013.

Lesmes, D. P. and Friedman, S. P.: Relationships between the electrical and hydrogeological properties of rocks and soils, in: Hydrogeophysics, edited by: Rubin, Y. and Hubbard, S. S., Springer, Dordrecht, 87–128, 2006.

Ma, R., McBratney, A., Whelan, B., Minasny, B., and Short, M.: Comparing temperature correction models for soil electrical conductivity measurement, Precis. Agric., 12, 55–66, 2011.

Massuel, S., Favreau, G., Descloitres, M., Le Troquer, Y., Albouy, Y., and Cappelaere, B.: Deep infiltration through a sandy alluvial fan in semiarid Niger inferred from electrical conductivity survey, vadose zone chemistry and hydrological modelling, Catena, 67, 105–118, 2006.

McDonnell, J. J.: Where does water go when it rains? Moving beyond the variable source area concept of rainfall–runoff response, Hydrol. Process., 17, 1869–1875, 2003.

McDonnell, J. J., Tanaka, T., Mitchell, M. J., and Ohte, N.: Hydrology and biogeochemistry of forested catchments, Hydrol. Process., 15, 1673–1674, 2001.

McDonnell, J. J., Sivapalan, M., Vaché, K., Dunn, S., Grant, G., Haggerty, R., Hinz, C., Hooper, R., Kirchner, J., Roderick, M. L., Selker, J., and Weiler, M.: Moving beyond heterogeneity and process complexity: a new vision for watershed hydrology, Water Resour. Res., 43, W07301, doi:10.1029/2006WR005467, 2007.

Michot, D., Benderitter, Y., Dorigny, A., Nicoullaud, B., King, D., and Tabbagh, A.: Spatial and temporal monitoring of soil water content with an irrigated corn crop cover using surface electrical resistivity tomography, Water Resour. Res., 39, SBH 14-1–SBH 14-20, 2003.

Miller, C. R., Routh, P. S., Brosten, T. R., and McNamara, J. P.: Application of time-lapse ERT imaging to watershed characterization, Geophysics, 73, G7–G17, 2008.

Moldenhauer, K.-M., Heller, K., Chifflard, P., Hübner, R., and Kleber, A.: Influence of cover beds on slope hydrology, in: Mid-Latitude Slope Deposits (Cover Beds), edited by: Kleber, A. and Terhorst, B., vol. 66 of Developments in Sedimentology, Elsevier, Amsterdam etc., 127–152, 2013.

Oldenborger, G. A., Knoll, M. D., Routh, P. S., and LaBrecque, D. J.: Time-lapse ERT monitoring of an injection/withdrawal experiment in a shallow unconfined aquifer, Geophysics, 72, F177–F188, 2007.

Perri, M. T., Cassiani, G., Gervasio, I., Deiana, R., and Binley, A.: A saline tracer test monitored via both surface and cross-borehole electrical resistivity tomography: comparison of time-lapse results, J. Appl. Geophys., 79, 6–16, 2012.

Popp, S., Altdorff, D., and Dietrich, P.: Assessment of shallow subsurface characterisation with non-invasive geophysical methods at the intermediate hill-slope scale, Hydrol. Earth Syst. Sci., 17, 1297–1307, doi:10.5194/hess-17-1297-2013, 2013.

Ramirez, A., Daily, W., LaBrecque, D. J., Owen, E., and Chesnut, D.: Monitoring an underground steam injection process using electrical resistance tomography, Water Resour. Res., 29, 73–87, 1993.

Robinson, D. A., Binley, A., Crook, N., Day-Lewis, F. D., Ferré, T. P. A., Grauch, V. J. S., Knight, R., Knoll, M. D., Lakshmi, V., Miller, R., Nyquist, J., Pellerin, L., Singha, K., and Slater, L.: Advancing process-based watershed hydrological research using near-surface geophysics: a vision for, and review of, electrical and magnetic geophysical methods, Hydrol. Process., 22, 3604–3635, 2008a.

Robinson, D. A., Campbell, C. S., Hopmans, J. W., Hornbuckle, B. K., Jones, S. B., Knight, R., Ogden, F., Selker, J., and Wendroth, O.: Soil moisture measurement for ecological and hydrological watershed-scale observatories: a review, Vadose Zone J., 7, 358–389, 2008b.

Robinson, D. A., Abdu, H., Lebron, I., and Jones, S. B.: Imaging of hill-slope soil moisture wetting patterns in a semi-arid oak savanna catchment using time-lapse electromagnetic induction, J. Hydrol., 416, 39–49, 2012.

Roy, A. and Apparao, A.: Depth of investigation in direct current methods, Geophysics, 36, 943–959, 1971.

Rücker, C. and Günther, T.: The simulation of Finite ERT electrodes using the complete electrode model, Geophysics, 76(4), F227–238, 2011.

Sauer, D., Scholten, T., and Felix-Henningsen, P.: Verbreitung und Eigenschaften periglaziärer Lagen im östlichen Westerwald in Abhängigkeit von Gestein, Exposition und Relief, Mitt. Dtsch. Bodenkdl. Ges., 96, 551–552, 2001.

Scholten, T.: Periglaziäre Lagen in Mittelgebirgslandschaften – Verbreitungssystematik, Eigenschaften und Bedeutung für den Landschaftswasser- und stoffhaushalt, in: Tagungsbeiträge IFZ-Workshop Ressourcensicherung in der Kulturlandschaft, Selbstverlag, Justus-Liebig-Universität, Gießen, 11–15, 1999.

Schön, J. H.: Physical Properties of Rocks: Fundamentals and Principles of Petrophysics, vol. 18 of Handbook of Geophysical Exploration: Seismic Exploration, Elsevier, Oxford, 2004.

Singha, K. and Gorelick, S. M.: Saline tracer visualized with three-dimensional electrical resistivity tomography: field-scale spatial moment analysis, Water Resour. Res., 41, W05023, doi:10.1029/2004WR003460, 2005.

Slater, L., Binley, A., Daily, W., and Johnson, R.: Cross-hole electrical imaging of a controlled saline tracer injection, J. Appl. Geophys., 44, 85–102, 2000.

Tilch, N., Uhlenbrook, S., Didszun, J., Wenninger, J., Kirnbauer, R., Zillgens, B., and Leibundgut, C.: Hydrologische Prozessforschung zur Hochwasserentstehung im Löhnersbach-Einzugsgebiet (Kitzbüheler Alpen, Österreich), Hydrol. Wasserbewirts., 50, 67–78, 2006.

Tromp-van Meerveld, H. J.: Hillslope hydrology: from patterns to processes, Ph.D. thesis, Oregon State University, Corvallis, 2004.

Tromp-van Meerveld, H. J. and McDonnell, J. J.: Assessment of multi-frequency electromagnetic induction for determining soil moisture patterns at the hillslope scale, J. Hydrol., 368, 56–67, 2009.

Uchida, T., McDonnell, J. J., and Asano, Y.: Functional intercomparison of hillslopes and small catchments by examining water source, flowpath and mean residence time, J. Hydrol., 327, 627–642, 2006.

Uhlenbrook, S.: Von der Abflussbildungsforschung zur prozessorientierten Modellierung – ein Review, Hydrol. Wasserbewirts., 49, 13–24, 2005.

Uhlenbrook, S., Didszun, J., and Wenninger, J.: Source areas and mixing of runoff components at the hillslope scale – a multitechnical approach, Hydrolog. Sci. J., 53, 741–753, 2008.

Vanderborght, J., Kemna, A., Hardelauf, H., and Vereecken, H.: Potential of electrical resistivity tomography to infer aquifer transport characteristics from tracer studies: a synthetic case study, Water Resour. Res., 41, W06013, doi:10.1029/2004WR003774, 2005.

Völkel, J., Leopold, M., Mahr, A., and Raab, T.: Zur Bedeutung kaltzeitlicher Hangsedimente in zentraleuropäischen Mittelgebirgslandschaften und zu Fragen ihrer Terminologie, Petermann. Geogr. Mitt., 146, 50–59, 2002a.

Völkel, J., Zepp, H., and Kleber, A.: Periglaziale Deckschichten in Mittelgebirgen – ein offenes Forschungsfeld, Ber. deuts. Landesk., 76, 101–114, 2002b.

Wenninger, J., Uhlenbrook, S., Tilch, N., and Leibundgut, C.: Experimental evidence of fast groundwater responses in a hillslope/floodplain area in the Black Forest Mountains, Germany, Hydrol. Process., 18, 3305–3322, 2004.

Zhou, B. and Dahlin, T.: Properties and effects of measurement errors on 2-D resistivity imaging surveying, Near Surf. Geophys., 1, 105–117, 2003.

Zhou, Q. Y., Shimada, J., and Sato, A.: Three-dimensional spatial and temporal monitoring of soil water content using electrical resistivity tomography, Water Resour. Res., 37, 273–285, 2001.

5

A high-resolution global-scale groundwater model

I. E. M. de Graaf[1]**, E. H. Sutanudjaja**[1]**, L. P. H. van Beek**[1]**, and M. F. P. Bierkens**[1,2]

[1]Department of Physical Geography, Faculty of Geosciences, Utrecht University, Utrecht, the Netherlands
[2]Unit Soil and Groundwater Systems, Deltares, Utrecht, the Netherlands

Correspondence to: I. E. M. de Graaf (i.e.m.degraaf@uu.nl)

Abstract. Groundwater is the world's largest accessible source of fresh water. It plays a vital role in satisfying basic needs for drinking water, agriculture and industrial activities. During times of drought groundwater sustains baseflow to rivers and wetlands, thereby supporting ecosystems. Most global-scale hydrological models (GHMs) do not include a groundwater flow component, mainly due to lack of geohydrological data at the global scale. For the simulation of lateral flow and groundwater head dynamics, a realistic physical representation of the groundwater system is needed, especially for GHMs that run at finer resolutions. In this study we present a global-scale groundwater model (run at 6′ resolution) using MODFLOW to construct an equilibrium water table at its natural state as the result of long-term climatic forcing. The used aquifer schematization and properties are based on available global data sets of lithology and transmissivities combined with the estimated thickness of an upper, unconfined aquifer. This model is forced with outputs from the land-surface PCRaster Global Water Balance (PCR-GLOBWB) model, specifically net recharge and surface water levels. A sensitivity analysis, in which the model was run with various parameter settings, showed that variation in saturated conductivity has the largest impact on the groundwater levels simulated. Validation with observed groundwater heads showed that groundwater heads are reasonably well simulated for many regions of the world, especially for sediment basins ($R^2 = 0.95$). The simulated regional-scale groundwater patterns and flow paths demonstrate the relevance of lateral groundwater flow in GHMs. Inter-basin groundwater flows can be a significant part of a basin's water budget and help to sustain river baseflows, especially during droughts. Also, water availability of larger aquifer systems can be positively affected by additional recharge from inter-basin groundwater flows.

1 Introduction

Groundwater is a crucial part of the global water cycle. It is the world's largest accessible source of fresh water and plays a vital role in satisfying basic needs of human society. It is a primary source for drinking water and supplies water for agriculture and industrial activities (Wada et al., 2014). During times of drought stored groundwater provides a buffer against water shortage and sustains baseflow to rivers and wetlands, thereby supporting ecosystems and biodiversity. However, in many parts of the world groundwater is abstracted at rates that exceed groundwater recharge, causing groundwater levels to drop while baseflow to rivers is no longer sustained (Konikow, 2011; Gleeson et al., 2012).

In order to understand variations in recharge and human water use affect groundwater head dynamics, lateral groundwater flow and groundwater surface water interactions should be included in global-scale hydrological models (GHMs), especially as these GHMs progressively move towards finer resolutions (Wood et al., 2012; Krakauer et al., 2014). Several studies (e.g. Bierkens and van den Hurk, 2007; Fan et al., 2007) have suggested that lateral groundwater flows can be important for regional climate conditions as they influence soil moisture and thus the water cycle and energy exchange between land and the lower atmosphere. Moreover, lateral groundwater flowing over catchment boundaries, i.e. inter-basin flows, can be a significant part of the water budget of a catchment, dependent on certain climate and geological conditions (Schaller and Fan, 2009). By supplementing the water budget, incoming inter-basin groundwater helps to sustain baseflows during droughts thereby increasing surface water availability for human water needs (de Graaf et al., 2014). Up to now, the current generation of GHMs typically does not include a lateral groundwater flow component mainly due to the lack of worldwide hydrogeological information (Gleeson et al., 2014). These data are available for parts of the developed world, but even there it is difficult to obtain data in a consistent manner. To cope with the unavailability of hydrogeological data Sutanudjaja et al. (2011) proposed the use of global data sets of surface lithology and elevation for aquifer parameterization. This method was tested by building a groundwater flow model for the Rhine–Meuse basin (30″ resolution) with promising results. Similarly, Vergnes et al. (2012) used global and European data sets to delimit the main aquifer basins for France (at 0.5° resolution) and parameterized these based on lithological information.

Recently, a pioneering study by Fan et al. (2013) presented a first ever high-resolution global groundwater table depth map. Their method, however, does not include hydrogeological information such as aquifer depths and transmissivities, but uses estimates from soil data. Also, the hydraulic connection between rivers and groundwater, which is the primary mode of drainage for groundwater in humid regions, is ignored. Moreover, their model requires calibration to head observations.

In this paper we present a global-scale groundwater model of an upper aquifer which is assumed to be unconfined. For the parameterization of the aquifer properties we relied entirely on available global lithological maps (Hartmann and Moosdorf, 2012) and databases on permeability (Gleeson et al., 2011). To overcome the lack of information about aquifer thickness worldwide, this is estimated based on extrapolation of available data from the USA. This can equally be extended to data-poor environments.

We forced the groundwater model with output from the global hydrological PCRaster Global Water Balance (PCR-GLOBWB) model (van Beek et al., 2011), specifically the net groundwater recharge and average surface water levels derived from routed channel discharge. This approach builds on earlier work by Sutanudjaja et al. (2011) and Sutanudjaja et al. (2014).

With this approach we were able to simulate groundwater heads of a upper unconfined aquifer, providing a first-order estimate of the spatial variability of water table heads as a function of climate and geology. In this paper we limit ourselves to a steady-state simulation as a prelude to transient simulations in forthcoming work. Also we did not yet perform a formal calibration of the model. We performed a sensitivity analysis using a Monte Carlo framework in which we ran the model with various hydrogeological parameter settings. Simulated groundwater heads from all realizations were evaluated against reported piezometer data and the parameter set with the highest coefficient of determination was used for further analysis. This resulted in a global map of average groundwater table depth in its natural state, i.e. in equilibrium with climate and without groundwater pumping. We simulated flow paths from the location of infiltration towards the location of drainage. Flow paths show areas where lateral

groundwater flows are important and inter-basin groundwater flows are significant and contribute to water availability in neighbouring watersheds. They also provide an indication of groundwater travel times.

Hereafter follows a description of the methods, in particular the parameterization of the upper aquifer, after which results of the sensitivity analysis and validation are presented. Next, the groundwater table depth map and flow path maps for Europe and Africa are presented. We end with conclusions and discussion.

2 Methods

2.1 General

The hydrological model of the terrestrial part of the world (excluding Greenland and Antarctica) developed in this study consists of two parts: (1) the dynamic land-surface model (PCR-GLOBWB) and (2) the steady-state groundwater model (MODFLOW). Both the land-surface model and groundwater model are run at 6′ resolution (approximately 11 km at the Equator). PCR-GLOBWB and MODFLOW are coupled offline where both models are run consecutively (Sutanudjaja et al., 2011).

2.1.1 Land-surface model

The PCR-GLOBWB model is a global hydrological model that simulates hydrological processes in and between two soil stores and one underlying linear groundwater store. For a detailed description of the PCR-GLOBWB model we refer to van Beek et al. (2011), and a summarized model description is given here. PCR-GLOBWB was run at 6′ resolution using a daily time step. Monthly climate data were taken from the CRU TS2.1 (Mitchell and Jones, 2005) with a spatial resolution of 0.5° and downscaled using the ERA-40 (Uppala et al., 2005) and ERA-Interim reanalysis (Dee et al., 2011) to obtain a daily climatic forcing see de Graaf et al., 2014 for a more detailed description of this forcing data set. Each grid cell contains a land surface that is represented by a vertical structured soil column comprising two soil layers (maximum depth 0.3 and 1.2 m, respectively), an underlying groundwater reservoir, and the overlying canopy. Sub-grid variability is included with regards to land cover (in this case using fractions of short and tall vegetation), soil conditions and topography. The model employs the improved Arno Scheme (Todini et al., 1996; Hagemann et al., 1999) to simulate variations in the fraction of saturated soil in order to quantify direct surface runoff. For each time step and for every grid cell the water balance of the soil column is calculated on the basis of the climatic forcing that imposes precipitation, reference-crop potential evapotranspiration and temperature. Actual evapotranspiration is calculated from reference-crop potential evaporation, time-varying crop factors and soil moisture conditions. Vertical exchange between the soil and

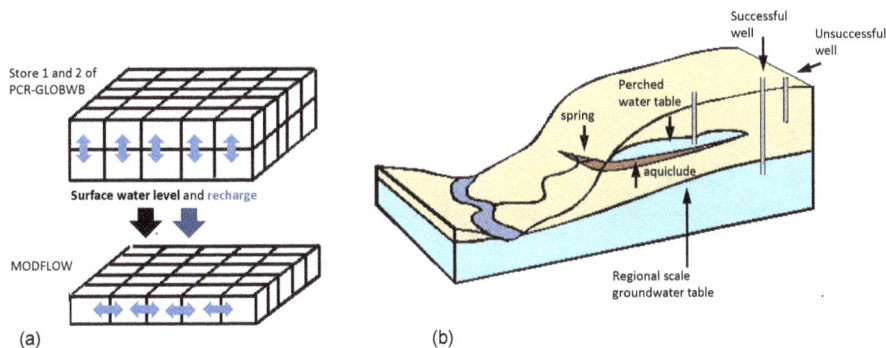

Figure 1. (a) Model structure used to couple the land-surface model PCR-GLOBWB with the groundwater model MODFLOW: first average annual net recharge and average annual channel discharge is calculated with PCR-GLOBWB. The latter is translated into surface water levels. Both recharge and surface water levels are used to force MODFLOW (Sutanudjaja et al., 2011). **(b)** Cross section illustrating the difference between the simulated regional-scale groundwater level and the perched groundwater levels that are often sampled.

groundwater occurs through percolation and capillary rise. Specific runoff from the soil column, comprising direct surface runoff, interflow and baseflow, is accumulated along the drainage network that consists of laterally connected surface water elements representing river channels, lakes or reservoirs. The accumulated runoff is routed to obtain discharge using the kinematic wave approximation of the Saint-Venant equations at a sub-daily time step. In the original version of PCR-GLOBWB no lateral groundwater flow is simulated. Groundwater flow within a cell is described as a linear store and recharge is simulated as percolation to the groundwater store, minus capillary rise from the groundwater store to the soil. However, in the current set-up, the capillary rise is disabled to force a one-way coupling from PCR-GLOBWB to MODFLOW.

2.1.2 Groundwater model

In this study the linear groundwater store of PCR-GLOBWB is replaced by a MODFLOW layer (McDonald and Harbaugh, 2000; Schmitz et al., 2009) simulating lateral groundwater flows and groundwater heads in a single-layer unconfined aquifer. Aquifer properties are prescribed and the MODFLOW layer is forced by outputs from PCR-GLOBWB, i.e. long-term averages of surface water levels and groundwater recharge (running period 1960–2010). Figure 1a illustrates the modelling strategy.

2.2 Estimating aquifer properties

Aquifer properties were initially based on two data sets; (1) the high-resolution global lithological map (GLiM) of Hartmann and Moosdorf (2012), and (2) the global permeability estimates of Gleeson et al. (2011).

The GLiM describes 15 lithology classes (see Table 1) (similar and expanding on Dürr et al., 2005). It is assumed that the lithological map represents the geology of the shallow subsurface accurately (Hartmann and Moosdorf, 2012).

For the global permeability map the lithology classes of Dürr et al. (2005) were combined to five hydrolithologies, representing broad lithologic categories with similar hydrogeological characteristics (Gleeson et al., 2011). In the GLiM, these hydrolithological units were subdivided further on the basis of texture in the case of unconsolidated and sedimentary rocks (Table 1). Gleeson et al. (2011) found that for all hydrolithologies permeability is representative for larger scales and that there is no discernible additional dependence of permeability on scale, with the exception of carbonates, most likely due to karst. The resulting map shows regional-scale permeability over the globe with the geometric mean permeability attributed to each hydrolithological unit. The geometric mean was obtained from calibrated permeabilities from groundwater models for units larger than 5 km in extent within 100 m depth. The polygons in the GLiM, delineating a hydrological unit, were subsequently gridded to 30″ (~ 1 km) and aggregated as the geometric mean at 6′ resolution.

To estimate aquifer transmissivities (kD in $\mathrm{m^2\,d^{-1}}$), aquifer thicknesses are required. Since no globally consistent data set on aquifer thickness is available, this was estimated using predominantly terrain attributes. Based on the assumption that unconfined productive aquifers coincide with sediment basins below river valleys, the distinction was made between (1) mountain ranges with negligible sediment thickness, consisting mainly of hard rock with secondary permeability and (2) sediment basins with thick sediment layers, presenting aquifers.

Aquifer thicknesses were then estimated as follows:

1. Mountain ranges and sediment basins were distinguished based on the difference between surface elevation and floodplain elevation within a cell. We used elevation data at 30″ from the HydroSHEDS data set to determine the floodplain elevation at 6′. First, for each 6′ cell we identified the lowest elevation at 30″ (maximum 144 values for a cell comprising only land

Table 1. Lithologic and hydrolithologic categories.

Lithologic categories[a]	Hydrolithologic categories[b]	$\log k\,\mu_{\mathrm{geo}}^{\mathrm{b}}$ [m^2]	σ^{b} [m^2]	$S_{\mathrm{y}}^{\mathrm{c}}$ [m m^{-1}]
Unconsolidated sediments	unconsolidated	−13.0	2.0	0.235
	c.g. unconsolidated	−10.9	1.2	0.360
	f.g. unconsolidated	−14.0	1.8	0.110
Siliciclastic sediments	siliciclastic sedimentary	−15.2	2.5	0.055
	c.g. siliciclastic sedimentary	−12.5	0.9	0.100
	f.g. siliciclastic sedimentary	−16.5	1.7	0.010
Mixed sedimentary rocks Carbonate sedimentary rocks Evaporites	carbonate	−11.8	1.5	0.140
Acid volcanic rocks Intermediate volcanic rocks Basic volcanic rocks	crystalline	−14.1	1.5	0.010
Acid plutonic rocks Intermediate plutonic rocks Basic plutonic rocks Pyroclastics Metamorphic	volcanic	−12.5	1.8	0.050
Water bodies Ice and glaciers	not assigned	−	−	−

[a] Hartmann and Moosdorf (2012). [b] Based on Gleeson et al. (2011), $\log k\,\mu_{\mathrm{geo}}$ is the geometric mean logarithmic permeability; σ is the standard deviation; f.g. and c.g. are fine grained and coarse grained respectively; [c] S_{y} is the storage coefficient, average per category.

area) and assigned this as the floodplain elevation for the entire cell (see Fig. 2, top panel). Next, the difference between surface elevation, also taken from the HydroSHEDS database, and floodplain elevation at 6′ was calculated. All cells with a floodplain elevation within 50 m b.s.l. (below the surface level) were assumed to form a sediment basin that constitutes an unconfined and relatively permeable aquifer (Fig. 2, top panel). These defined sediment basins included 70 % of the unconsolidated sediments mapped in the GLiM. The sediment basins consist of 56 % unconsolidated sediments, 25 % consolidated sediments and 19 % metamorphic or plutonic rocks. The latter are mainly found over the old cratons of Africa and the flat, recently glaciated areas of Laurasia.

2. By definition basins are linked to sedimentary environments in fluvial systems and deltas. Sediments are deposited perpendicular to the main gradient (constituting the transversal axis of the basin), with grain size and volumes decreasing at greater distance away from the transversal axis. Grain size also decreases along the transversal axis, distinguishing proximal (near the source of sediment) and distal parts. We assumed that gradation in grain size is captured in the GLiM but differentiation in depth is not. Instead, we used relative elevation as a measure of proximity to the river measured

along the transversal axis and as an indicator of the associated depth. We standardized the relative elevation and used this to define the distribution of aquifer thickness using a log-normal distribution, assuming thickness is non-negative and positively skewed. In more detail, we used the following procedure.

First, for each cell location x belonging to the sediment basins, a measure expressing the relative difference between land-surface elevation and floodplain elevation was calculated:

$$F'(x) = 1 - \frac{F(x) - F_{\min}}{F_{\max} - F_{\min}}, \qquad (1)$$

where $F(x)$ is the difference of surface and floodplain elevation at location x. $F_{\min}$ and $F_{\max}$ are the minimal and maximal value, corresponding to a difference between land-surface and floodplain elevation of 0 and 50 m respectively (following from the method to distinguish sediment basins from mountain ranges). This measure leads to a thinning layer further from the river towards the edge of the sediment basin (Fig. 2, bottom panel). $F'(x)$ can be seen as the likelihood of finding a thick sedimentary aquifer at a particular location x. A map of the spatial distribution of $F'(x)$ is given in the Supplement (Fig. S1).

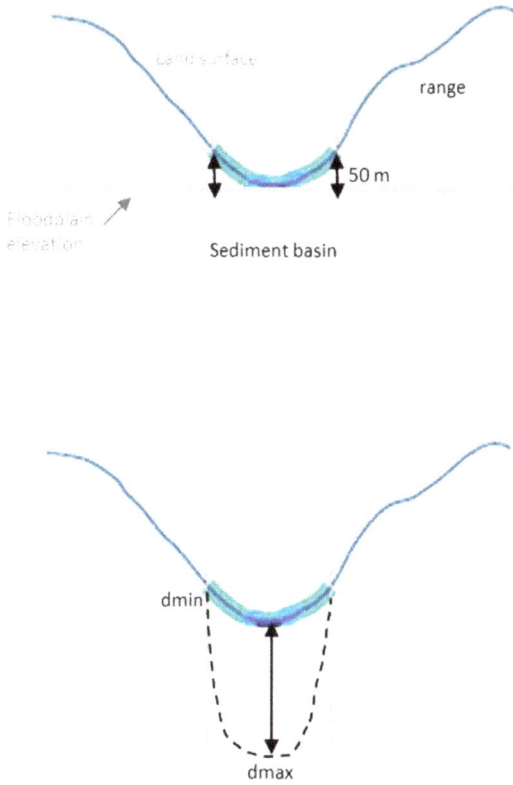

Figure 2. Top panel: definition of sediment basins and mountain ranges, based on terrain attributes (land-surface elevation and floodplain elevation). Bottom panel: estimation of aquifer thickness.

Second, the associated z-score is then calculated as

$$Z(x) = G^{-1}(F'(x)), \tag{2}$$

where $G^{-1}()$ is the inverse of the standard normal distribution.

3. Next, statistics on the thickness of unconsolidated sediments were obtained from available regional-scale groundwater studies in the USA (e.g. Central Valley, California, Faunt et al., 2009; Mississippi basin, Clark and Hart, 2009; in total six studies were used). As a measure of difference between aquifer systems, for each study the average thickness was determined resulting in a range of average thickness between 50 and 500 m. We assume that the thicknesses of the delineated sediments basins correspond with the total thickness of the upper aquifer and that this thickness is log-normally distributed. Therefore, this distribution is described using the average thickness of the ln-transformed thickness $\overline{\ln D}$. This $\overline{\ln D}$ is chosen uniformly over the globe and sampled from the range of thicknesses:

$$\overline{\ln D} = U(50;\ 500). \tag{3}$$

Moreover, as a measure of variation in thickness within aquifers systems, an average coefficient of variation was determined from the same USA regional groundwater studies. The coefficient of variation of the ln-transformed thickness $\mathrm{CV}_{\overline{\ln D}}$ was fixed when calculating the global distribution of aquifer thickness.

4. For each realization a spatial distribution of aquifer thickness is generated, assuming a log-normal distribution with random average $\overline{\ln D}$, sampled from $U(50;\ 500)$ with a fixed coefficient of variation $\mathrm{CV}_{\overline{\ln D}}$, and using the standard normal ordinate $Z(x)$ that is based on the topographical controls within each delineated basin:

$$Y(x) = \overline{\ln D}(1 + \mathrm{CV}_{\overline{\ln D}} Z(x)), \tag{4}$$

$$D(x) = e^{Y(x)}. \tag{5}$$

Therefore, in this representation $Y(x)$ and $D(x)$ are random because $\overline{\ln D}$ is random, while spatial variation is determined in $Z(x)$ reflecting the likelihood of a thick aquifer. Average aquifer thicknesses was simulated randomly from $U(50, 500)$ resulting in 100 equally likely maps of aquifer thickness. The result of the best performing run (selected after evaluation to groundwater head data) is presented in Fig. 3a.

Transmissivities were calculated using the estimated aquifer thickness. To estimate permeability at greater depth we combined the concept of exponentially decreasing permeability of the continental crust with depth (Ingebritsen and Manning, 1999) with data on near-surface permeability from Gleeson et al. (2011). The permeability decline with depth is prescribed by the sediment–bedrock profile at a location, which depends strongly on terrain slope; the steeper the land, the thinner the weathered layer and the sharper the decrease in permeability with depth. This is expressed through the e-folding depth α (range and global spatial distribution of e-folding depth is given in Figs. S2 and S3 and taken from Miguez-Macho et al., 2008). With near-surface permeability k_0 (m d^{-1}, from Table 1) the transmissivity $T(x)$ (m^2 d^{-1}) over the aquifer depth $D(x)$ (m) can then be calculated as

$$T(x) = \int_0^{D(x)} k_0 e^{\frac{-z}{\alpha}} \mathrm{d}z. \tag{6}$$

As Eq. (6) is an exponential function, permeabilities will approximate zero at greater depth.

It is assumed that conductivities are horizontally homogeneous within a hydrolithological class. The globally calculated transmissivities are presented in Fig. 3b. Note that for mountain ranges low permeabilities are calculated that represent the permeabilities of the bedrock, thereby neglecting weathered regolith soils with high permeabilities that develop on the more gentle slopes. As a result perched water

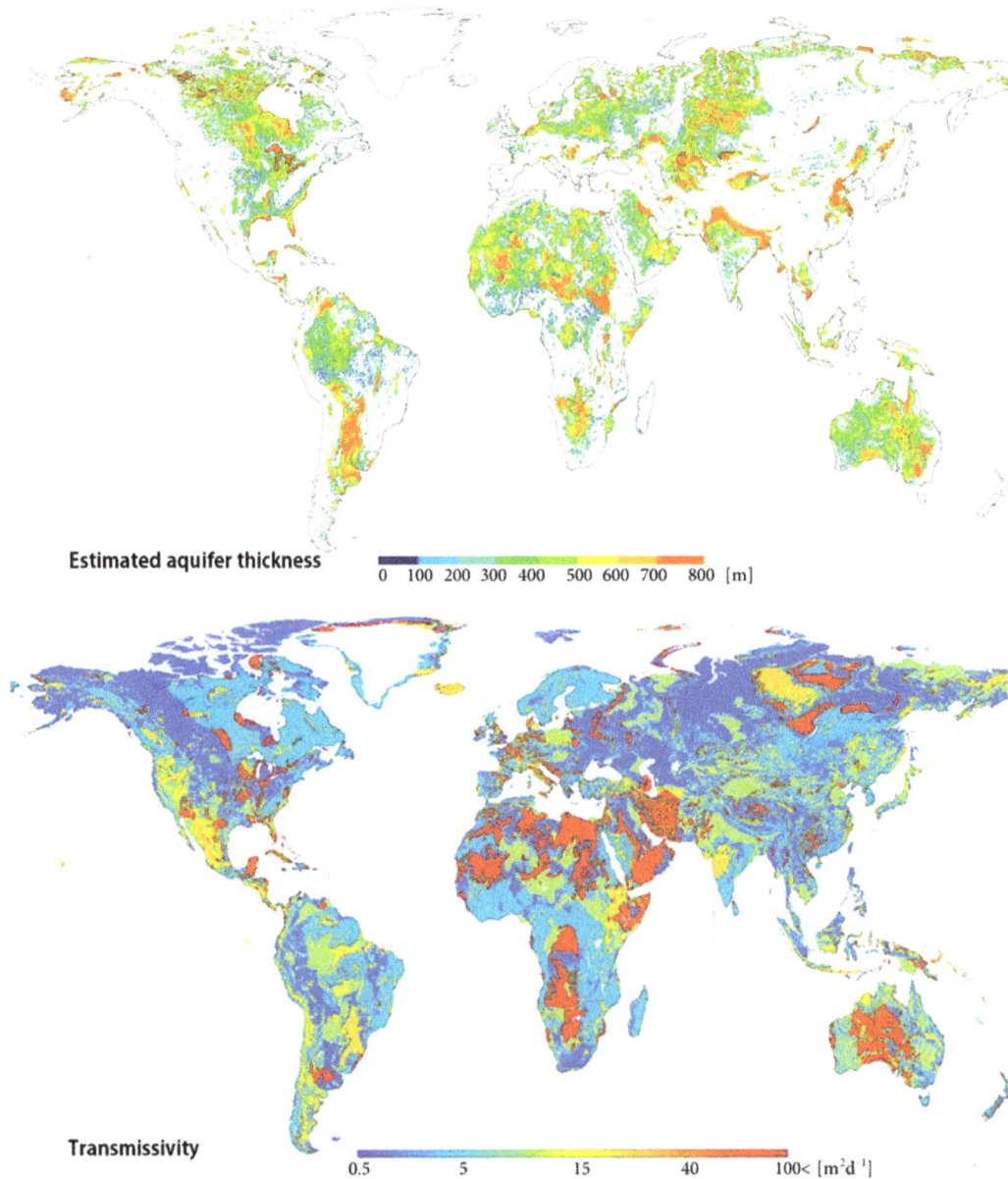

Figure 3. Calculated aquifer thicknesses and transmissivities.

tables that develop in these soils are not included in the simulated lateral groundwater flow (illustrated in Fig. 1b). Instead,
runoff associated with these perched water tables is taken
care of in the land-surface model (PCR-GLOBWB) as storm
flow or interflow from the second soil reservoir. It should be
recognized that our MODFLOW model is built at 6′ and cells
thus have different length units. To account for this a spatially
variable anisotropy factor can be introduced. We have not yet
implemented this option, but will do so in future work.

2.3 Boundary conditions, recharge and drainage levels

For large lakes and the ocean a Dirichlet boundary condition
was used. For the ocean the groundwater head was set at 0 m,

water levels of the lakes were set at elevation levels provided
by the HydroSHEDS digital elevation map.

The steady-state groundwater recharge, shown in Fig. 4
and obtained from PCR-GLOBWB as the long-term average
(for 1960–2000), was used as input for the recharge package of MODFLOW. In the MODFLOW calculation, the input
value of recharge is multiplied by the MODFLOW cell di-
mension to get a volume per unit time, $L^3 T^{-1}$. However, the
input coming from our hydrological model is calculated for
a geographic projected cell, and thus varying surface areas.
For this reason we modified our recharge input as follows:

$$RCH_{inp} = RCH_{act} \times \frac{A_{cell}}{A_{MF}}, \tag{7}$$

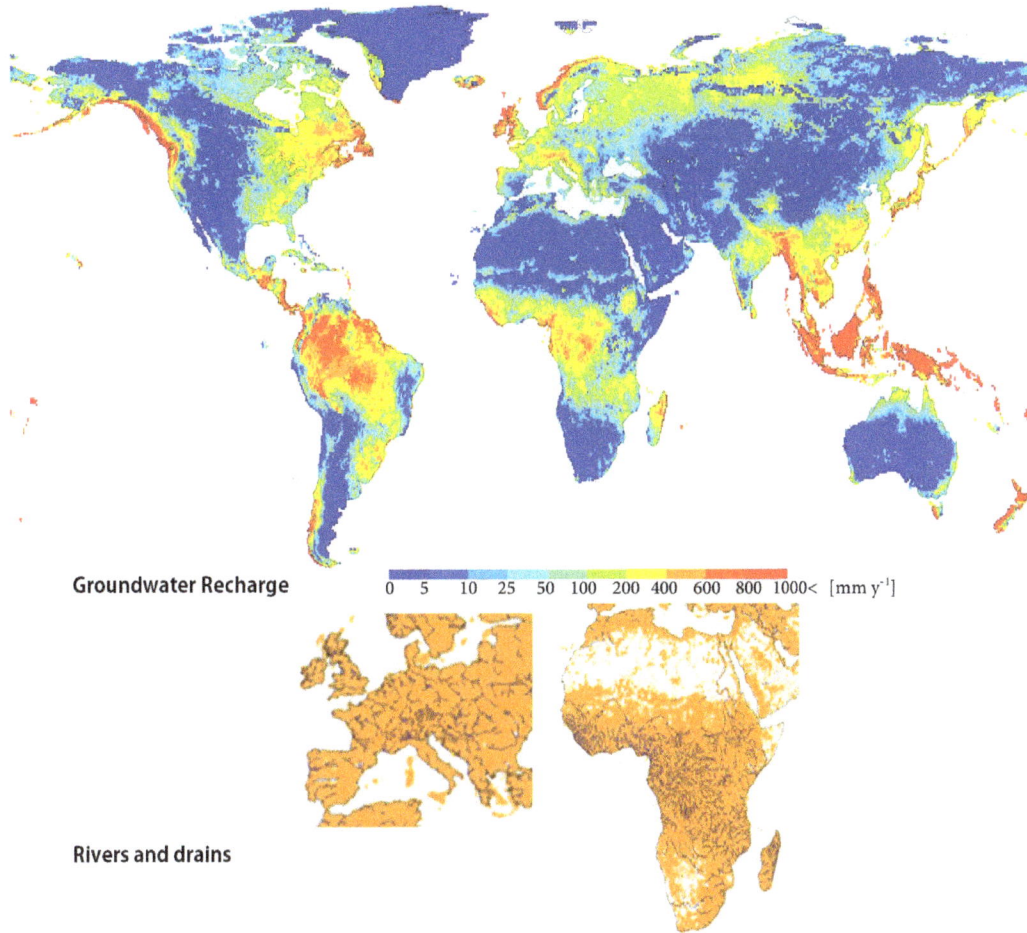

Figure 4. Top panel: steady-state recharge input as obtained from averaging PCR-GLOBWB recharge output over the period 1957–2002. Bottom panel: the hydrography of the imposed stream network is superimposed: large rivers, with widths > 10 m (blue) and smaller rivers, with widths < 10 m (orange).

where A_{cell} is the cell area of the projected cell, A_{MF} is the cell area of the MODFLOW cell, RCH_{act} is the groundwater recharge coming from PCR-GLOBWB (Fig. 4) and RCH_{inp} is the modified input for the MODFLOW calculation.

We used the MODFLOW river (RIV) and drain (DRN) packages to incorporate interactions between groundwater bodies and the surface water. We distinguished three levels of groundwater–surface water interactions: (1) large rivers, with a width > 10 m, (2) smaller rivers, with a width < 10 m and (3) springs and streams higher up in river valleys.

1. For the larger rivers the interactions are governed by actual groundwater heads and surface water levels. The latter can be obtained from the long-term average naturalized river discharge, $\overline{Q_{chn}}$ (calculated by PCR-GLOBWB), by using assumed channel properties. These are channel width, W_{chn} (L), channel depth, D_{chn} (L), Manning roughness coefficient, n (L$^{-1/3}$ T^{-1}) and channel longitudinal slope, Sl (–).

The channel width was calculated using Lacey's formula (Lacey, 1930)

$$W_{chn} \approx P_{bkfl} = 4.8 \times Q_{bkfl}^{0.5}, \qquad (8)$$

where P_{bkfl} (m) is the wetted perimeter, Q_{bkfl} is the long-term averaged natural bankfull discharge (m^3 s^{-1}) and 4.8 (s$^{0.5}$ m$^{-0.5}$) is an empirical factor derived from the relationship between discharge and channel geometry (Savenije, 2003). In large natural braided rivers P_{bkfl} is slightly larger than W_{chn}. The bankfull discharge was calculated from the simulated river discharges and occurs, as a rule of thumb, every 1.5 years. Combining Lacey's formula with Manning's formula (Manning, 1891) assuming a rectangular channel gives the channel depth:

$$D_{\text{chn}} = \left(\frac{n \times Q_{\text{bkfl}}^{0.5}}{4.8 \times \text{SI}^{0.5}} \right)^{\frac{3}{5}}. \qquad (9)$$

By subtracting D_{chn} from surface elevation we estimated the bottom elevation of the river bed RBOT (m). The average river head HRIV (m) was subsequently calculated from the long-term average naturalized river discharge $\overline{Q_{\text{chn}}}$ using the Manning formula

$$\text{HRIV} = \text{RBOT} + \left(\frac{n \times \overline{Q_{\text{chn}}^{0.5}}}{B_{\text{chn}} \times \text{SI}^{0.5}} \right)^{\frac{3}{5}}. \qquad (10)$$

The RBOT and the HRIV were used as input for the RIV package in MODFLOW to calculate the flow between the river and aquifer: Q_{riv} (m³ s⁻¹). If the head in the cell connected to the river drops below the bottom of the river bed, water enters the groundwater system from the river at a constant rate. If the head is above the bottom of the river bed, water will either enter or leave the aquifer system depending on whether the head is above or below the river head. Q_{riv} is positive when water from the river enters the aquifer and is calculated as follows:

$$Q_{\text{riv}} = \begin{cases} c \times (\text{HRIV} - h) & \text{if } h > \text{RBOT} \\ c \times (\text{HRIV} - \text{RBOT}) & \text{if } h \leq \text{RBOT}, \end{cases} \qquad (11)$$

where h is groundwater head (m), and c is a conductance (m² d⁻¹) calculated as

$$c = \frac{1}{\text{BRES}} \times P_{\text{chn}} \times L_{\text{chn}}, \qquad (12)$$

where BRES is bed resistance (d, taken 1 day here), L_{chn} (m) is the channel length (approximated the diagonal cell length) and P_{chn} is the wetted perimeter (approximated by W_{chn}). The river package was used only for cells with large rivers, i.e. $W_{\text{chn}} \geq 10\,\text{m}$.

2. To simulate smaller rivers, $W_{\text{chn}} < 10\,\text{m}$, the DRN package was used. Water can only leave the groundwater system through the drain when head rises above the drainage level which was taken equal to the surface elevation, DEM. The drainage Q_{drn} (m³ d⁻¹) is then calculated as follows:

$$Q_{\text{drn}} = \begin{cases} c \times (\text{DEM} - h) & \text{if } h > \text{DEM} \\ c \times 0 & \text{if } h \leq \text{DEM}. \end{cases} \qquad (13)$$

Figure 4 shows larger rivers and active smaller rivers for Europe and Africa.

3. Q_{riv} and Q_{drn} quantify flow between streams and aquifer and are the main components of the baseflow

Q_{bf} which is negative when water flows into the river. However, at 6′ resolution the main stream is insufficient to represent truthfully all locations within a cell where groundwater levels intersect the terrain and additional drainage is needed to represent local sags, springs, and streams higher up in valleys in mountainous areas. To resolve this issue, we assumed that groundwater above the floodplain level can be tapped by local springs (illustrated in Fig. 1b) which were presented by means of a linear storage–outflow relationship. To be consistent with the RIV and DRN packages, this term is also negative when water is drained. Thus, total groundwater drainage was simulated as

$$Q_{\text{bf}} = (Q_{\text{riv}} + Q_{\text{drn}}) - (J S_{3,\text{flp}}), \qquad (14)$$

where $S_{3,\text{flp}}$ (m) is the groundwater storage above the floodplain as obtained from PCR-GLOBWB and J (d⁻¹) is a recession coefficient parameterized based on Kraaijenhof van der Leur (1958):

$$J = \frac{\pi T}{4 S_y L^2}, \qquad (15)$$

where T (m² d⁻¹) is the transmissivity as used in the groundwater model, S_y is the storage coefficient assumed for each hydrolithological category (see Table 1) and L is the average distance between streams and rivers as obtained from the stream density (see van Beek et al., 2011).

2.4 Sensitivity analysis of aquifer properties and recharge

In groundwater modelling the transmissivity and groundwater recharge are important parameters and subject to large uncertainty. In this study we investigated the sensitivity of the model outcome to changes in the aquifer parameters (conductivity, thickness) and recharge.

For each parameter a Monte Carlo simulation of 100 samples was performed. This simulation followed a log-normal distribution for layer thickness and saturated conductivity. For groundwater recharge a normal distribution was used. For layer thickness, mean and standard deviation were obtained by combining several case studies of the USA and extrapolating this globally (see Sect. 2.2, Eqs. 1–5). Means and standard deviations of saturated conductivities per hydrolithological class were taken from Gleeson et al. (2011) (see Table 1). Mean and standard deviations for groundwater recharge were taken for the PCR-GLOBWB sensitivity study of Wada et al. (2014).

The variation in groundwater depth caused by changing one parameter was evaluated by calculating maps coefficients of variation (presented in Fig. 5). To obtain the uncertainty from the combination of these parameters, for each parameter 10 evenly distributed quantiles were determined and combined into 1000 parameter sets with which to run the model.

Again variation in groundwater depth was evaluated by calculating maps of coefficients of variation.

2.5 Validation of groundwater heads

Simulated groundwater depths were validated against a compilation of reported piezometer data (Fan et al., 2013). The average of the reported data was used if more than one observation was available within a $6'$ cell, giving a total of 65 303 cells with observations worldwide (of the total 6 480 000 cells). The water table head, instead of depth, was evaluated as it measures the potential energy that drives flow and is therefore physically more meaningful. The coefficient of determination (R^2) and regression coefficient (α) were calculated for every run (results presented in Fig. 6). Residuals (res) were calculated as simulated heads minus observed heads and maps are presented (Fig. 7).

2.6 Simulating flow paths

Particle tracking, using MODPATH (Pollock, 1994), was included to track flow paths and estimate travel times of groundwater flows. For this simulation cell-to-cell flux densities, defined as the specific discharge per unit of cross sectional area, were used. A flow path is computed by tracking the particle from one cell to the next until it reaches a boundary or sink. It shows the path through the subsoil that the groundwater follows from the location of infiltration towards the location of drainage. In our case the particle was stopped when it reached the ocean, a lake or the local drainage (rivers or drains). It provides insights in regional-scale groundwater movements and groundwater age, indicating areas where lateral groundwater flows are significant and inter-basin groundwater flows are important. The latter positively affects water budgets of neighbouring catchments or recharges the larger aquifer systems, thereby increasing water availability of these neighbouring catchments. Results are presented for Europe and Africa, showing paths and travel times (Fig. 9).

3 Results and discussion

3.1 Sensitivity analysis

Figure 5 shows the coefficient of variation (CV) of calculated groundwater depths with changing parameter settings for saturated conductivity, aquifer thickness, and recharge. Overall CVs are small, less than 1. Higher CVs are found for the Sahara and Australian desert, where recharge is low, transmissivities are high, and groundwater levels become disconnected from the surface. This emphasizes the influence of regional-scale lateral flow in these areas. Higher variations are also found for areas with shallow groundwater tables and higher transmissivities and recharge, like the Amazon and Indus Basin.

Figures of CVs of simulated groundwater depths resulting from changing a single parameter only are presented in the Supplement of this paper (Fig. S4). The CVs from changing saturated conductivity are almost similar to the total CV, illustrating that saturated conductivity is the predominant control of groundwater depth. This is expected as the standard deviation of saturated conductivity is large for several hydrolithological classes (Table 1), changing saturated conductivity by orders of magnitude. In general a higher-saturated conductivity leads to lower water tables and more significant regional groundwater flow, and vice versa.

The other two parameters, aquifer thickness and groundwater recharge, are of lower importance. Although different thicknesses do change transmissivities, impact on calculated groundwater depths is small. Also, the effect of changing groundwater recharge is small. This is the direct result of the small relative uncertainty compared to hydraulic conductivity. Beside this, drainage is self-limiting; as recharge increases, the water table rises and the hydraulic gradient is steepened, accelerating drainage and lowering the water table. This dampens the water table sensitivity to recharge uncertainties.

3.2 Validation of groundwater simulated heads

Simulated groundwater heads were compared to piezometer observations. A scatter plot of the best performing run (after changing three parameters) is presented in Fig. 6 and spatial patterns are presented in Fig. 7. It should be mentioned here that for most regions of the world no observation data are available (see Fig. S5) or are incomplete (i.e. no elevation measurement). While interpreting the results it should be noted that observation locations are biased towards river valleys, coastal ribbons and the areas where productive aquifers occur. Also, observations are taken at a certain moment in time, and thus are liable to seasonal effects and drawdown as a result of abstractions, while simulated groundwater heads represent the steady-state average. Beside this, for the mountain ranges it is likely that observations are located in small mountain valleys with shallow local water tables, partly from infiltrating streams. Our grid resolution is too coarse to capture these small-scale features. Also, occasionally observations of perched water tables in hill slopes are included. These perched groundwater tables are not described by our large-scale groundwater model (simulating the regional-scale groundwater; see Fig. 1b), but captured in the land-surface model as interflow.

For all runs the computed coefficient of determination (R^2) was calculated and found to fall between 0.75 and 0.87. For the 10 best performing runs it ranges between 0.85 and 0.87. The scatter of the best performing run is given in Fig. 6. The presented scatter and statistics of R^2 and regression coefficient α in Fig. 6 show that the model performance is good. However, the scatter shows a strong underestimation of groundwater heads, meaning that simulated ground-

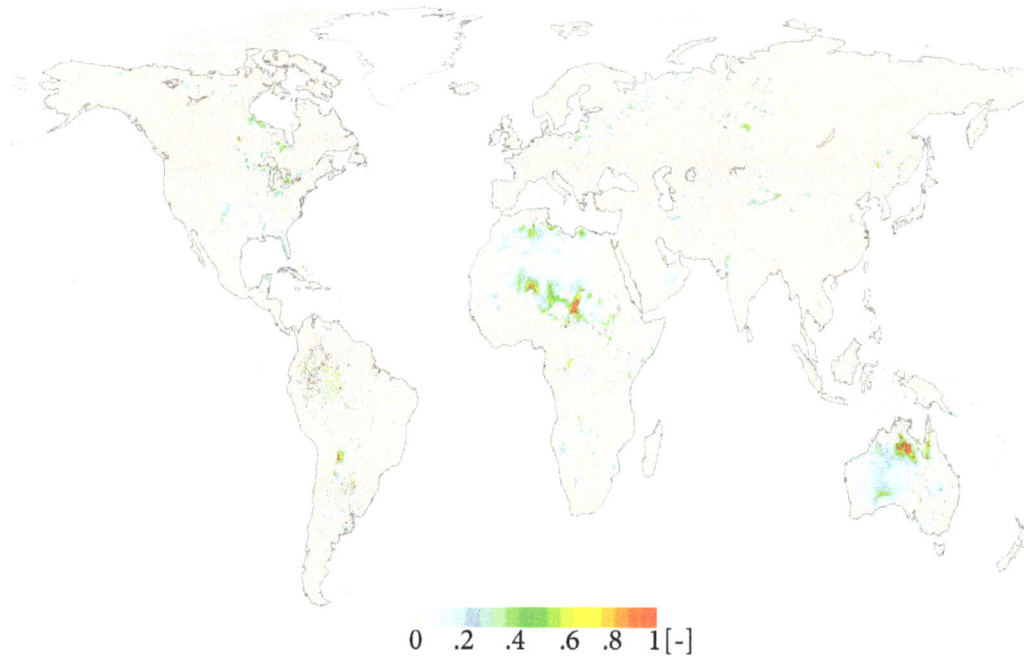

Figure 5. Coefficient of variation of groundwater depth of 1000 runs with different parameter settings for aquifer thickness, saturated conductivity and groundwater recharge.

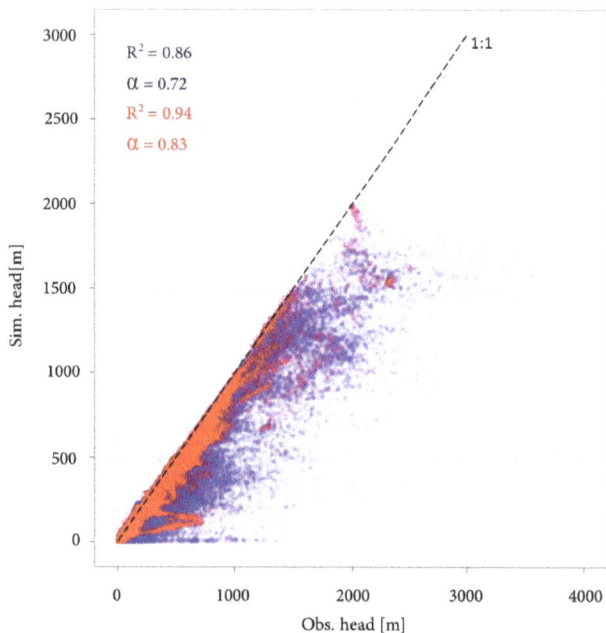

Figure 6. Scatter plot of observed heads against simulated heads for sediment basins (red) and mountain ranges (blue).

water tables are too deep compared with the observations. This appears especially for higher elevated areas (also shown in Fig. 7). This underestimation is expected as most likely for higher elevated areas shallow local water tables are sampled which are not captured by our model as a result of the limited grid resolution. Therefore, we evaluated R^2 and α for mountain ranges and sediment basins separately shown in Fig. 6 in blue and red respectively. The R^2 for the sediment basins is slightly better than for mountain ranges, but in general water table elevations here are still underestimated. The R^2 for sediment basins ranges between 0.90 and 0.95, and for the 10 best performing runs between 0.945 and 0.946. For the lower elevations (approx. 0–500 m) a small overestimations of heads can be seen as well.

In Fig. 7 the spatial distribution of the residuals of groundwater heads and corresponding histograms are shown for Europe and the USA. The figures confirm the above-stated conclusion that heads are generally underestimated compared to the observations. The largest underestimations area found for higher elevated areas, such as the Rocky Mountains. Groundwater heads are best estimated for lower flatter areas, like the Mississippi embayment and the Netherlands. The histograms show that larger residuals are found for areas where groundwater levels are deeper, and smaller residuals are found for more shallow groundwater levels. This shows that, although absolute values of groundwater heads are underestimated, the general pattern of deep and shallow groundwater is well captured by the model.

3.3 Global groundwater depth map

Figure 8 shows for the best performing run the simulated steady-state groundwater table depths at its natural state (without pumping), in metres below the land surface (result of the best performing run).

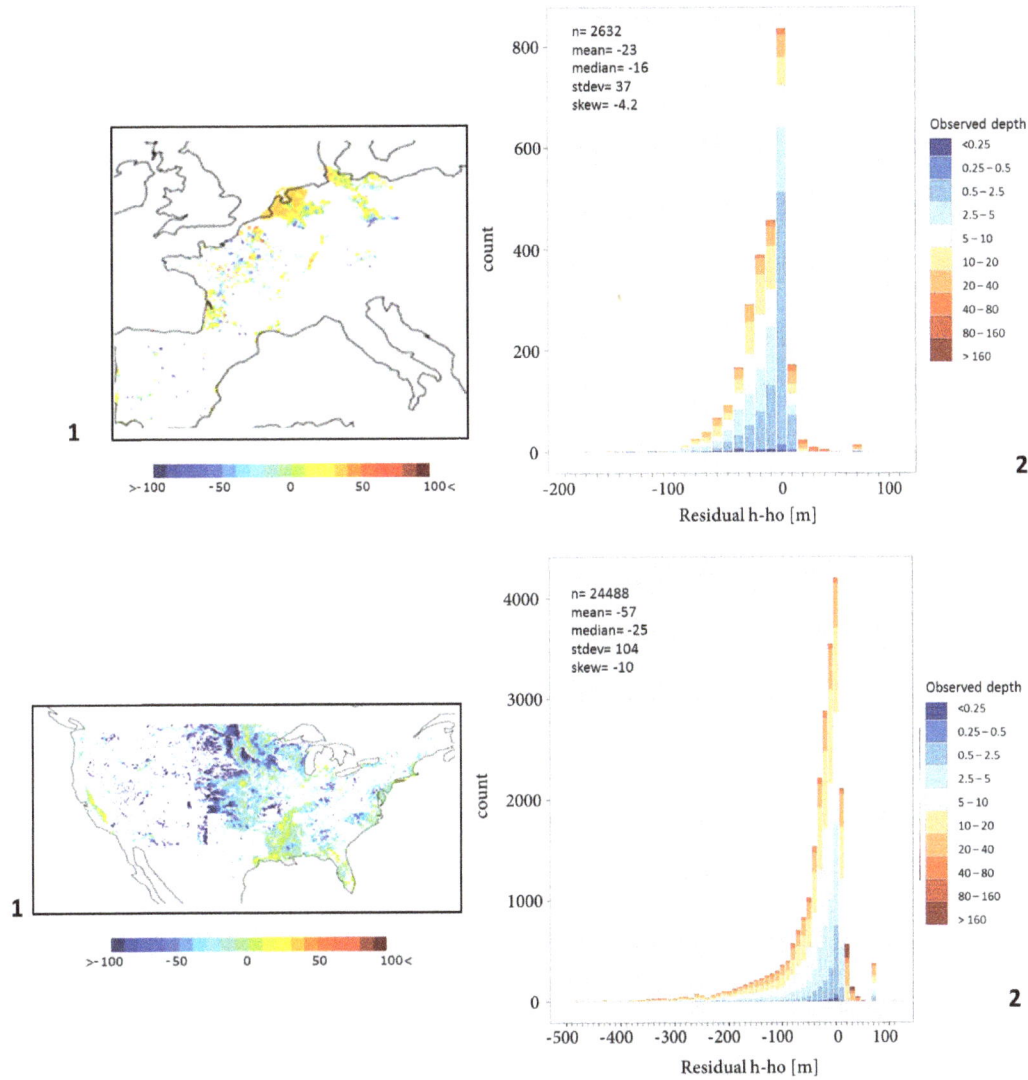

Figure 7. Maps of residuals for NW Europe and USA. (**1**) Residuals and (**2**) histograms of residuals. Each bar in the histogram is clustered based on observed on classified groundwater depths.

General patterns in water table depths can be identified. At the global scale, sea level is the main control of groundwater depth. Throughout the entire coastal ribbon shallow groundwater tables occur. These areas expand where flat coastal plains meet the sea, including major river basins such as the Mississippi, the Indus and large wetlands. At the regional scale, recharge is the main control in combination with scale topography. For regions with high groundwater recharge rates, shallow groundwater tables are simulated, for example the tropical swamps of the Amazon. The influence of the regional topography is also evident in the central Amazon and for the flat lowlands of South America as these regions receive water from elevated areas.

Regions with low recharge rates correspond with deep groundwater where groundwater head gets disconnected from the local topography. The great deserts stand out

(hyper-arid regions dotted in Fig. 8). Also for the mountain ranges of the world deep groundwater tables are simulated. As stated before, small local valleys with higher local groundwater tables are not captured by the model due to the used grid resolution. The mountainous regions where local and perched water tables are likely to occur are masked in the figure with a semi-transparent layer.

3.4 Groundwater flow paths and travel time

Figure 9 shows the simulated large-scale flow paths for Europe and Africa where different colours indicate the simulated travel times. These figures show both short and long inter-basin flow paths, which are stopped when they reach the local drainage, such as a lake or the ocean. Long flow paths are for example found in eastern Europe, where flow paths

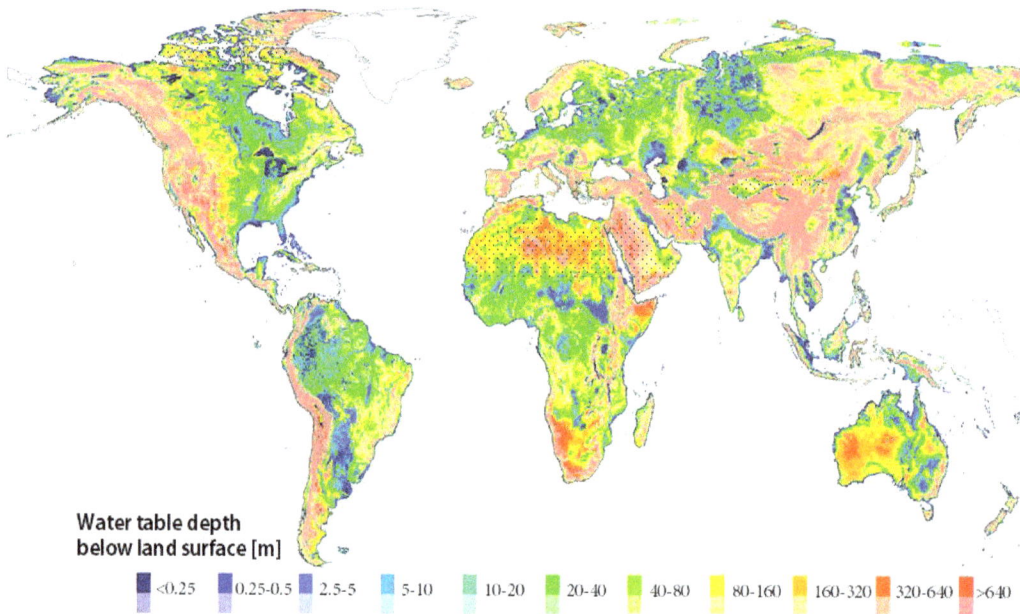

Figure 8. Simulated water table depth in metres below the land surface. The result of a steady-state natural run, using the best estimated parameter set. Semi-transparent colours indicate deep groundwater regions where most likely more shallow perched and local water tables not captured by the model. Dotted areas indicate hyper-arid zones, distinguished at the grid resolution. White areas indicate no-data values.

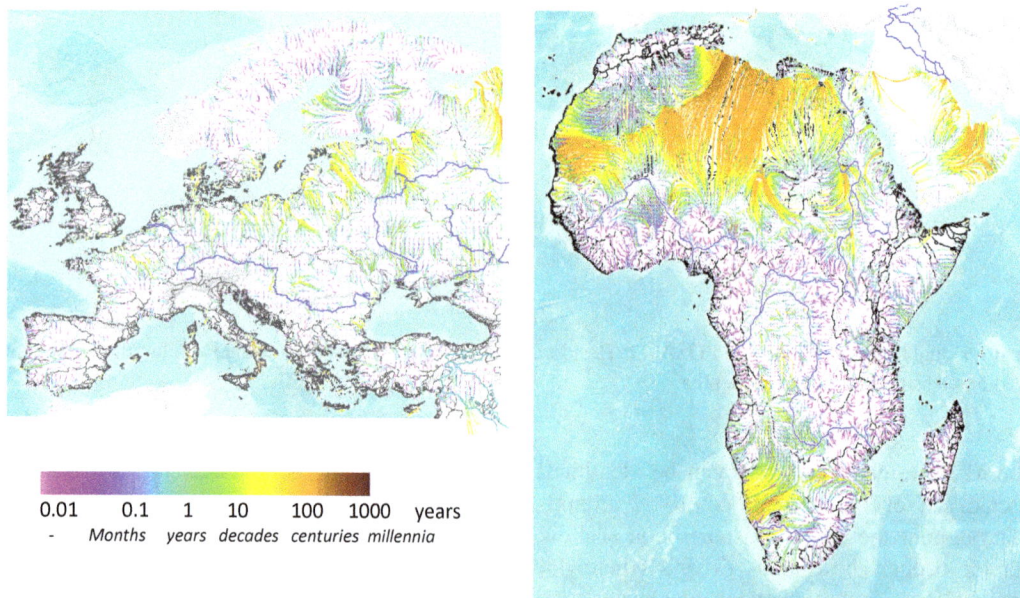

Figure 9. Flow paths simulated for NW Europe and Africa, underlain by river basin boundaries, overlain with major rivers.

cross several catchment boundaries and end as submarine groundwater discharge. Also, inter-basin flow paths recharge larger aquifer systems, such is the case for the upper-Danube aquifer system. For Africa long inter-basin flow paths are evident for the desert area as well. The flow path simulations show that especially for sediment areas, inter-basin groundwater flow is important and significant at least at longer timescales. It should be noted that these larger-scale flow paths are associated with the scale of the model. Obviously, superimposed on these areas sub-regional to local-scale flow paths of shallow groundwater systems exist that are not captured by our model (Toth, 1963).

4 Conclusions

In this paper a global-scale groundwater model of an upper unconfined aquifer layer is presented. A feasible and relatively simple method is introduced to overcome the limited information available for aquifer parameterization; available global data sets for lithology and saturated conductivity were used such that the parameterization method can be expanded to data-poor environments.

By applying this method we are able to produce a global picture of water table depths at fine resolution (6′) within good accuracy in many part of the world, especially for sediment basins ($R^2 = 0.95$ and $\alpha = 0.84$). The sediment basins are specific areas of interest, as these include the major aquifer systems of the world (e.g. Indus, Ganges, High Plains). For the higher and steeper terrain, groundwater depths are in general overestimated compared to observations (simulated depths deeper than observed); as a result, perched water tables on hillsides are not included in the groundwater model but are present in the observations. Additionally, the model resolution and the aquifer property estimation are still too coarse to capture shallow water tables in small sediment pockets in small mountain valleys.

The results presented in this study confirm the relevance of taking lateral groundwater flow into account in global-scale hydrological models. Short and long flow paths, also over catchment boundaries as inter-basin flow paths, are simulated. The latter can be of major importance as it provides additional recharge to a catchment and thereby helps to sustain river baseflow in times of droughts, supporting ecosystems and wetlands and increasing surface water availability for human water use. Also, inter-basin groundwater flows can act as additional recharge to large aquifer systems, thereby increasing water availability in these aquifers.

Obviously the model presented here must be considered as a first-order attempt towards global groundwater modelling and consequently has a number of limitations that prevent it from simulating groundwater dynamics completely truthfully.

Firstly, the model simulates a natural dynamic steady state; it does not provide any information about groundwater fluctuations caused by climate (seasonal and inter-annual) or human water use. Obviously, as we have estimated specific yield as well, extension to transient simulations is straightforward and will be attempted in a future study.

Secondly, only one unconfined layer is modelled here, while in reality, multi-layered aquifers including unconsolidated and consolidated layers can be present. Before we can include human groundwater use globally, these multi-layered aquifers should be included in the model as this holds vital information on the accessibility and quality of global groundwater resources. However, the information on these aspects is sparse and incomplete.

Thirdly, capillary rise of the water table into the soil has not yet been implemented, although several studies have pointed out that it can affect soil moisture, evaporation or even precipitation (e.g. Bierkens and van den Hurk, 2007; Fan et al., 2013; Lam et al., 2011). Further, there is no dynamic interaction between groundwater and surface water, as the drainage level of rivers does not change over time.

That being said, our model has the ability to capture the large-scale distribution of groundwater levels and as such can serve as a starting point leading to a tool to assess groundwater level fluctuations and their sensitivity to human water intervention and climate.

The next step of this work will be to expand the current aquifer schematization with multi-layered and confined aquifer systems. The model will become transient and fully coupled to the land-surface model in order to incorporate capillary rise to the soil moisture and link river dynamics with groundwater. Human water use will be included as well. The goal will be to represent the impact of human water use on groundwater dynamics and river discharges. It will show where and when limits of groundwater abstractions will be reached. This is vital information needed to ensure sustainable and efficient groundwater use, particularly for semi-arid regions where groundwater demand will intensify due to the increase of drought frequency and duration, combined with population growth, expansion of irrigation areas and rising standard of living.

Acknowledgements. The authors thank Peter Vermeulen (Deltares) for his assistance with iMOD which we used for the particle tracking. The authors also would like to thank the three reviewers, Nir Krakauer (The City College of New York), Mary Hill (USGS) and one anonymous reviewer for their constructive and thoughtful suggestions on an earlier version of this paper. This study was funded by the Netherlands Organization for Scientific Research (NWO) in the project Planetary Boundaries Fresh Water Cycle.

Edited by: H.-J. Hendricks Franssen

References

Beven, K. J. and Kirkby, M. J.: Considerations in the development and validation of a simple physically based, variable contributing area model of catchment hydrology, in: Surface and subsurface hydrology, Proc. Fort Collins 3rd international hydrology symposium, July 1977, Fort Collins, 23–36, 1979.

Bierkens, M. F. P. and van den Hurk, B. J. J. M.: Groundwater convergence as a possible mechanism for multi-year persistence in rainfall, Geophys. Res. Lett., 34, L02402, doi:10.1029/2006GL028396, 2007.

Clark, B. and Hart, R.: The Mississippi Embayment Regional Aquifer Study (MERAS): Documentation of a groundwater-flow model contructed to assess water Availability in the Mississippi Embayment, Tech. Rep. 5172, US Geological Survey Scientific Investigations Report, US Geological Survey, Reston, Virginia, 2009.

de Graaf, I. E. M., van Beek, L. P. H., Wada, Y., and Bierkens, M. F. P.: Dynamic attribution of global water demand to surface water and groundwater resources: effects of abstractions and return flows on river discharges, Adv. Water Resour., 64, 21–33, doi:10.1016/j.advwatres.2013.12.002, 2014.

Dee, D., Uppala, S., Simmons, A., Berrisford, P., Poli, P., Kobayashi, S., Andrae, U., Balmaseda, M., Balsamo, G., Bauer, P., Bechtold, P., Beljaars, A., van de Berg, L., Bidlot, J., Bormann, N., Delsol, C., Dragani, R., Fuentes, M., Geer, A., Haimberger, L., Healy, S., Hersbach, H., Hólm, E., Isaksen, L., Kållberg, P., Köhler, M., Matricardi, M., Mcnally, A., Monge-Sanz, B., Morcrette, J.-J., Park, B.-K., Peubey, C., de Rosnay, P., Tavolato, C., Thépaut, J.-N., and Vitart, F.: The ERA-Interim reanalysis: configuration and performance of the data assimilation system, Q. J. Roy. Meteorol. Soc., 137, 553–597, 2011.

Dürr, H. H., Meybeck, M., and Dürr, S. H.: Lithologic composition of the Earth's continental surfaces derived from a new digital map emphasizing riverine material transfer, Global Biogeochem. Cy., 19, GB4S10, doi:10.1029/2005GB002515, 2005.

Fan, Y., Miguez-Macho, G., Weaver, C. P., Walko, R., and Robock, A.: Incorporating water table dynamics in climate modeling: 1. Water table observations and equilibrium water table simulations, J. Geophys. Res.-Atmos., 112, D10125, doi:10.1029/2006JD008111, 2007.

Fan, Y., Li, H., and Miguez-Macho, G.: Global patterns of groundwater table depth, Science, 339, 940–943, doi:10.1126/science.1229881, 2013.

Faunt, C. C. (Ed.): Groundwater availability of the Central Valley Aquifer, US Geological Survey Professional Paper 1766, US Geological Survey, California, p. 255, 2009.

Gleeson, T., Smith, L., Moosdorf, N., Hartmann, J., Dürr, H. H., Manning, A. H., van Beek, L. P. H., and Jellinek, A. M.: Mapping permeability over the surface of the Earth, Geophys. Res. Lett., 38, L02401, doi:10.1029/2010GL045565, 2011.

Gleeson, T., Wada, Y., Bierkens, M. F. P., and van Beek, L. P. H.: Water balance of global aquifers revealed by groundwater footprint, Nature, 488, 197–200, doi:10.1038/nature11295, 2012.

Gleeson, T., Moosdorf, N., Hartmann, J., and van Beek, L. P. H.: A glimpse beneath earth's surface: GLobal HYdrogeological MaPS (GLHYMPS) of permeability and porosity, Geophys. Res. Lett., 41, 3891–3898, doi:10.1002/2014GL059856, 2014.

Hagemann, S., Botzet, M., Dümenil, L., and Machenhauer, B.: Derivation of global GCM boundary conditions from 1 km land use satellite data, Rep. 289, Tech. Rep., Max Planck Institude for Meteorology, Hamburg, Germany, 1999.

Hartmann, J. and Moosdorf, N.: The new global lithological map database GLiM: a representation of rock properties at the Earth surface, Geochem. Geophy. Geosy., 13, Q12004, doi:10.1029/2012GC004370, 2012.

Ingebritsen, S. and Manning, C.: Geological implications of a permeability-depth curve for the continental crust, Geology, 27, 1107–1110, doi:10.1130/0091-7613(1999)027<1107:GIOAPD>2.3.CO;2, 1999.

Konikow, L. F.: Contribution of global groundwwater depletion since 1900 to sea-level rise,geophysical research letters, 38, L17401, doi:10.1029/2011GL048604, 2011.

Kraaijenhof van der Leur, D.: A study of non-steady grounground flow with special reference to a reservoir coefficient, de ingenieur, 70, 87–94, 1958.

Krakauer, N. Y., Li, H., and Fan, Y.: Groundwater flow across spatial scales: importance for climate modeling, Environ. Res. Lett., 9, 034003, doi:10.1088/1748-9326/9/3/034003, 2014.

Lacey, G.: Stable cchannel in alluvium, P. I. Civ. Eng., 229, 259–292, 1930.

Lam, A., Karssenberg, D., van den Hurk, B. J. J. M., and Bierkens, M. F. P.: Spatial and temporal connections in groundwater contribution to evaporation, Hydrol. Earth Syst. Sci., 15, 2621–2630, doi:10.5194/hess-15-2621-2011, 2011.

Manning, R.: On the flow of water in open cchannel and pipes, Transactions of Institution of Civil Engineers of Ireland, Dublin, 161–207, 1891.

McDonald, M. G. and Harbaugh, A. W.: MODFLOW-2000, the US Geological Survey modular ground-water model – User guide to modularmodular concepts and the Ground-Water Flow Process, Tech. rep., US Geological Survey, Reston, Virginia, 2000.

Miguez-Macho, G., Li, H., and Fan, Y.: Simulated water table and soil moisture climatology over north america, B. Am. Meteorol. Soc., 89, 663–672, doi:10.1175/BAMS-89-5-663, 2008.

Mitchell, T. D. and Jones, P. D.: An improved method of constructing a database of monthly climate observations and associated high-resolution grids, Int. J. Climatol., 25, 693–712, doi:10.1002/joc.1181, 2005.

Pollock, D.: User's Guide for MODPATH/MODPATH-PLOT, version 3: A particle tracking post-processing package for MODFLOW, the US Geological Survey finite-difference groundwater flow model, Tech. rep., US Geological Survey Open-File Report 94-464, US Geological Survey, Reston, Virginia, 1994.

Savenije, H. H.: The width of a bankfull channel; Lacey's formula explained, J. Hydrol., 276, 176–183, doi:10.1016/S0022-1694(03)00069-6, 2003.

Schaller, M. F. and Fan, Y.: River basins as groundwater exporters and importers: implications for water cycle and climate modeling, J. Geophys. Res.-Atmos., 114, D04103, doi:10.1029/2008JD010636, 2009.

Schmitz, O., Karssenberg, D., van Deursen, W., and Wesseling, C.: Linking external components to spatio-temporal modeling framework: coupling MODLFW and PCRaster, Environ. Modell. Softw., 24, 1088–1099, doi:10.1016/j.envsoft.2009.02.018, 2009.

Summerfield, M. and Hulton, N.: Natural controls of fluvial denudation rates in major world drainage basins, J. Geophys. Res., 99, 13871–13883, 1994.

Sutanudjaja, E. H., van Beek, L. P. H., de Jong, S. M., van Geer, F. C., and Bierkens, M. F. P.: Large-scale groundwater modeling using global datasets: a test case for the Rhine-Meuse basin, Hydrol. Earth Syst. Sci., 15, 2913–2935, doi:10.5194/hess-15-2913-2011, 2011.

Sutanudjaja, E. H., van Beek, L. P. H., de Jong, S. M., van Geer, F. C., and Bierkens, M. F. P.: Calibrating a large-extent high-resolution coupled groundwater-land surface model using soil moisture and discharge data, Water Resour. Res., 50, 687–705 doi:10.1002/2013WR013807, 2014.

Todini, T.: The arno rainfall-runoff model, J. Hydrol., 175, 339–382, 1996.

Toth, J.: A theoretical analysis of groundwaterflow in small drainage basins, J. Geophys. Res., 68, 2156–2202, doi:10.1029/JZ068i016p04795, 1963.

Uppala, S. M., Kållberg, P. W., Simmons, A. J., Andrae, U., Bechtold, V. D. C., Fiorino, M., Gibson, J. K., Haseler, J., Hernandez, A., Kelly, G. A., Li, X., Onogi, K., Saarinen, S., Sokka, N., Allan, R. P., Andersson, E., Arpe, K., Balmaseda, M. A., Beljaars, A. C. M., Berg, L. V. D., Bidlot, J., Bormann, N., Caires, S., Chevallier, F., Dethof, A., Dragosavac, M., Fisher, M., Fuentes, M., Hagemann, S., Hólm, E., Hoskins, B. J., Isaksen, L., Janssen, P. A. E. M., Jenne, R., Mcnally, A. P., Mahfouf, J.-F., Morcrette, J.-J., Rayner, N. A., Saunders, R. W., Simon, P., Sterl, A., Trenberth, K. E., Untch, A., Vasiljevic, D., Viterbo, P., and Woollen, J.: The ERA-40 re-analysis, Q. J. Roy. Meteorol. Soc., 131, 2961–3012, doi:10.1256/qj.04.176, 2005.

van Beek, L. P. H., Wada, Y., and Bierkens, M. F. P.: Global monthly water stress: 1. Water balance and water availability, Water Resour. Res., 47, W07517, doi:10.1029/2010WR009791, 2011.

Vergnes, J.-P., Decharme, B., Alkama, R., Martin, E., Habets, F., and Douville, H.: A simple groundwater scheme for hydrological and climate applications: description and offline evaluation over france, J. Hydrometeorol., 13, 1149–1171, doi:10.1175/JHM-D-11-0149.1, 2012.

Wada, Y., Wisser, D., and Bierkens, M. F. P.: Global modeling of withdrawal, allocation and consumptive use of surface water and groundwater resources, Earth Syst. Dynam., 5, 15–40, doi:10.5194/esd-5-15-2014, 2014.

Wood, E. F., Roundy, J. K., Troy, T. J., van Beek, R., Bierkens, M., Blyth, E., de Roo, A., Döll, P., Ek, M., Famiglietti, J., Gochis, D., van de Giesen, N., Houser, P., Jaffe, P., Kollet, S., Lehner, B., Lettenmaier, D. P., Peters-Lidard, C. D., Sivapalan, M., Sheffield, J., Wade, A. J., and Whitehead, P.: Reply to comment by Keith, J. Beven and Hannah, L. Cloke on "Hyperresolution global land surface modeling: Meeting a grand challenge for monitoring Earth's terrestrial water", Water Resour. Res., 48, W01802, doi:10.1029/2011WR011202, 2012.

6

Scalable statistics of correlated random variables and extremes applied to deep borehole porosities

A. Guadagnini[1,2], S. P. Neuman[1], T. Nan[1], M. Riva[1,2], and C. L. Winter[1]

[1]Department of Hydrology and Water Resources, University of Arizona, Tucson, Arizona 85721, USA
[2]Dipartimeno di Ingegneria Civile e Ambientale, Politecnico di Milano, Piazza L. Da Vinci 32, 20133 Milan, Italy

Correspondence to: T. Nan (tcnan.nju@gmail.com)

Abstract. We analyze scale-dependent statistics of correlated random hydrogeological variables and their extremes using neutron porosity data from six deep boreholes, in three diverse depositional environments, as example. We show that key statistics of porosity increments behave and scale in manners typical of many earth and environmental (as well as other) variables. These scaling behaviors include a tendency of increments to have symmetric, non-Gaussian frequency distributions characterized by heavy tails that decay with separation distance or lag; power-law scaling of sample structure functions (statistical moments of absolute increments) in midranges of lags; linear relationships between log structure functions of successive orders at all lags, known as extended self-similarity or ESS; and nonlinear scaling of structure function power-law exponents with function order, a phenomenon commonly attributed in the literature to multifractals. Elsewhere we proposed, explored and demonstrated a new method of geostatistical inference that captures all of these phenomena within a unified theoretical framework. The framework views data as samples from random fields constituting scale mixtures of truncated (monofractal) fractional Brownian motion (tfBm) or fractional Gaussian noise (tfGn). Important questions not addressed in previous studies concern the distribution and statistical scaling of extreme incremental values. Of special interest in hydrology (and many other areas) are statistics of absolute increments exceeding given thresholds, known as peaks over threshold or POTs. In this paper we explore the statistical scaling of data and, for the first time, corresponding POTs associated with samples from scale mixtures of tfBm or tfGn. We demonstrate that porosity data we analyze possess properties of such samples and thus follow the theory we proposed. The porosity data are of additional value in revealing a remarkable cross-over from one scaling regime to another at certain lags. The phenomena we uncover are of key importance for the analysis of fluid flow and solute as well as particulate transport in complex hydrogeologic environments.

1 Introduction

Hydrogeologic variables such as log permeability are known to vary with scales of measurement, observation, domain of investigation, spatial correlation and resolution (Neuman and Di Federico, 2003). The statistics of these and diverse environmental (as well as earth, financial, astrophysical, biological and many other) variables are likewise known to vary with scale. This is especially true of statistics characterizing spatial and/or temporal increments of these variables. Symptoms of such statistical scaling include irregular spatial variability, persistence or antipersistence of increments (large and small values tending to either persist or alternate rapidly in space and/or time); tendency of increments to have symmetric, non-Gaussian frequency distributions characterized by heavy tails that often decay with separation distance or lag; power-law scaling of sample structure functions (statistical moments of absolute increments) in midranges of lags, with breakdown in power-law scaling at small and/or large lags; linear relationships between log structure functions of successive orders at all lags, also known as extended self-similarity or ESS; and nonlinear scaling of structure function power-law exponents with function order. The traditional interpretation of these widely documented behaviors has been based on the concept of multifractals. This, however, does not explain observed breakdown in power-law scaling at small and large lags or extended power-law scaling (Neuman et al., 2013, and references therein).

Of special concern are the statistics of extremes, which have received much attention among hydrologists (Katz et al., 2002) and others concerned with a wide range of phenomena including snow avalanches on mountain slopes (Ancey, 2012); rupture events associated with the propagation of cracks or sliding along faults in brittle materials including rock failure, landslides and earthquakes (Amitrano, 2012; Lei, 2012; Main and Naylor, 2012) as well as volcanic eruptions, landslides, wildfires and floods (Sachs et al., 2012; Schoenberg and Patel, 2012; Süveges and Davison, 2012); demographic and financial crises (Akaev et al., 2012; Janczura and Weron, 2012); neuronal avalanches and coherence potentials in the mammalian cerebral cortex (de Arcangelis, 2012; Plenz, 2012); citations of scientific papers (Golosovsky and Solomon, 2012); and distributions of city sizes (Pisarenko and Sornette, 2012). Extreme values cluster around heavy tails of data frequency distributions which are often modeled as stretched exponential, lognormal or power functions. There is growing evidence that these frequency distributions, as well as other geospatial and/or temporal statistics of many data, vary with scale. A key related question concerns the scale dependence of frequency distributions (typically generalized extreme value or GEV in the case of block extrema and generalized Pareto distribution or GPD in the case of peaks over thresholds or POTs, e.g., Embrechts et al., 1997) and statistics of extremes at the tails of the original data distributions (e.g., Riva et al., 2013a).

In this paper we explore the statistical scaling of variables and, for the first time, corresponding POTs using as an example neutron porosity data and their POTs from six deep boreholes in three different depositional environments. These data are of interest because, as we show below, (a) they possess statistics that scale in manners typical of many earth, environmental and other variables and (b) reveal a remarkable cross-over from one scaling regime to another at certain separation distances or lags. The phenomena we uncover vis-à-vis neutron porosity data, and corresponding extremes, are of critical importance for the analysis of fluid flow and solute as well as particulate transport in complex hydrogeologic environments. This is so because spatial variability of porosity controls fluid flow velocity distributions in geologic media and has an impact on solute and particulate concentration dynamics. Extreme values of porosity are particularly relevant to depositional processes responsible for the development of preferential flow paths through heterogeneous porous and fractured media. Neutron porosity logs are widely used to characterize stratigraphic sequences and the geostatistical description of geological structures of lithotypes in multilayer systems of aquifers and aquitards (e.g., Barrash and Reboulet, 2004; Tronicke and Holliger, 2005). Combined with laboratory-determined particle size distributions, porosity data may allow one to infer spatial distribu-

tions (see review of Vuković and Soro, 1992) and covariances (Riva et al., 2014) of hydraulic conductivity.

Statistical scaling of hydrogeological data such as permeability or hydraulic conductivity has been studied amongst others by Painter (2001), Meerschaert et al. (2004), Kozubowski et al. (2006), Siena et al. (2012, 2014), Riva et al. (2013b, 2013c), and Guadagnini et al. (2012, 2013, 2014). Whereas research in the subsurface hydrology literature has not addressed specifically the distribution and statistical scaling of extreme incremental values, spatial correlations between values significantly in excess of the mean have been studied vis-à-vis variables such as transmissivity and their relevance to transport processes has been highlighted. Sanchez-Vila et al. (1996) conjectured that observed scale dependence of transmissivities estimated from large-scale pumping tests could be related to strong connectivity between regions of elevated transmissivity, as opposed to spatial persistence of average or low transmissivity values. Spatial correlation of extreme conductivity values was examined for the first time by Gómez-Hernández and Wen (1998). In these authors' opinion the standard multi-Gaussian assumption was not consistent with observed short solute travel times resulting from fast spatially connected pathways. Connectivity of high permeability zones thus became an important concept underlying some modern interpretations of effective conductivity and solute travel time (see for example Meier et al., 1998; Wen and Gómez-Hernández, 1998; Western et al., 2001; Fogg et al., 2000; Zinn and Harvey, 2003; Knudby and Carrera, 2005, 2006; Knudby et al., 2006; Nield, 2008, and references therein). The above ideas have motivated the development of multipoint geostatistical methods of analysis such as those described in a recent special issue of the journal *Mathematical Geosciences* on 20 years of multipoint statistics (e.g., Renard and Mariethoz, 2014; Mariethoz and Renard, 2014, and references therein).

Notably, attempts by hydrologists to investigate the manner in which statistics of extremes vary with scale have centered almost exclusively on peak rainfall intensities and stream flows. Whereas some have found statistical measures of rainfall extremes to exhibit linear (sometimes termed simple) scaling (Menabde et al., 1999; Garcia-Bartual and Schneider, 2001; De Michele et al., 2001) under at least some conditions (Burlando and Rosso, 1996; Veneziano and Furcolo, 2002; Yu et al., 2004), most authors describe them by means of nonlinear (often called multiscaling) models (Burlando and Rosso, 1996; Veneziano and Furcolo, 2002; Castro et al., 2004; Langousis and Veneziano, 2007; Mohymont and Demarée, 2006). Statistical measures of peak stream flows were considered by Javelle et al. (1999), Menabde and Sivapalan (2001) and Rigon et al. (2011) to scale linearly. Work on the scaling of GEVs and/or GPDs associated with extreme rainfall and/or stream flow was reported amongst others by Nguyen et al. (1998), Menabde et al. (1999), Menabde and Sivapalan (2001), Willems (2000), Trefry et al. (2005), Veneziano et al. (2009) and Veneziano and Yoon (2013).

The general tendency has been to interpret linear scaling as a manifestation of monofractal behavior analogous to that of fractional Brownian motion (fBm) or fractional Gaussian noise (fGn). Nonlinear scaling has commonly been attributed to multifractal behavior, a viewpoint espoused originally by Schertzer and Lovejoy (1987) and expanded on recently by Veneziano and Yoon (2013).

Work by our group has demonstrated theoretically (Neuman, 2010, 2011; Guadagnini and Neuman, 2011; Siena et al., 2012; Neuman et al., 2013), computationally (Guadagnini et al., 2012; Neuman et al., 2013) and on the basis of varied pedological, hydrological and hydrogeological data (Siena et al., 2012, 2014; Riva et al., 2013b, c; Guadagnini et al., 2012, 2013, 2014) that statistical scaling behaviors of the kind traditionally attributed to multifractals can be interpreted more simply and consistently by viewing the data as samples from stationary sub-Gaussian random fields subordinated to truncated fBm (tfBm) or fGn (tfGn). Such sub-Gaussian fields are scale mixtures of stationary Gaussian fields with random variances (Andrews and Mallows, 1974; West, 1987) that we model as being lognormal or Lévy-stable (Samorodnitsky and Taqqu, 1994). In this sense our approach bears partial relationship to cascades of Gaussian-scale mixtures that Ebtehaj and Foufoula-Georgiou (2010) use to reproduce coherent structures and extremes of precipitation reflectivity images in the wavelet domain.

Our analysis suggests that, quantitatively, the statistics of neutron porosity increments and their POTs at intra-layer vertical separation scales (or lags) differ from those at inter-layer scales. Qualitatively, however, the statistics of porosity increments at each of these two scales behave in a manner that the literature would typically associate with multifractals. This behavior includes all statistical scaling symptoms described above. Our alternative interpretation of the data allows us to obtain maximum likelihood (ML) estimates of all parameters characterizing the underlying truncated sub-Gaussian fields at both intra- and inter-layer scales. Most importantly, we offer what appears to be the first data-driven exploration (following a synthetic study of outliers by Riva et al., 2013a) of how statistics of POTs associated with such families of sub-Gaussian fields vary with scale.

2   Source of neutron porosity data

As stated in Sect. 1, we illustrate and explore our approach on neutron porosity data from six deep vertical boreholes in three different depositional environments. These are part of a broader set of geophysical logs from the same boreholes, previously described and analyzed within a multifractal framework by Dashtian et al. (2011), provided to us courtesy of Professor Muhammad Sahimi, University of Southern California. Three of the wells (numbered here 1, 2 and 3) are drilled in the Maroon field within which a gas drive is used to produce oil and natural gas, Wells 4 and 5 in the Ahwaz

oil field, and Well 6 in the Tabnak gas field. The Maroon and Ahwaz fields in southwestern Iran, and the Tabnak field in southern Iran, have distinct geologies. Whereas carbonate rock content is highest in the Tabnak and lowest in the Maroon and Ahwaz fields, the opposite is true about sandstone content. Though we do not have information about the relative geographic locations of the six wells, we note that Dashtian et al. (2011) analyzed data from each well independently of those from the remaining five wells. We do the same on the assumption that distances between the wells are sufficiently large to allow treating data from each well as being statistically independent of the rest.

3   Theoretical basis and method of inference

Summary information about the available neutron porosity (P) data is listed in Table 1. As the sampling interval between available values in Well 6 is half of that in Wells 1–5, we disregard every other measurement in analyzing these data, leaving a total of 4267 values. Most of our analysis concerns increments in recorded P values at various separation distances or lags, s, in each well. Lags are taken to be integer multiples, $s = s_n \times \Delta z$, of the vertical spacing, $\Delta z = 0.1524\,\text{m}$, between recorded values.

As stated in Sec. 1, we view the data as samples from stationary sub-Gaussian random fields subordinated to truncated fBm (tfBm) or fGn (tfGn). Sub-Gaussian random variables, defined in Appendix A following standard statistical terminology (e.g., Samorodnitsky and Taqqu, 1994), are scale mixtures of Gaussian variables with random variances. We consider two sub-Gaussian variables, one α-stable with Gaussian variances that are $\alpha/2$-stable, and another normal–lognormal (NLN) variable with lognormal Gaussian variances. There is no physical basis for their choice, just as there usually is no such basis for working with the Gaussian distribution. Lévy-stable (or α-stable) probability distributions are frequently employed due to their ability to interpret heavy tails displayed by empirical distributions of data. While convenient in this sense, this model has the drawback of being associated with densities with diverging moments of order larger than α, notably the variance (e.g., Neuman et al., 2013, and references therein). The use of a lognormal subordinator provides us with the ability to represent tailing behaviors reasonably well with the additional benefit that associated densities possess finite moments of all orders. Regardless of this choice, our approach is compatible with diverse types of subordinators. Using ML we compare the ability of the above two subordinators to (a) capture critical distributional features of our data and (b) yield reliable parameters of the underlying sub-Gaussian random fields.

Statistical scaling of the data is analyzed in part on the basis of sample structure functions, $S_N^q(s_n)$, of order q. Structure functions are moments of order q of absolute increments (e.g., Frisch, 1995). The corresponding sample moments are

Table 1. Summary information about available neutron porosity (P) data.

Reservoir	Well #	Sampling interval (m)	Min P (%)	Max P (%)	Mean P (%)	Standard deviation SD (%)	Number of data points used
Maroon	1	0.1524	0	46.04	14	6.4	3567
(MN)	2	0.1524	0*	74.29	17.27	9.98	4049
	3	0.1524	0*	37.6	15.72	8.54	2945
	1 + 2 + 3	0.1524	0*	74.29	15.74	8.62	10 561
Ahwaz	4	0.1524	0	36.01	16.47	6.82	3882
(AZ)	5	0.1524	0	47.91	16.05	8.35	6949
Tabnak (TBK)	6	0.0762**	0	96.9	9.28	13.2	4267

* These, being negative and very close to zero, were set equal to zero. ** We disregard every other measurement in analyzing these data.

constructed with $N(s_n)$ absolute increments at normalized (by Δz) lags s_n:

$$S_N^q(s_n) = \frac{1}{N(s_n)} \sum_{j=1}^{N(s_n)} |\Delta P_j(s_n)|^q, \tag{1}$$

where $\Delta P_j(s_n)$ is the jth increment of P values separated by lag s_n. The variable P is said exhibits power-law scaling if $S_N^q(s_n) \propto s_n^{\xi(q)}$, where the power or scaling exponent, $\xi(q)$, depends solely on the order q. The exponent is estimated through linear fits of $\log(S_N^q)$ to $\log(s_n)$ within the range of lags where such linear behavior is indicated. We refer to this approach of assessing and quantifying power-law scaling as method of moments.

As shown by Neuman et al. (2013, and references therein), another way to assess the dependence of scaling exponents $\xi(q)$ on q is through ESS or extended power-law scaling. ESS is an empirical approach originally introduced by Benzi et al. (1993a, b, 1996) to widen the range of lags over which velocities in fully developed turbulence scale according to Eq. (1). The approach calls for plotting the S_N^{q+1} vs. S_N^q for various q values and quantifying the resulting linear dependence between them (see Neuman et al., 2013, and references therein). In this work we apply both methods to available neutron porosity data.

To estimate parameters characterizing the distribution of the underlying (Gaussian) tfBm or tfGn, we consider the zero-mean tfBm $G'(x; \lambda_l, \lambda_u)$ defined by Di Federico and Neuman (1997) as a Gaussian random function of space having variance

$$\sigma_G^2(\lambda_l, \lambda_u) = \sigma_G^2(\lambda_u) - \sigma_G^2(\lambda_l), \tag{2}$$

variogram or semi-structure function of second order

$$\gamma_G(s; \lambda_l, \lambda_u) = \gamma_G(s; \lambda_u) - \gamma_G(s; \lambda_l), \tag{3}$$

and integral autocorrelation scale

$$I(\lambda_l, \lambda_u) = \frac{2H}{1+2H} \frac{\lambda_u^{1+2H} - \lambda_l^{1+2H}}{\lambda_u^{2H} - \lambda_l^{2H}}, \tag{4}$$

where for $m = l, u$,

$$\sigma_G^2(\lambda_m) = A\lambda_m^{2H}/2H, \tag{5}$$

$$\gamma_G(s; \lambda_m) = \sigma_G^2(\lambda_m)\rho(s/\lambda_m), \tag{6}$$

A is a coefficient, H is a Hurst scaling exponent and s is lag. The tfBm variogram $\gamma_G(s; \lambda_l, \lambda_u)$ is a weighted integral of variograms characterizing stationary Gaussian fields, or modes, having integral scales λ and variances $\sigma^2(\lambda) = A\lambda^{2H}/2H$, between lower and upper cutoff scales, λ_l and λ_u, respectively. Here we consider modes having Gaussian variograms in which case

$$\rho(s/\lambda_m) = \left[1 - \exp\left(-\frac{\pi}{4}\frac{s^2}{\lambda_m^2}\right) + \left(\frac{\pi}{4}\frac{s^2}{\lambda_m^2}\right)^H \right.$$
$$\left. \Gamma\left(1 - H, \frac{\pi}{4}\frac{s^2}{\lambda_m^2}\right) \right] 0 < H < 1, \tag{7}$$

where $\Gamma(\cdot, \cdot)$ is the incomplete gamma function. In the limits $\lambda_l \to 0$ and $\lambda_u \to \infty$, $\gamma_G(s; \lambda_l, \lambda_u)$ tends to a power variogram (PV) $\gamma^2(s) = Bs^{2H}$ where $B = A(\pi/4)^{2H/2}\Gamma(1 - 2H/2)/2H$, Γ being the gamma function. The stationary tfBm $G'(x; \lambda_l, \lambda_u)$ thus tends to nonstationary fBm, $G'(x; 0, \infty)$, the stationary increments of which, $\Delta G(x, x + s; 0, \infty)$, form fGn. It follows that when $\lambda_u < \infty$, $\gamma_G(s; \lambda_l, \lambda_u)$ is a truncated power variogram (TPV) characterizing a (stationary) truncated version of fBm (tfBm).

We treat neutron porosity increments in each borehole as a sample from a zero-mean random field, $\Delta Y(x, x + s; \lambda_l, \lambda_u)$, subordinated to tfBm according to (see Appendix A)

$$\Delta Y(x, x + s; \lambda_l, \lambda_u) = W^{1/2}\Delta G(x, x + s; \lambda_l, \lambda_u), \tag{8}$$

where $s \geq 0$ is lag and the subordinator, W, is a non-negative random variable independent of ΔG (and of G'). As stated

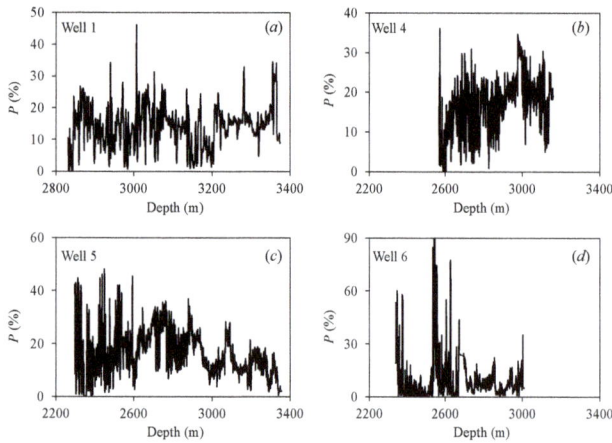

Figure 1. Variation of neutron porosity (P) with depth in Wells 1 (Maroon field), 4–5 (Ahwaz field) and 6 (Tabnak field).

above, we allow W to be Lévy-stable or lognormal. Appendix A explains that, in the first case, W is $\alpha/2$-stable, totally skewed to the right of zero (hence non-negative) with scale parameter $\sigma_S = \left(\frac{\cos\pi\alpha}{4}\right)^{2/\alpha}$, unit skewness and zero shift. The corresponding univariate pdf (probability density function) of $\Delta Y (x, x + s; \lambda_\mathrm{l}, \lambda_\mathrm{u})$ is symmetric α-stable with zero skewness and shift. The pdf possesses heavy, power-law tails. In the second case $W^{1/2} = e^V$, where V is zero-mean Gaussian with variance $\sigma_V^2 = (2 - \alpha)^2$. This renders $W^{1/2} \equiv 1$ when $\alpha = 2$ and its pdf increasingly skewed to the right as α diminishes. The corresponding univariate NLN pdf of $\Delta Y (x, x + s; \lambda_\mathrm{l}, \lambda_\mathrm{u})$ possesses heavier tails than the exponential tails of the Gaussian to which NLN tends asymptotically as α increases toward 2. Whereas α-stable variables do not possess finite moments of order $\geq \alpha$, all moments of NLN variables are finite. Parameters of the variogram characterizing the underlying Gaussian field are estimated through ML model calibration, as detailed in Sect. 7 for the two types of subordinators we consider.

4   Frequency distributions of neutron porosity data

Figure 1 shows how the neutron porosity data vary with depth in Wells 1, 4, 5 and 6. Frequency distributions of deviations, $P' = P - P_\mathrm{a}$, from average values, P_a, in Wells 1, 4 and 6 are plotted on arithmetic and semi-logarithmic scales in Fig. 2. The empirical frequency distributions exhibit sharp peaks, asymmetry and slight bimodality. Also shown in Fig. 2 are ML fits of a Gaussian and two sub-Gaussian pdfs to the empirical frequency distributions. Figure 1 shows that neutron porosity values in Well 6 exhibit greater variability than in other wells. This could be due to a larger carbonate content in formations penetrated by Well 6 than in those penetrated by other wells (see Sect. 2), rendering the former more heterogeneous than the rest.

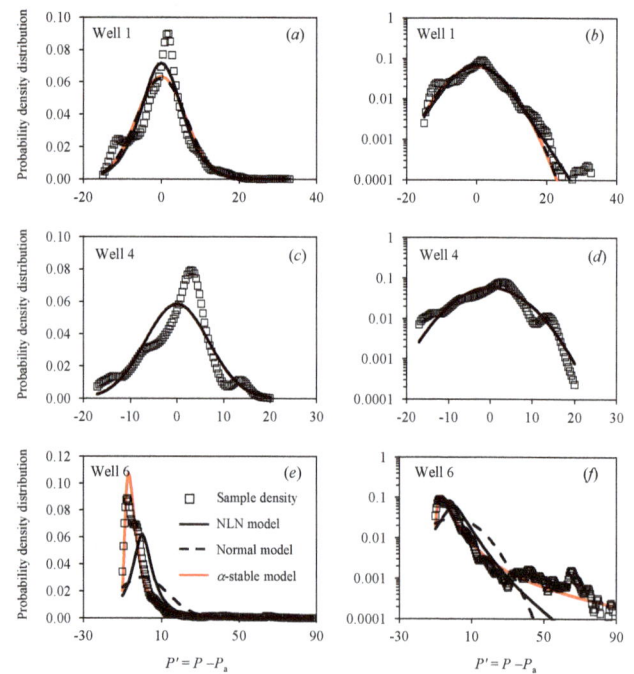

Figure 2. Frequency distributions on arithmetic and semi-logarithmic scales of $P' = P - P_\mathrm{a}$ in (**a, b**) Well 1 (Maroon field), (**c, d**) Well 4 (Ahwaz field), and (**e, f**) Well 6 (Tabnak field). Also shown are ML fits of Gaussian (dashed), α-stable (solid red), and NLN (black solid) pdfs.

ML fits to Gaussian and α-stable pdfs is accomplished with a code developed by Nolan (2001) and to NLN using a code we have written in Matlab. The quality of these fits is variable; in the case of Well 1, the NLN model is seen to fit the empirical frequency distribution slightly better than do the other two models but, in the case of Well 6, the α-stable model is seen to be best and Gaussian model worst. Formal Kolmogorov–Smirnov (KS), χ^2 and Shapiro–Wilk tests conducted on some of the data tend to reject the Gaussian model at a significance level of 0.05.

5   Frequency distributions of neutron porosity increments

Rather than presenting results in terms of lag s we report them below in terms of normalized (by Δz) integer values, s_n. Figure 3 shows how increments $\Delta P (s_\mathrm{n})$ at three different normalized lags ($s_\mathrm{n} = 1, 32, 1024$) vary with sequential (integer) vertical position in Wells 1 (Maroon field), 4 (Ahwaz field) and 6 (Tabnak field).

Frequency distributions of $\Delta P (s_\mathrm{n})$ at the same three lags in Wells 1 and 4 are plotted on semi-logarithmic scale in Fig. 4. The empirical frequency distributions exhibit pronounced symmetry with sharp peaks and heavy tails, which decay toward Gaussian shapes as lags increase. At all lags, the empirical frequency distributions of increments are rep-

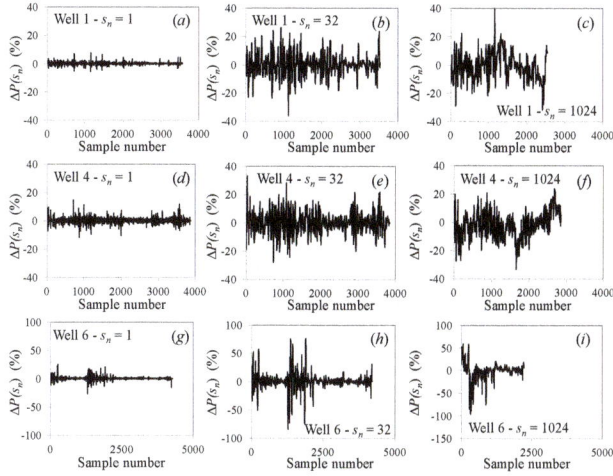

Figure 3. Increments $\Delta P(s_n)$ of P at normalized lags $s_n = 1$ ($s = 0.15\,\text{m}$), 32 ($s = 4.80\,\text{m}$), and 1024 ($s = 153.60\,\text{m}$) versus sequential (integer) vertical position in **(a–c)** Well 1 (Maroon field), **(d–f)** Well 4 (Ahwaz field), and **(g–i)** Well 6 (Tabnak field).

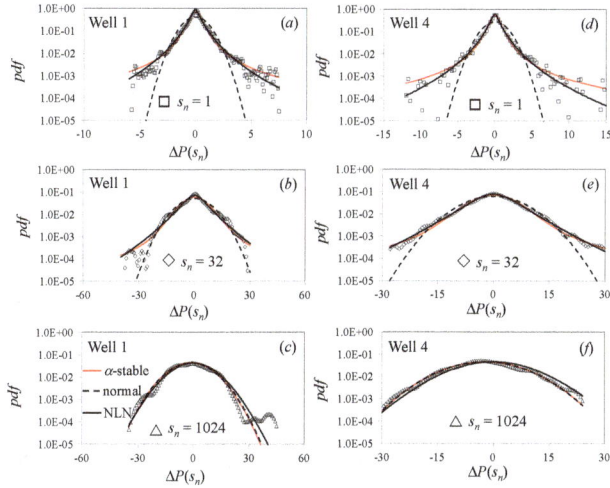

Figure 4. Frequency distributions of increments $\Delta P(s_n)$ of P at normalized lags $s_n = 1$ ($s = 0.15\,\text{m}$), 32 ($s = 4.80\,\text{m}$), and 1024 ($s = 153.60\,\text{m}$) in **(a–c)** Well 1 (Maroon field) and **(d–f)** Well 4 (Ahwaz field). Also shown are ML fits of Gaussian (dashed), α-stable (solid, red), and NLN (black, solid) pdfs.

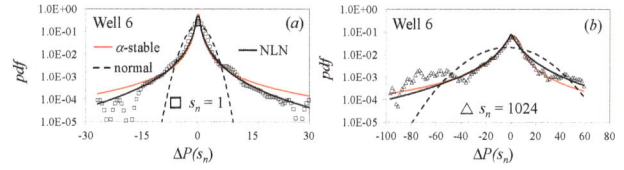

Figure 5. Frequency distributions of increments $\Delta P(s_n)$ of P at normalized lags $s_n = 1$ ($s = 0.15\,\text{m}$) and 1024 ($s = 153.60\,\text{m}$) in Well 6 (Tabnak field). Also shown are ML fits of Gaussian (dashed), α-stable (solid, red), and NLN (black, solid) pdfs.

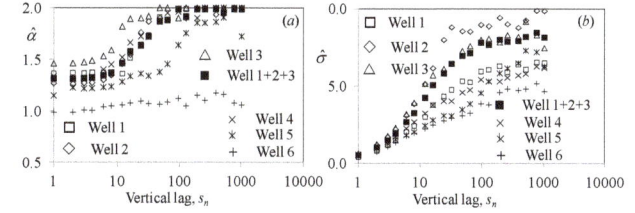

Figure 6. ML estimates $\hat{\alpha}$ and $\hat{\sigma}$ of stability and scale parameters, respectively, characterizing α-stable distribution models of increments $\Delta P(s_n)$ of P in all wells versus normalized lag.

in Well 6 are represented equally well by α-stable and NLN models.

Figure 6 shows how estimates $\hat{\alpha}$ and $\hat{\sigma}$ of stability and scale parameters, respectively, characterizing α-stable distribution models (see Appendix A) of neutron porosity increments in all wells vary with normalized lag. Estimates $\hat{\alpha}$ of the stability index, α, in Wells 1–3 (Maroon field) and 4–5 (Ahwaz field) exceed 1 and increase asymptotically toward 2 with increasing lag, confirming that the increments become Gaussian at large lags. In Well 6 (Tabnak field) $\hat{\alpha}$ fluctuates around a value that exceeds 1 by a small amount. Estimates $\hat{\sigma}$ of the scaling index σ, which measures the width of the α-stable distribution, first increase with lag and then stabilize in all wells. All these behaviors are consistent with sub-Gaussian random fields associated with α-stable subordinators; whether or not α does or does not grow with lag depends on how these fields are generated (see Riva et al., 2013c; Neuman et al., 2013). We do not show but note that parameters of NLN distribution models fitted to the increments also vary with lag in a way that renders them asymptotically Gaussian at large lags, with the exception of Well 6.

6 Statistical scaling of neutron porosity increments

Next we analyze the scaling behavior of sample structure functions, $S_N^q(s_n)$, of order q defined in Eq. (1). Figure 7 shows how such structure functions of orders $q = 0.5$, 1.0 and 2.0 vary with s_n in Wells 1 (Maroon) and 6 (Tabnak). Log–log regression lines fitted to the data separately at vertical distance scales $s_n < 10$ and $s_n > 12$ suggest, at relatively high levels of confidence (coefficients of determination, R^2,

resented quite closely by α-stable and NLN models fitted to them by ML. Negative log likelihood (NLL) measures of best fit associated with these two models as well as values of the Kashyap (1982) information criterion, KIC, demonstrate that they fit the empirical frequency distributions equally well (not shown). The same is true for all increments in all other wells. Frequency distributions of $\Delta P(s_n)$ plotted for two normalized lags in Well 6 (Fig. 5) are likewise symmetric with sharp peaks and heavy tails which, however, do not decay with lag. Empirical frequency distributions of $\Delta P(s_n)$

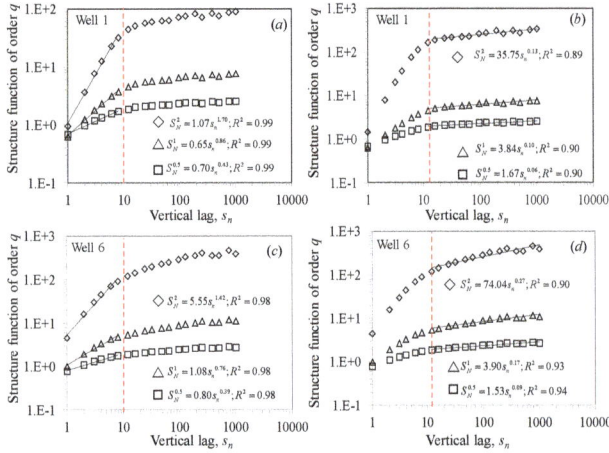

Figure 7. $S_N^q(s_n)$ versus normalized lag for $q = 0.5$, 1.0, and 2.0 in Wells 1 (Maroon) and 6 (Tabnak). Red dashed line demarcates breaks in power-law scaling regimes. Logarithmic-scale regression lines and corresponding power-law relations between $S_N^q(s_n)$ and s_n are given in (**a**) for Well 1 at $s_n < 10$, (**b**) Well 1 at $s_n > 12$, (**c**) Well 6 at $s_n < 10$, and (**d**) Well 6 at $s_n > 12$.

Table 2. Method of moments estimates of H for porosity increments at $s_n < 10$ (denoted by subscript w) and $s_n > 12$ (subscript b).

Well	$\hat{H}_w$	$\hat{H}_b$
1 (Maroon field)	0.86	0.10
2 (Maroon field)	0.87	0.08
3 (Maroon field)	0.85	0.11
4 (Ahwaz field)	0.70	0.11
5 (Ahwaz field)	0.66	0.16
6 (Tabnak field)	0.75	0.17

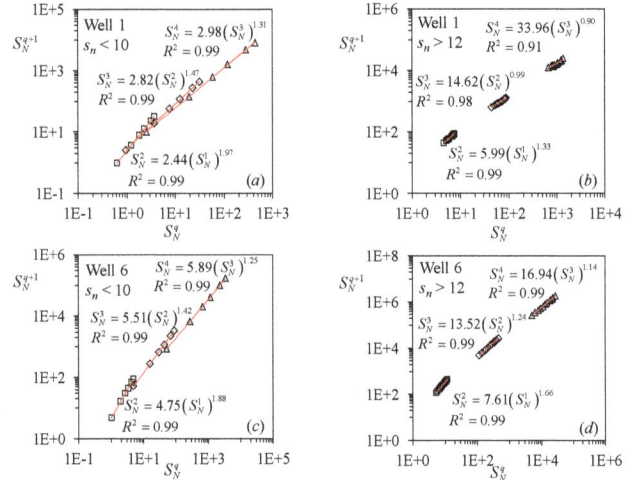

Figure 8. S_N^{q+1} vs. S_N^q for $q = 1$, 2 and 3 in Wells 1 (Maroon) and 6 (Tabnak). Logarithmic-scale regression lines and corresponding power-law relations between S_N^{q+1} and S_N^q are given in (**a**) for Well 1 at $s_n < 10$, (**b**) Well 1 at $s_n > 12$, (**c**) Well 6 at $s_n < 10$, and (**d**) Well 6 at $s_n > 12$.

ranging from 0.98 to 0.99 at $s_n < 10$ and from 0.89 to 0.99 at $s_n > 12$), that $S_N^q(s_n)$ varies as a power of s_n in each of these two scale ranges. Power-law exponents are larger at small ($s_n < 10$) than at large ($s_n > 12$) lags. We thus have a crossover between two diverse power-law regimes at distance scales 1.5–1.8 m delineated in Fig. 7 by a dashed red line. We interpret the power-law scaling of $S_N^q(s_n)$ with $s_n < 10$ representing variability within, and that at $s_n > 12$ variability between, sedimentary layers at each site. Similar dual power-law scaling behavior is exhibited by structure functions of increments from Wells 2–5 (not shown). The identification of layers of diverse geomaterials is related to depositional processes which take place over time in any sedimentary basin of the kind we deal with here. Dashtian et al. (2011) concluded that these formations are layered based on complete suites of well logs at each of the three sites. We note further that a similar dual-scaling phenomenon has recently been reported by Siena et al. (2014) vis-á-vis porosities and specific surface areas imaged using X-ray computer microtomography throughout a millimeter-scale block of Estaillades limestone, at a spatial resolution of 3.3 μm, as well as Lagrangian velocities computed by solving the Stokes equation in the sample pore space.

Following the most recent examples of Guadagnini et al. (2013, 2014) we use the method of moments described in Sect. 3 to obtain estimates, $\hat{H}_w$ and $\hat{H}_b$, of Hurst scaling exponents, H_w and H_b, characterizing the within- and between-layers scaling behaviors of neutron porosity increments, respectively, in each well. $\hat{H}_w$ and $\hat{H}_b$ are set equal to the slopes, $\xi_w(q = 1)$ and $\xi_b(q = 1)$, of regression lines fitted to $S_N^1(s_n)$ on log–log scale at $s_n < 10$ and $s_n > 12$, respectively. Values of these estimates are listed, for all six wells,

in Table 2. As $\hat{H}_w > 1/\hat{\alpha}$ and $\hat{H}_b \ll 1/\hat{\alpha}$ in all cases, we conclude that whereas intra-layer variability is persistent (large values tend to follow large values and small values tend to follow small values), inter-layer variability is strongly antipersistent (small and large values tend to alternate rapidly). The latter is likely a manifestation of strong variations in environments responsible for the deposition of alternating sedimentary layers.

As no theory other than ours (Siena et al., 2012; Neuman et al., 2013) is known to explain extended self-similarity (ESS) of variables that do not necessarily satisfy Burger's equation (Chakraborty et al., 2010), demonstrating that $\Delta P(s_n)$ satisfy ESS is akin to verifying that these data conform to our theoretical scaling framework. That this is indeed the case becomes evident upon examining the high-confidence ($R^2 = 0.91$–0.99) straight-line relationships between log S_N^{q+1} and log S_N^q, and corresponding power-law relationships between S_N^{q+1} and S_N^q, at $s_n < 10$ and $s_n > 12$ in Fig. 8 for $q = 1$, 2 and 3 in Wells 1 (Maroon) and 6 (Tabnak). Similar ESS relationships hold (not shown) in Wells 2–5.

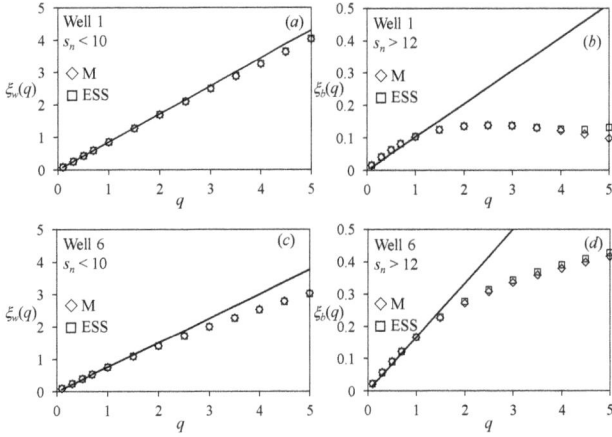

Figure 9. $\xi_w(q)$ and $\xi_b(q)$ evaluated as functions of q by the method of moments (M) and ESS in **(a)** Well 1 at $s_n < 10$, **(b)** Well 1 at $s_n > 12$, **(c)** Well 6 at $s_n < 10$, and **(d)** Well 6 at $s_n > 12$.

Our next step is to compute functional relationships between power exponents $\xi_w(q)$ and $\xi_b(q)$, and the order q, of structure functions that scale as power laws of lag. In the method of moments these powers are the slopes of regression lines fitted to log–log plots of $S_N^q(s_n)$ versus s_n, such as those depicted in Fig. 7. In the case of ESS we use $\xi_w(q = 1)$ and $\xi_b(q = 1)$, determined by the method of moments, as reference values for the sequential computation of $\xi_w(q)$ and $\xi_b(q)$ at $q > 1$ based on known power-law relationships between S_N^{q+1} and S_N^q, such as those given in Fig. 8. Corresponding plots of $\xi_w(q)$ and $\xi_b(q)$ as functions of q, evaluated by the method of moments and ESS in Wells 1 and 6 at $s_n < 10$ and $s_n > 12$, are presented in Fig. 9. Results obtained by the two methods are, for the most part, very similar. With the exception of $\xi_w(q)$ at $s_n < 10$ in Wells 1, 2, and 3 (Maroon field), in all cases (including those corresponding to Wells 2–5, which we do not show) $\xi_w(q)$ and $\xi_b(q)$ delineate convex functions that fall below straight lines having slopes $\hat{H}_w$ and $\hat{H}_b$, respectively, which pass through the origin. Tradition has it that whereas such straight lines are characteristic of monofractal (self-affine, additive) random fields, nonlinear variations of power exponents such as those exhibited by $\xi_w(q)$ and $\xi_b(q)$ in Fig. 9 are symptomatic of (multiplicative) multifractals. Yet we have seen that the data in this paper conform to a statistical scaling theory in which the underlying random fields are subordinated to truncated versions of monofractal fBm or fGn. As we have previously demonstrated theoretically (Neuman, 2010, 2011; Neuman et al., 2013) and computationally (Guadagnini et al., 2012), nonlinear scaling of such data is nothing but a random artifact of sampling from similar fields.

7 Estimation of variogram parameters

We saw that our analysis supports treating the neutron porosity data from each well as a random sample from a stationary sub-Gaussian random field subordinated to tfBm or tfGn. Our previous ML fits of univariate α-stable and NLN pdf models to neutron porosity increments in each well have yielded estimates of all distributional parameters characterizing these models. We also found the data to exhibit different modes of scaling at $s_n < 10$ and $s_n > 12$ and obtained estimates of H for each of these two ranges of lags. All that remains to fully characterize the multivariate random fields, $\Delta Y(x, x + s; \lambda_l, \lambda_u)$, which we take to underlie the incremental data, is to estimate the parameters A, λ_l and λ_u (and, optionally, H) of TPVs corresponding to $s_n < 10$ and $s_n > 12$. We do so next for each of the two subordinators we consider.

Assuming first that neutron porosity increments in each well are α-stable, one can estimate the scale parameter $\sigma(s; \lambda_l, \lambda_u)$ of their distribution at any lag, s, from the theoretical relationship (Samorodnitsky and Taqqu, 1994)

$$\hat{\sigma}(s; \lambda_l, \lambda_u) = \sqrt{\gamma_G(s; \lambda_l, \lambda_u)}. \qquad (9)$$

Here we employ this relationship separately for normalized lag ranges $s_n < 10$ and $s_n > 12$. We saw earlier that structure functions of neutron porosity data in both lag ranges, including second-order structure functions, can be closely represented in each well by power laws. In other words, the TPVs within these lag ranges are effectively PVs. We recall that this happens in the limits as λ_l and λ_u tend, respectively, to zero and infinity. We note further that λ_l should be a fraction of the measurement scale. In our case, the measurement scale can be considered as smaller than the 0.15 m data resolution scale (in Well 6 data resolution is 0.07 m). When compared to the much larger length scale of each borehole (on the order of 10^3 m), λ_l is negligibly small and can be disregarded. Accordingly, we set $\lambda_l = 0$ and λ_u to a sufficiently large number to ensure that the TPV $\gamma_G(s; \lambda_l, \lambda_u)$ reduces, within both working lag ranges, to the PV $\gamma(s) = Bs^{2H}$. Then, in a manner analogous to that outlined most recently by Guadagnini et al. (2013, 2014), we obtain ML estimates $\hat{A}$ of A in two ways: once by adopting corresponding method-of-moment estimates $\hat{H}_w$ and $\hat{H}_b$ from Table 2 and once by estimating the latter jointly with A. Both sets of estimates are obtained upon fitting the theoretical PV $\gamma(s) = Bs^{2H}$ to sample scale parameters $\hat{\sigma}(s_n)$ such as those plotted versus s_n in Fig. 7b. The fits are depicted graphically in Fig. 10 for Wells 1 and 6. The corresponding parameter estimates and 95 % confidence limits are listed, for all wells and both lag ranges, in Table 3. The two sets of estimates lie within each other's 95 % confidence intervals, implying that they are equally reliable.

Next we consider the case where neutron porosity increments in each well are NLN. Due to finiteness of all (statistical) moments associated with this model, structure functions of order $q = 2$ in Fig. 7 coincide with twice the variogram of

Table 3. Estimates $\hat{A}$ of A given estimates $\hat{H}$ of H from Table 2, and joint estimates $\hat{A}$ and $\hat{H}$, of PVs with associated 95 % confidence limits (in parenthesis) for all wells at $s_n < 10$ and $s_n > 12$ in the case of α-stable subordinator.

Data source	$\hat{A}$ estimated using $\hat{H}$ from Table 2		Joint estimates $\hat{A}$ and $\hat{H}$	
	$\hat{H}$	$\hat{A}$	$\hat{H}$	$\hat{A}$
Well 1 $s_n < 10$	0.86	0.06 (0.05; 0.07)	0.87 (0.78; 0.97)	0.05 (0.02; 0.13)
Well 1 $s_n > 12$	0.10	2.12 (1.84; 2.45)	0.14 (0.10; 0.20)	2.00 (1.66; 2.43)
Well 2 $s_n < 10$	0.87	0.12 (0.11; 0.13)	0.91 (0.86; 0.96)	0.08 (0.04; 0.16)
Well 2 $s_n > 12$	0.08	5.14 (4.48; 5.90)	0.10 (0.06; 0.16)	5.27 (4.56; 6.08)
Well 3 $s_n < 10$	0.85	0.16 (0.14; 0.17)	0.89 (0.82 0.96)	0.11 (0.05; 0.23)
Well 3 $s_n > 12$	0.11	4.02 (3.60; 4.49)	0.09 (0.06; 0.14)	4.02 (3.59; 4.51)
Well 4 $s_n < 10$	0.70	0.21 (0.19; 0.24)	0.76 (0.70; 0.83)	0.16 (0.11; 0.23)
Well 4 $s_n > 12$	0.11	1.80 (1.67; 1.94)	0.13 (0.11; 0.16)	1.74 (1.59; 1.90)
Well 5 $s_n < 10$	0.66	0.18 (0.15; 0.23)	0.70 (0.53; 0.93)	0.15 (0.06; 0.37)
Well 5 $s_n > 12$	0.16	1.36 (1.13; 1.65)	0.25 (0.22; 0.30)	0.84 (0.64; 1.11)
Well 6 $s_n < 10$	0.75	0.09 (0.08; 0.11)	0.81 (0.70; 0.94)	0.06 (0.03; 0.14)
Well 6 $s_n > 12$	0.17	0.86 (0.78; 0.94)	0.18 (0.15; 0.22)	0.80 (0.66; 0.96)

Table 4. Estimates $\hat{C}$ of C given estimates $\hat{H}$ of H from Table 2, and joint estimates $\hat{C}$ and $\hat{H}$, of PVs with associated 95 % confidence limits (in parenthesis) for all wells at $s_n < 10$ and $s_n > 12$ in the case of lognormal subordinator.

Data source	$\hat{C}$ estimated using $\hat{H}$ from Table 2		Joint estimates $\hat{C}$ and $\hat{H}$	
	$\hat{H}$	$\hat{C}$	$\hat{H}$	$\hat{C}$
Well 1 $s_n < 10$	0.86	0.52 (0.46; 0.58)	0.85 (0.75; 0.96)	0.53 (0.40; 0.70)
Well 1 $s_n > 12$	0.10	13.22 (12.36; 14.13)	0.07 (0.05; 0.08)	17.88 (15.44; 20.70)
Well 2 $s_n < 10$	0.87	1.35 (1.18; 1.53)	0.84 (0.74; 0.96)	1.43 (1.07; 1.92)
Well 2 $s_n > 12$	0.08	39.31 (36.17; 42.72)	0.04 (0.03; 0.07)	55.61 (45.31; 68.24)
Well 3 $s_n < 10$	0.85	0.87 (0.76; 1.00)	0.83 (0.72; 0.95)	0.91 (0.67; 1.25)
Well 3 $s_n > 12$	0.11	19.96 (18.30; 21.77)	0.09 (0.06; 0.12)	24.88 (18.72; 33.06)
Well 4 $s_n < 10$	0.70	1.09 (0.92; 1.31)	0.65 (0.52; 0.80)	1.23 (0.85; 1.80)
Well 4 $s_n > 12$	0.11	10.02 (9.48; 10.59)	0.08 (0.07; 0.09)	13.01 (11.66; 14.52)
Well 5 $s_n < 10$	0.66	1.59 (1.35; 1.88)	0.61 (0.50; 0.75)	1.78 (1.25; 2.53)
Well 5 $s_n > 12$	0.16	8.69 (7.73; 9.76)	0.09 (0.08; 0.11)	16.05 (13.83; 18.61)
Well 6 $s_n < 10$	0.76	2.52 (2.15; 2.95)	0.71 (0.60; 0.84)	2.77 (1.98; 3.89)
Well 6 $s_n > 12$	0.17	26.90 (24.45; 29.58)	0.14 (0.11; 0.17)	37.02 (27.90; 49.11)

neutron porosity. As shown in Appendix A, the variogram of $Y'(x; \lambda_l, \lambda_u)$ is given by

$$\gamma_Y (s_n; \lambda_l, \lambda_u) = \left(\mu_w^2 + \sigma_w^2\right) \gamma_G (s_n; \lambda_l, \lambda_u), \qquad (10)$$

where μ_w and σ_w^2 are defined in (Eq. A1). We replace Eq. (10) by $\gamma_Y (s) = C s^{2H}$ and fit the latter by ML to second-order sample structure functions of porosity increments in each well, separately for $s_n < 10$ and $s_n > 12$. Joint estimates of C and H for each range of lags, as well as ML estimates of C based on method-of-moment estimates $\hat{H}_w$ and $\hat{H}_b$ from Table 2, together with associated 95 % confidence intervals, are listed in Table 4. Corresponding best fits are depicted graphically in Fig. 11. Here again the two sets of estimates

lie within each other's 95 % confidence intervals, implying that they are equally reliable.

8   Frequency distributions of peaks over thresholds

Extreme value analyses of randomly varying data typically concern block maxima (BM) and/or POTs. The number of neutron porosity increments, $\Delta P(s_n)$, available to us at any normalized lag at any well are insufficient to conduct a statistically meaningful analysis of BM. For this reason, and for the fact that POTs provide a higher resolution of maxima than do BM, we focus in this paper exclusively on the former. In way of illustration we consider absolute increments $|\Delta P(s_n)|$ to constitute POTs whenever they exceed

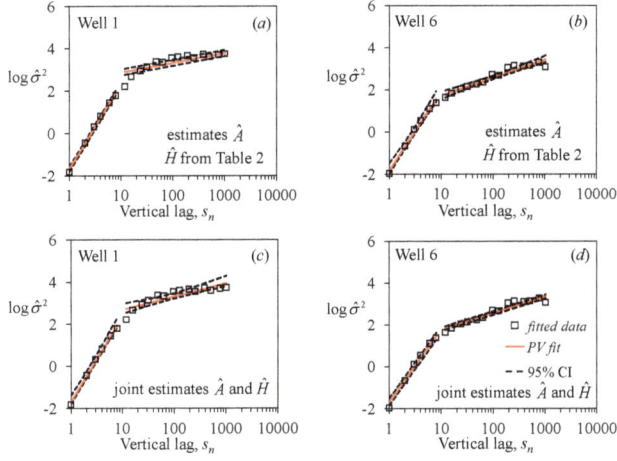

Figure 10. Sample scale parameter square $\hat{\sigma}^2 (s_n)$ as functions of s_n (squares), ML fitted PVs (solid lines) and 95 % confidence limits (broken curves) in Wells 1 and 6 based on (**a, b**) estimates $\hat{A}$ given estimates $\hat{H}$ from Table 2 and (**c, d**) joint estimates of $\hat{A}$ and $\hat{H}$.

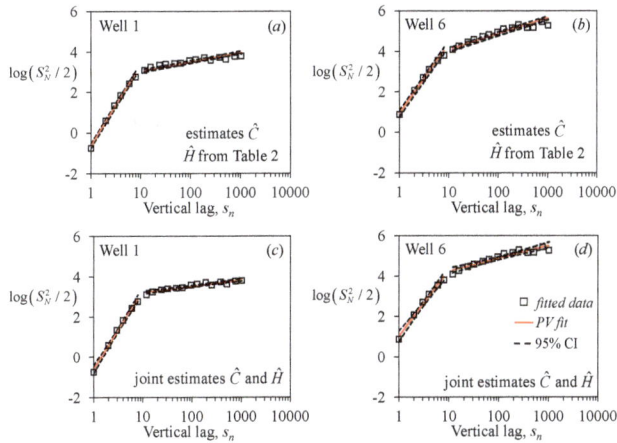

Figure 11. Sample structure functions, $S_N^2 (s_n)$, of order $q = 2$ as functions of s_n (squares), ML fitted PVs (solid lines) and 95 % confidence limits (broken curves) in Wells 1 and 6 based on (**a, b**) estimates $\hat{C}$ given estimates $\hat{H}$ from Table 2 and (**c, d**) joint estimates of $\hat{C}$ and $\hat{H}$.

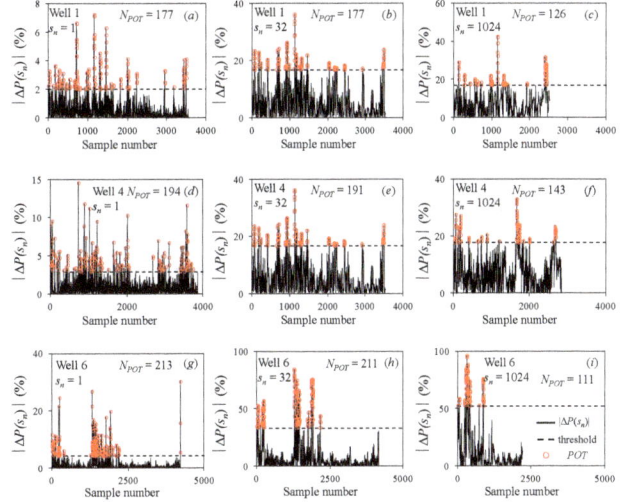

Figure 12. POTs of absolute increments $|\Delta P (s_n)|$ at normalized lags $s_n = 1, 32$, and 1024 versus sequential (integer) vertical positions in (**a–c**) Well 1 (Maroon), (**d–f**) Well 4 (Ahwaz), and (**g–i**) Well 6 (Tabnak).

Table 5. POT sample sizes and Kolmogorov–Smirnov p values associated with three lags in various wells.

s_n	Well no.	No. of samples	No. of POT samples	p value (KS test)
1	1	3566	177	0.240
	2	4048	202	0.994
	3	2944	147	0.706
	4	3881	194	0.437
	5	6948	208	0.970
	6	4265	213	0.788
32	1	3535	177	0.612
	2	4017	201	0.199
	3	2913	146	0.394
	4	3850	191	0.426
	5	6917	208	0.313
	6	4203	210	0.215
1024	1	2543	126	0.089
	2	3025	151	0.530
	3	1921	96	0.928
	4	2858	143	0.473
	5	5925	178	0.072
	6	2219	111	0.590

a non-negative threshold, u_t, equal to the 95 % quantile of $|\Delta P (s_n)|$ values in a sample. This renders about 5 % of all sampled $|\Delta P (s_n)|$ values of POTs. Figure 12 identifies POTs associated with sequences of porosity increments depicted in Fig. 3.

In each well, sample autocorrelation of non-overlapping neutron porosity increments at diverse normalized lags diminishes rapidly with the number, n, of these normalized increments (not shown), in line with theoretical expressions (18)–(20) of Neuman (2010). We expect autocorrelations between POTs to be weaker, possibly justifying a representation of their frequency distributions by GPDs (see Appendix B) which, theoretically, apply to independent identi-

cally distributed (iid) variables. To test this, we plot in Fig. 13 quantile–quantile (Q–Q) plots of GPD fits to frequency distributions of POTs identified in Fig. 12. Included in Fig. 13 are 95 % confidence intervals of these fits and p values of KS goodness-of-fit tests. A list of POT sample sizes and p values associated with the same three lags in all wells is provided in Table 5. The p value is the probability of obtaining given data when a null hypothesis is true. As all p values in

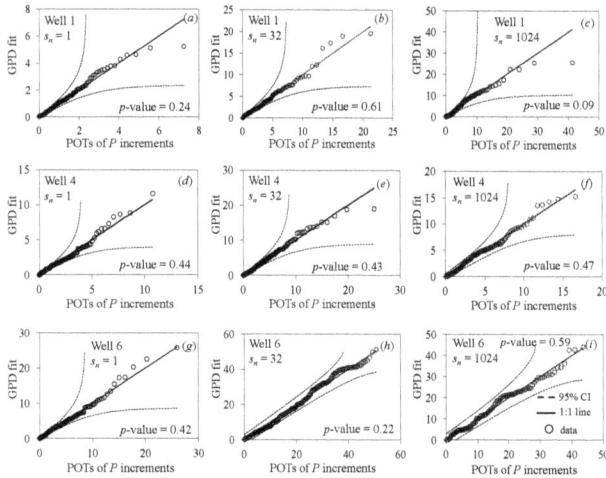

Figure 13. Quantile–quantile plots of GPD fits to frequency distributions of POTs of *porosity* increments at normalized lag $s_n = 1$, 32 and 1024 in (**a–c**) Well 1 (Maroon), (**d–f**) Well 4 (Ahwaz), and (**g–i**) Well 6 (Tabnak). Also shown are a line of unit slope (solid), 95 % confidence intervals (dashed), and p values of Kolmogorov–Smirnov tests.

Table 6. Method of moments estimates of H for POTs at $s_n < 10$ (denoted by subscript w) and $s_n > 12$ (subscript b).

Well	$\hat{H}_w$	$\hat{H}_b$
1 (Maroon field)	0.84	0.02
2 (Maroon field)	0.83	0.0001
3 (Maroon field)	0.80	0.06
4 (Ahwaz field)	0.61	0.03
5 (Ahwaz field)	0.60	0.02
6 (Tabnak field)	0.71	0.11

Table 5 exceed 0.05, one cannot reject (at a significance level of 0.05) the null hypothesis that all POTs have GPDs.

Figure 14 shows variations of best fit GPD shape (ξ_{POT}, governing the tail behavior of the distribution) and scale (σ_{POT}, governing the spread of the distribution) parameters with normalized lag, and corresponding 95 % uncertainty bounds, in the same wells as in Fig. 13. With the exception of Well 6 in which ξ_{POT} first diminishes with lag and then stabilizes, this parameter fluctuates but does not vary systematically with lag. The same applies to the shape parameter of each fitted GPD. However, σ_{POT} in all wells increases as a power of lag before stabilizing at larger lags, as does the scale parameter of α-stable distributions fitted to all neutron porosity increments in Fig. 6b.

9 Statistical scaling of peaks over thresholds

We end our analysis by exploring the scaling behavior of q-order sample structure functions of POT in absolute increments $|\Delta P_{POT,j}(s_n)|$. Following Eq. (1), these sample struc-

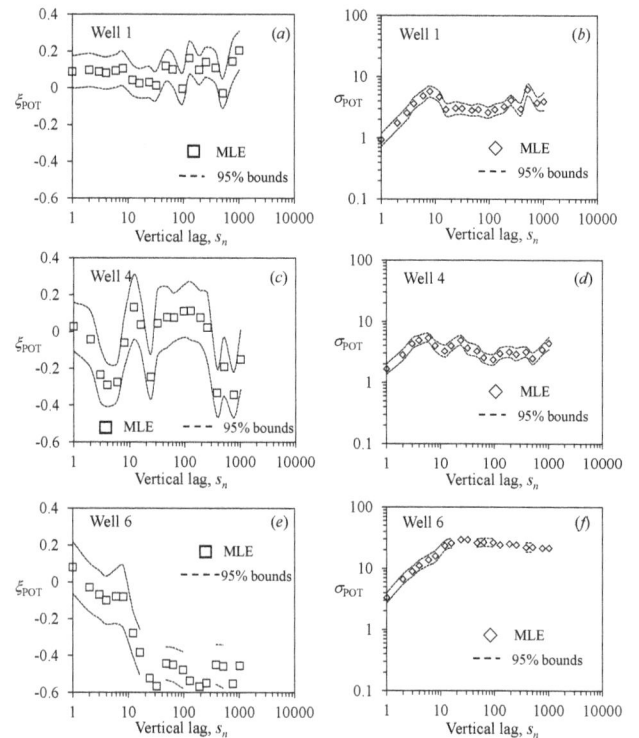

Figure 14. Variations of best fit GPD shape (ξ_{POT}) and scale (σ_{POT}) parameters with normalized lag in (**a, b**) Well 1 (Maroon), (**c, d**) Well 4 (Ahwaz), and (**e, f**) Well 6 (Tabnak). Also shown are 95 % uncertainty bounds.

ture functions are defined as

$$S^q_{N_{POT}}(s_n) = \frac{1}{N_{POT}(s_n)} \sum_{j=1}^{N_{POT}(s_n)} \left| \Delta P_{POT,j}(s_n) \right|^q, \qquad (11)$$

where $N_{POT}(s_n)$ is the number of POTs at normalized lag s_n. We do so as we did earlier for all increments, according to the methodology summarized in Sect. 3. Figure 15 depicts variations of $S^q_{N_{POT}}(s_n)$ with normalized lag for $q = 0.5, 1.0$, and 2.0 in Wells 1 (Maroon) and 6 (Tabnak). A red dashed line in the figure demarcates cross-over between two diverse power-law scaling regimes at $s_n < 10$ and $s_n > 12$. Included in Fig. 15 are logarithmic-scale regression lines and corresponding power-law relations between $S^q_{N_{POT}}(s_n)$ and s_n in each well and scaling regime. The scaling behavior in Fig. 15 is similar to that shown previously for all (unfiltered) porosity increments in Fig. 7. Corresponding estimates of Hurst exponent are listed in Table 6; these too differ little from those obtained earlier for all porosity increments (Table 2) with the exception of estimates $\hat{H}_b$ which are consistently lower than those associated with unfiltered increments. Like the latter (Fig. 8), POTs exhibit ESS at all lags in the scaling intervals $s_n < 10$ and $s_n > 12$ (not shown).

Our final step is to compute functional relationships between power exponents $\xi_w(q)$ and $\xi_b(q)$, and the order q, of POT structure functions that scale as power-laws of lag. We

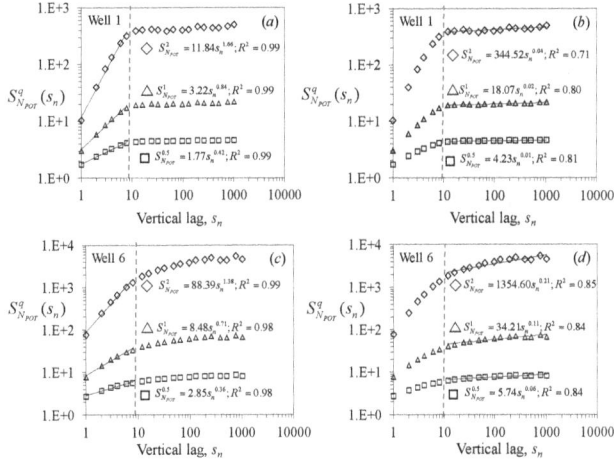

Figure 15. $S^q_{N_{POT}}(s_n)$ versus normalized lag for $q = 0.5$, 1.0, and 2.0 in Wells 1 (Maroon) and 6 (Tabnak). Red dashed line demarcates breaks in power-law scaling regimes. Logarithmic-scale regression lines and corresponding power-law relations between $S^q_{N_{POT}}(s_n)$ and s_n are given in (**a**) for Well 1 at $s_n < 10$, (**b**) Well 1 at $s_n > 12$, (**c**) Well 6 at $s_n < 10$, and (**d**) Well 6 at $s_n > 12$.

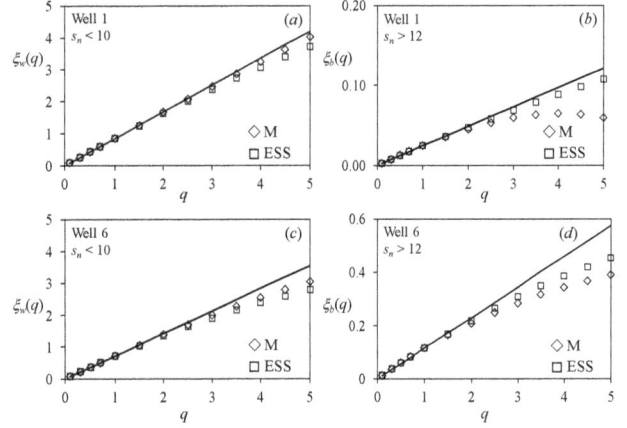

Figure 16. $\xi_w(q)$ and $\xi_b(q)$ evaluated for POTs as functions of q by M and ESS in (**a**) Well 1 at $s_n < 10$, (**b**) Well 1 at $s_n > 12$, (**c**) Well 6 at $s_n < 10$, and (**d**) Well 6 at $s_n > 12$.

do so as we did previously for unfiltered porosity increments. Corresponding plots of $\xi_w(q)$ and $\xi_b(q)$ as functions of q, evaluated by the method of moments and ESS in Wells 1 and 6 at $s_n < 10$ and $s_n > 12$, are presented in Fig. 16. Results obtained by the two methods are again, for the most part, very similar. Similar behavior has been shown by us elsewhere (Guadagnini et al., 2012) to be consistent with increments sampled from random fields subordinated to tfBm or tfGn.

10 Conclusions

After showing that neutron porosity data from six deep boreholes in three geologic environments have statistical scaling properties characteristic of samples from scale mixtures of truncated fractional Brownian motion (tfBm) or fractional Gaussian noise (tfGn), we used these data to explore the statistical behavior of extreme porosity increments, the absolute values of which exceed certain thresholds. We expect our results to hold for many earth, environmental and other variables that were shown elsewhere to possess similar statistical scaling properties. These results include the following:

1. The frequency distributions of neutron porosities in any well, or group of wells in any one of the three geologic environments, are non-Gaussian with sharp peaks, asymmetry and slight bimodality.

2. The frequency distributions of neutron porosity increments in any well, or group of wells at one of the three sites, are zero-mean symmetric with heavy tails that decay with increasing vertical separation distance or lag. At all lags, the distributions are represented closely by

either α-stable or normal–lognormal probability density models that tend to Gaussian with increasing lag.

3. Order q structure functions of absolute neutron porosity increments grow approximately as positive powers $\xi_w(q)$ of normalized lag, s_n, at $s_n < 10$ and as much smaller positive powers, $\xi_b(q)$, of s_n at $s_n > 12$. We interpret this dual power-law scaling to represent within- or intra-layer variability at $s_n < 10$ and between- or inter-layer variability at $s_n > 12$. Values of $\xi_w(q = 1)$ and $\xi_b(q = 1)$ provide method-of-moment estimates of Hurst exponents H_w and H_b for these two power-law scaling ranges, respectively.

4. Structure functions of absolute neutron porosity increments exhibit extended self similarity (ESS) at all normalized lags within both power-law scaling ranges, $s_n < 10$ and $s_n > 12$.

5. Values of power-law exponents $\xi_w(q)$ and $\xi_b(q)$ associated with absolute neutron porosity data, computed by the method of moments and by ESS, are for the most part very similar. Whereas such nonlinear scaling of power-law exponents has traditionally been viewed as a hallmark of multifractality (or, more recently, of fractional Laplace motion), we find the neutron porosity data in this paper to behave in a way fully consistent with that of samples from sub-Gaussian random fields subordinated to truncated (monofractal, self-affine, Gaussian) fractional Brownian motion or fractional Gaussian noise. The latter is the only view known to be theoretically consistent with ESS in the case of data, such as those considered here, that do not necessarily satisfy Burger's equation.

6. Our method of interpretation allows one to fully characterize the sub-Gaussian random field that underlies a

given set of data by estimating the parameters of corresponding (generally truncated) power variograms.

7. The autocorrelation of neutron porosity increments diminishes rapidly with the number, n, of non-overlapping increments in a separation distance (lag). This helps explain why sample distributions of peaks over thresholds (POTs, taken here to be absolute increments which exceed their 95 % quantile) are described reasonably well by a generalized Pareto distribution (GPD) model, which in theory applies to iid extrema. Whereas GPD shape parameter estimates do not show systematic variations with lag except in one well, corresponding estimates of GPD shape parameters tend to increase as a power of small lags and stabilize at larger lags. The same happens with scale parameters of α-stable distributions fitted to all (unfiltered) neutron porosity increments.

8. In all other respects, POTs show statistical scaling very similar to that of unfiltered increments. Estimates of POT Hurst exponents are very close to those obtained for unfiltered increments, with the exception of $\hat{H}_b$, that are consistently lower than those associated with unfiltered increments. Such nonlinear scaling is consistent with our method of interpreting the data. To our knowledge, this is the first documented example of POT statistical scaling interpreted on the basis of sub-Gaussian theory. We are not aware of any known theoretical reason why statistics of POT increments would necessarily scale in a manner similar to that of their parent population, as they do here.

Appendix A

Let $\Delta Y(x, x+s) = W^{1/2} \Delta G(x, x+s)$, where x is a spatial (or temporal) coordinate, $s \geq 0$ is lag, $W^{1/2}$ is a random variable acting as subordinator, and ΔG is a zero-mean Gaussian random field of increments with pdf (probability density function) $f_{\Delta G}(\Delta g)$ and variance $\sigma_{\Delta G}^2$ dependent on lag, ΔG and $W^{1/2}$ being statistically independent of each other at all lags. In this paper we consider W to be either Lévy-stable or lognormal.

In the first case (e.g., Samorodnitsky and Taqqu, 1994) W is $\alpha/2$-stable, totally skewed to the right of zero (hence nonnegative) with scale parameter $\sigma_S = \left(\cos \frac{\pi \alpha}{4}\right)^{2/\alpha}$, unit skewness and zero shift. The corresponding pdf of ΔY is symmetric α-stable with zero skewness and shift. In the second case we follow Neuman (2011) and Guadagnini et al. (2012) by setting $W^{1/2} = e^V$, where V is zero-mean Gaussian with variance $\sigma_V^2 = (2-\alpha)^2$, yielding the following respective mean and variance expressions for $W^{1/2}$,

$$\mu_w = \exp(\sigma_V^2/2) \text{ and } \sigma_w^2 = \exp\left(\sigma_V^2\right)\left[\exp\left(\sigma_V^2\right) - 1\right]. \tag{A1}$$

Correspondingly, the pdf of ΔY is

$$f_{\Delta Y}(\Delta y) = \int_{-\infty}^{\infty} \frac{1}{|u|} f_U(u) f_{\Delta G}\left(\frac{\Delta y}{u}\right) du, \tag{A2}$$

where $U = W^{1/2}$, and $u = w^{1/2}$. Since $U = W^{1/2} > 0$ one has

$$f_{\Delta Y}(\Delta y) = \int_0^{\infty} \frac{1}{u} f_U(u) f_{\Delta G}\left(\frac{\Delta y}{u}\right) du. \tag{A3}$$

As $\Delta G \sim N(0, \sigma_{\Delta G}^2)$ and $U = W^{1/2} \sim \ln N(0, \sigma_V^2)$, Eq. (A3) becomes

$$f_{\Delta Y}(\Delta y) = \frac{1}{2\pi \sigma_V} \int_0^{\infty} \frac{1}{u^2} e^{-\frac{\Delta y^2}{2u^2}} \cdot e^{-\frac{(\ln u - \ln \sigma_{\Delta G})^2}{2\sigma_V^2}} du. \tag{A4}$$

This is the normal–lognormal (NLN) pdf we refer to in the text. In it $\sigma_{\Delta G}$ plays the role of a scale parameter, and σ_V of a shape factor. Letting $\sigma_V \to 0$ is tantamount to letting Eq. (A4) converge to a normal density $f_{\Delta Y}(\Delta y) = \frac{1}{\sqrt{2\pi}\sigma_{\Delta G}} e^{-\frac{(\Delta y)^2}{2\sigma_{\Delta G}^2}}$. The larger is σ_V the heavier are the tails and the sharper is the peak of the NLN distribution. Fitting Eq. (A4) by maximum likelihood (ML) to sample frequency distributions of ΔY allows one to estimate $\sigma_{\Delta G}^2$ and σ_V^2, which in turn allows one to estimate μ_w and σ_w^2 according to Eq. (A1). The variance of ΔY is $\sigma_{\Delta Y}^2 = \left(\mu_w^2 + \sigma_w^2\right)\sigma_{\Delta G}^2$

and the variogram of Y' is

$$\gamma_Y(s) = \frac{1}{2} E\left[(\Delta Y(x,s))^2\right] = E\left[\left(W^{1/2}\right)^2\right]$$
$$\cdot \frac{1}{2} E\left[(\Delta G(x,s))^2\right] = \left(\sigma_w^2 + \mu_w^2\right) \gamma_G(s), \tag{A5}$$

where $\gamma_G(s)$ is the variogram of G'. Once μ_w and σ_w^2 have been estimated by maximum likelihood on the basis of ΔY data as described above, fitting Eq. (A5) to corresponding second-order sample structure functions allows one to estimate all parameters of $\gamma_G(s)$.

In case G' has a power variogram, $\gamma_G(s) = Bs^{2H}$, of the kind we consider in the manuscript so does Y,

$$\gamma_Y(s) = \left(\sigma_w^2 + \mu_w^2\right) \gamma_G(s) = Cs^{2H}, \tag{A6}$$

where C is a coefficient. Fitting Eq. (A6) to second-order sample structure functions of corresponding increments allows one to estimate C and H.

Appendix B

In this work empirical distributions of POTs (peaks over thresholds) of absolute neutron porosity increments at normalized lag s_n, $|\Delta P(s_n)|$, are shown to fit well-known two-parameter generalized Pareto distributions (GPDs). A GPD is described in terms of the following cumulative distribution function (CDF):

$$H(y) = 1 - (1 + y\xi_{POT}/\sigma_{POT})^{-1/\xi_{POT}},$$
$$y = |\Delta P(s_n)| - u_t > 0 \tag{B1}$$

where ξ_{POT} and σ_{POT} are the shape and scale parameters, respectively, governing tail behavior and spread of the distribution; and u_t is the predetermined threshold. Equation (B1) reduces to a Pareto (type-II) distribution when $\xi_{POT} > 0$, an exponential distribution when $\xi_{POT} = 0$ and a generalized beta distribution (of the first kind) when $\xi_{POT} < 0$ (Arnold, 2008).

Acknowledgements. This work was supported in part through a contract between the University of Arizona and Vanderbilt University under the Consortium for Risk Evaluation with Stakeholder Participation (CRESP) III, funded by the US Department of Energy. Funding from MIUR (Italian Ministry of Education, Universities and Research (PRIN2010-11 project: Innovative methods for water resources under hydro-climatic uncertainty scenarios) is also acknowledged. We thank Professor Muhammad Sahimi, University of Southern California, for having generously shared with us borehole geophysical log data, some of which we analyze in this paper.

Edited by: J. Carrera

References

Akaev, A., Sadovnichy, V., and Korotayev, A.: On the dynamics of the world demographic transition and financial-economic crises forecasts, Eur. Phys. J.-Special Topics, 205, 355–373, doi:10.1140/epjst/e2012-01578-2, 2012.

Amitrano, D.: Variability in the power-law distributions of rupture events, Eur. Phys. J.-Special Topics, 205, 199–215, doi:10.1140/epjst/e2012-01571-9, 2012.

Ancey, C.: Are there "dragon-kings" events (i.e. genuine outliers) among extreme avalanches?, Eur. Phys. J.-Special Topics, 205, 117–129, doi:10.1140/epjst/e2012-01565-7, 2012.

Andrews, D. F. and Mallows, C. L.: Scale Mixtures of Normal Distributions, J. Roy. Stat. Soc. B Met., 36, 99–102, 1974.

Arnold, B. C.: Pareto and Generalized Pareto Distributions, in: Modeling Income Distributions and Lorenz Curves, edited by: Chotikapanich, D., Springer Science & Business Media, New York NY, 119–145, 2008.

Barrash, W. and Reboulet, E. C.: Significance of porosity for stratigraphy and textural composition in subsurface coarse fluvial deposits, Boise Hydrogeophysical Research Site, Geol. Soc. Am. Bull., 116, 1059–1073, doi:10.1130/B25370.1, 2004.

Benzi, R., Ciliberto, S., Baudet, C., Chavarria, G. R., and Tripiccione, R.: Extended self-similarity in the dissipation range of fully developed turbulence, Europhys. Lett., 24, 275–279, 1993a.

Benzi, R., Ciliberto, S., Tripiccione, R., Baudet, C., Massaioli, F., and Succi, S.: Extended self-similarity in turbulent flows, Phys. Rev. E, 48, R29–R32, 1993b.

Benzi, R., Biferale, L., Ciliberto, S., Struglia, M. V., and Tripiccione, R.: Generalised scaling in fully developed turbulence, Phys. D, 96, 162–181, 1996.

Burlando, P. and Rosso, R.: Scaling and multiscaling models of depth-duration-frequency curves for storm precipitation, J. Hydrol., 187, 45–64, doi:10.1016/S0022-1694(96)03086-7, 1996.

Castro, J. J., Carsteanu, A. A., and Flores, C. G.: Intensity-duration-area-frequency functions for precipitation in a multi-fractal framework, Physica A, 338, 206–210, doi:10.1016/j.physa.2004.02.043, 2004.

Chakraborty, S., Frisch, U., and Ray, S. S.: Extended self-similarity works for the Burgers equation and why, J. Fluid Mech., 649, 275–285, doi:10.1017/S0022112010000595, 2010.

Dashtian, H., Jafari, G. R., Sahimi, M., and Masihi, M.: Scaling, multifractality, and long-range correlations in well log data of large-scale porous media, Physica A, 390, 2096–2111, doi:10.1016/j.physa.2011.01.010, 2011.

de Arcangelis, L.: Are dragon-king neuronal avalanches dungeons for self-organized brain activity?, Eur. Phys. J.-Special Topics, 205, 243–257, doi:10.1140/epjst/e2012-01574-6, 2012.

De Michele, C., Kottegoda, N. T., and Rosso, R.: The derivation of areal reduction factor of storm rainfall from its scaling properties, Water Resour. Res., 37, 3247–3252, doi:10.1029/2001WR000346, 2001.

Di Federico, V. and Neuman, S. P.: Scaling of random fields by means of truncated power variograms and associated spectra, Water Resour. Res., 33, 1075–1085, doi:10.1029/97WR00299, 1997.

Ebtehaj, M. and Foufoula-Georgiou, E.: Orographic signature on multiscale statistics of extreme rainfall: A storm-scale study, J. Geophys. Res.-Atmos., 115, D23112, doi:10.1029/2010JD014093, 2010.

Embrechts, P., Mikosch, T., and Klüppelberg, C.: Modelling Extremal Events For Insurance and Finance, Springer-Verlag, London, UK, 1997.

Fogg, G. E., Carle, S. F., and Green, C.: Connected-network paradigm for the alluvial aquifer system, in: Theory, Modeling, and Field Investigation in Hydrogeology: A Special Volume in Honor of Shlomo P. Neuman's 60th Birthday, edited by: Zhang, D. and Winter, C. L., Geological Society of America Special Paper 348, Boulder, Colorado, 25–42, 2000.

Frisch, U.: Turbulence, Cambridge University Press, Cambridge, 1995.

Garcia-Bartual, R. and Schneider, M.: Estimating maximum expected short-duration rainfall intensities from extreme convective storms, Phys. Chem. Earth Pt. B, 26, 675–681, doi:10.1016/S1464-1909(01)00068-5, 2001.

Golosovsky, M. and Solomon, S.: Runaway events dominate the heavy tail of citation distributions, Eur. Phys. J.-Special Topics, 205, 303–311, doi:10.1140/epjst/e2012-01576-4, 2012.

Gómez-Hernández, J. J. and Wen, X.-H.: To be or not to be multi-Gaussian. A reflection on stochastic hydrogeology, Adv. Water Resour., 21, 47–61, doi:10.1016/S0309-1708(96)00031-0, 1998.

Guadagnini, A. and Neuman, S. P.: Extended power-law scaling of self-affine signals exhibiting apparent multifractality, Geophys. Res. Lett., 38, L13403, doi:10.1029/2011gl047727, 2011.

Guadagnini, A., Neuman, S. P., and Riva, M.: Numerical investigation of apparent multifractality of samples from processes subordinated to truncated fBm, Hydrol. Proc., 26, 2894–2908, doi:10.1002/Hyp.8358, 2012.

Guadagnini, A., Neuman, S. P., Schaap, M. G., and Riva, M.: Anisotropic statistical scaling of vadose zone hydraulic property estimates near Maricopa, Arizona, Water Resour. Res., 49, 8463–8479, doi:10.1002/2013wr014286, 2013.

Guadagnini, A., Neuman, S. P., Schaap, M. G., and Riva, M.: Anisotropic statistical scaling of soil and sediment texture in a stratified deep vadose zone near Maricopa, Arizona, Geoderma, 214, 217–227, doi:10.1016/j.geoderma.2013.09.008, 2014.

Janczura, J. and Weron, R.: Black swans or dragon-kings? A simple test for deviations from the power law, Eur. Phys. J.-Special Topics, 205, 79–93, doi:10.1140/epjst/e2012-01563-9, 2012.

Javelle, P., Gresillon, J. M., and Galea, G.: Discharge-duration-frequency curve modelling for floods and scale invariance, Comptes Rendus De L Academie Des Sciences Serie Ii Fascicule a-Sciences De La Terre Et Des Planetes, 329, 39–44, doi:10.1016/S1251-8050(99)80225-0, 1999.

Kashyap, R. L.: Optimal choice of AR and MA parts in autoregressive moving average models, IEEET Pattern. Anal. 4, 99–104, 1982.

Katz, R. W., Parlange, M. B., and Naveau, P.: Statistics of extremes in hydrology, Adv. Water Resour., 25, 1287–1304, doi:10.1016/S0309-1708(02)00056-8, 2002.

Knudby, C. and Carrera, J.: On the relationship between indicators of geostatistical, flow and transport connectivity, Adv. Water Resour. 28, 405–421, doi:10.1016/j.advwatres.2004.09.001, 2005.

Knudby, C. and Carrera, J.: On the use of apparent hydraulic diffusivity as an indicator of connectivity, J. Hydrol., 329, 377–389, doi:10.1016/j.jhydrol.2006.02.026, 2006.

Knudby, C., Carrera, J., Bumgardner, J. D., and Fogg, G. E.: Binary upscaling – the role of connectivity and a new formula, Adv. Water Resour. 29, 590–604, 2006.

Kozubowski, T. J., Meerschaert, M. M., and Podgorski, K.: Fractional Laplace motion, Adv. Appl. Probab. 38, 451–464, doi:10.1239/aap/1151337079, 2006.

Langousis, A. and Veneziano, D.: Intensity-duration-frequency curves from scaling representations of rainfall, Water Resour. Res., 43, W02422, doi:10.1029/2006wr005245, 2007.

Lei, X.: Dragon-Kings in rock fracturing: Insights gained from rock fracture tests in the laboratory, Eur. Phys. J.-Special Topics, 205, 217–230, doi:10.1140/epjst/e2012-01572-8, 2012.

Main, I. and Naylor, M.: Extreme events and predictability of catastrophic failure in composite materials and in the Earth, Eur. Phys. J.-Special Topics, 205, 183–197, doi:10.1140/epjst/e2012-01570-x, 2012.

Mariethoz, G. and Renard, P.: Special Issues on 20 years of multiple-point statistics: part 2, Math. Geosci., 46, 517–518, doi:10.1007/s11004-014-9545-y, 2014.

Meerschaert, M. M., Kozubowski, T. J., Molz, F. J., and Lu, S.: Fractional Laplace model for hydraulic conductivity, Geophys. Res. Lett., 31, L08501, doi:10.1029/2003GL019320, 2004.

Meier, P. M., Carrera, J., and Sanchez-Vila, X.: An evaluation of Jacob's method for the interpretation of pumping tests in heterogeneous formations, Water Resour. Res., 34, 1011–1025, doi:10.1029/98WR00008, 1998.

Menabde, M. and Sivapalan, M.: Linking space-time variability of river runoff and rainfall fields: a dynamic approach, Adv. Water Resour., 24, 1001–1014, doi:10.1016/S0309-1708(01)00038-0, 2001.

Menabde, M., Seed, A., and Pegram, G.: A simple scaling model for extreme rainfall, Water Resour. Res., 35, 335–339, doi:10.1029/1998wr900012, 1999.

Mohymont, B. and Demarée, G. R.: Intensity-duration-frequency curves for precipitation at Yangambi, Congo, derived by means of various models of Montana type, Hydrolog. Sci. J., 51, 239–253, doi:10.1623/hysj.51.2.239, 2006.

Neuman, S. P.: Apparent/spurious multifractality of absolute increments sampled from truncated fractional Gaussian/Levy noise, Geophys. Res. Lett., 37, L09403, doi:10.1029/2010gl043314, 2010.

Neuman, S. P.: Apparent multifractality and scale-dependent distribution of data sampled from self-affine processes, Hydrol. Process., 25, 1837–1840, doi:10.1002/Hyp.7967, 2011.

Neuman, S. P. and Di Federico, V.: Multifaceted nature of hydrogeologic scaling and its interpretation, Rev. Geophys., 41, 1014, doi:10.1029/2003RG000130, 2003.

Neuman, S. P., Guadagnini, A., Riva, M., and Siena, M.: Recent advances in statistical and scaling analysis of earth and environmental variables, in: Advances in Hydrogeology, edited by: Mishra, P. K. and Kuhlman, K. L., Springer, New York, 2013.

Nguyen, V. T. V., Nguyen, T. D., and Wang, H.: Regional estimation of short duration rainfall extremes, Water Sci. Technol., 37, 15–19, doi:10.1016/S0273-1223(98)00311-4, 1998.

Nield, D. A.: Connectivity and Effective Hydraulic Conductivity, Transp. Porous Med., 74, 129–132, doi:10.1007/s11242-007-9185-5, 2008.

Nolan, J.: Maximum likelihood estimation of stable parameters, in: Lévy Processes: Theory and Applications, edited by: Barndorff-Nielsen, O., Mikosch, T., and Resnick, S., Birkhauser, Boston, 2001.

Painter, S.: Flexible scaling model for use in random field simulation of hydraulic conductivity, Water Resour. Res. 37, 1155–1163, 2001.

Pisarenko, V. F. and Sornette, D.: Robust statistical tests of Dragon-Kings beyond power law distributions, Eur. Phys. J.-Special Topics, 205, 95–115, doi:10.1140/epjst/e2012-01564-8, 2012.

Plenz, D.: Neuronal avalanches and coherence potentials, Eur. Phys. J.-Special Topics, 205, 259–301, doi:10.1140/epjst/e2012-01575-5, 2012.

Renard, P. and Mariethoz, G.: Special Issues on 20 years of multiple-point statistics: part 1, Math. Geosci., 46, 129–131, doi:10.1007/s11004-014-9524-3, 2014.

Rigon, R., D'Odorico, P., and Bertoldi, G.: The geomorphic structure of the runoff peak, Hydrol. Earth Syst. Sci., 15, 1853–1863, doi:10.5194/hess-15-1853-2011, 2011.

Riva, M., Neuman, S. P., and Guadagnini, A.: On the identification of Dragon Kings among extreme-valued outliers, Nonlin. Processes Geophys., 20, 549–561, doi:10.5194/npg-20-549-2013, 2013a.

Riva, M., Neuman, S. P., Guadagnini, A., and Siena, M.: Anisotropic scaling of berea sandstone log air permeability statistics, Vadose Zone J., 12, doi:10.2136/Vzj2012.0153, 2013b.

Riva, M., Neuman, S. P., and Guadagnini, A.: Sub-Gaussian model of processes with heavy-tailed distributions applied to air permeabilities of fractured tuff, Stoch. Env. Res. Risk A., 27, 195–207, doi:10.1007/s00477-012-0576-y, 2013c.

Riva, M., Sanchez-Vila, X., and Guadagnini, A.: Estimation of spatial covariance of log conductivity from particle size data, Water Resour. Res., 50, 5298–5308, doi:10.1002/2014WR015566, 2014.

Sachs, M., Yoder, M., Turcotte, D., Rundle, J., and Malamud, B.: Black swans, power laws, and dragon-kings: Earthquakes, volcanic eruptions, landslides, wildfires, floods, and SOC models, Eur. Phys. J.-Special Topics, 205, 167–182, doi:10.1140/epjst/e2012-01569-3, 2012.

Samorodnitsky, G. and Taqqu, M. S.: Stable Non-Gaussian Random Processes, Chapman and Hall, New York, 1994.

Sanchez-Vila, X., Carrera, J., and Girardi, J. P.: Scale effects in transmissivity, J. Hydrol., 183, 1–22, doi:10.1016/S0022-1694(96)80031-X, 1996.

Schertzer, D. and Lovejoy, S.: Physical Modeling and Analysis of Rain and Clouds by Anisotropic Scaling Multiplicative Processes, J. Geophys. Res.-Atmos., 92, 9693–9714, doi:10.1029/Jd092id08p09693, 1987.

Schoenberg, F. and Patel, R.: Comparison of Pareto and tapered Pareto distributions for environmental phenomena, Eur. Phys. J.-Special Topics, 205, 159–166, doi:10.1140/epjst/e2012-01568-4, 2012.

Siena, M., Guadagnini, A., Riva, M., and Neuman, S. P.: Extended power-law scaling of air permeabilities measured on a block of tuff, Hydrol. Earth Syst. Sci., 16, 29–42, doi:10.5194/hess-16-29-2012, 2012.

Siena, M., Guadagnini, A., Riva, M., Bijeljic, B., Pereira Nunes, J. P., and Blunt, M. J.: Statistical scaling of pore-scale Lagrangian velocities in natural porous media, Phys. Rev. E, 90, 023013, doi:10.1103/PhysRevE.90.023013, 2014.

Süveges, M. and Davison, A.: A case study of a "Dragon-King": The 1999 Venezuelan catastrophe, Eur. Phys. J.-Special Topics, 205, 131–146, doi:10.1140/epjst/e2012-01566-6, 2012.

Trefry, C. M., Watkins, D. W., and Johnson, D.: Regional rainfall frequency analysis for the state of Michigan, J. Hydrol. Eng., 10, 437–449, doi:10.1061/(Asce)1084-0699(2005)10:6(437), 2005.

Tronicke, J. and Holliger, K.: Quantitative integration of hydrogeophysical data: Conditional geostatistical simulation for characterizing heterogeneous alluvial aquifers, Geophysics, 70, H1–H10, doi:10.1190/1.1925744, 2005.

Veneziano, D. and Furcolo, P.: Multifractality of rainfall and scaling of intensity-duration-frequency curves, Water Resour. Res., 38, 1306, doi:10.1029/2001WR000372, 2002.

Veneziano, D. and Yoon, S.: Rainfall extremes, excesses, and intensity-duration-frequency curves: A unified asymptotic framework and new nonasymptotic results based on multifractal measures, Water Resour. Res., 49, 4320–4334, doi:10.1002/wrcr.20352, 2013.

Veneziano, D., Langousis, A., and Lepore, C.: New asymptotic and preasymptotic results on rainfall maxima from multifractal theory, Water Resour. Res., 45, W11421, doi:10.1029/2009wr008257, 2009.

Vuković, M. and Soro, A.: Determination of hydraulic conductivity of porous media from grain-size composition, Water Resources Publications, ISBN:0-918334-77-2, 1992.

Wen, X.-H. and Gómez-Hernández, J. J.: Numerical modeling of macrodispersion in heterogeneous media – a comparison of multi-Gaussian and non-multi-Gaussian models, J. Contam. Hydrol., 30, 129–156, doi:10.1016/S0169-7722(97)00035-1, 1998.

West, M.: On scale mixtures of normal distributions, Biometrika, 74, 646–648, doi:10.1093/biomet/74.3.646, 1987.

Western, A. W., Blöschl, G., and Grayson, R. B.: Toward capturing hydrologically significant connectivity in spatial patterns, Water Resour. Res., 37, 83–97, doi:10.1029/2000WR900241, 2001.

Willems, P.: Compound intensity/duration/frequency-relationships of extreme precipitation for two seasons and two storm types, J. Hydrol., 233, 189–205, doi:10.1016/S0022-1694(00)00233-X, 2000.

Yu, P. S., Yang, T. C., and Lin, C. S.: Regional rainfall intensity formulas based on scaling property of rainfall, J. Hydrol., 295, 108–123, doi:10.1016/j.jhydrol.2004.03.003, 2004.

Zinn, B. and Harvey, C. F.: When good statistical models of aquifer heterogeneity go bad: A comparison of flow, dispersion, and mass transfer in connected and multivariate Gaussian hydraulic conductivity fields, Water Resour. Res., 39, 1051, doi:10.1029/2001WR001146, 2003.

7

A framework for assessing hydrological regime sensitivity to climate change in a convective rainfall environment: a case study of two medium-sized eastern Mediterranean catchments, Israel

N. Peleg[1], E. Shamir[2], K. P. Georgakakos[2,3], and E. Morin[4]

[1]Hydrology and Water Resources Program, Hebrew University of Jerusalem, Givat Ram, Jerusalem 91904, Israel
[2]Hydrologic Research Center, San Diego, California, USA
[3]Scripps Institution of Oceanography, University of California San Diego, California, USA
[4]Department of Geography, Hebrew University of Jerusalem, Jerusalem 91905, Israel

Correspondence to: N. Peleg (nadav.peleg@mail.huji.ac.il)

Abstract. A modeling framework is formulated and applied to assess the sensitivity of the hydrological regime of two catchments in a convective rainfall environment with respect to projected climate change. The study uses likely rainfall scenarios with high spatiotemporal resolution that are dependent on projected changes in the driving regional meteorological synoptic systems. The framework was applied to a case study in two medium-sized Mediterranean catchments in Israel, affected by convective rainfall, by combining the HiReS-WG rainfall generator and the SAC-SMA hydrological model. The projected climate change impact on the hydrological regime was examined for the RCP4.5 and RCP8.5 emission scenarios, comparing the historical (beginning of the 21st century) and future (mid-21st-century) periods from three general circulation model simulations available from CMIP5. Focusing on changes in the occurrence frequency of regional synoptic systems and their impact on rainfall and streamflow patterns, we find that the mean annual rainfall over the catchments is projected to be reduced by 15 % (outer range 2–23 %) and 18 % (7–25 %) for the RCP4.5 sand RCP8.5 emission scenarios, respectively. The mean annual streamflow volumes are projected to be reduced by 45 % (10–60 %) and 47 % (16–66 %). The average events' streamflow volumes for a given event rainfall depth are projected to be lower by a factor of 1.4–2.1. Moreover, the streamflow season in these ephemeral streams is projected to be shorter by 22 % and 26–28 % for the RCP4.5 and RCP8.5, respectively. The amplification in reduction of streamflow volumes relative to rainfall amounts is related to the projected reduc-

tion in soil moisture, as a result of fewer rainfall events and longer dry spells between rainfall events during the wet season. The dominant factors for the projected reduction in rainfall amount were the reduction in occurrence of wet synoptic systems and the shortening of the wet synoptic systems durations. Changes in the occurrence frequency of the two dominant types of the regional wet synoptic systems (active Red Sea trough and Mediterranean low) were found to have a minor impact on the total rainfall.

1 Introduction

Rainfall spatial and temporal variability plays a major role in the hydrological response of catchments affecting runoff timing, streamflow volume, and peak discharge (Morin et al., 2006; Morin and Yakir, 2014; Paschalis et al., 2014; Rozalis et al., 2010; Singh, 1997). The role of rainfall variability is even greater in climate regimes where a substantial portion of rainfall is convective and runoff is sensitive to this type of rainfall (Morin and Yakir, 2014; Peleg and Morin, 2014; Rozalis et al., 2010; Smith et al., 2000; Yakir and Morin, 2011). The hydrological response was found to be sensitive to rainfall spatial variability in very small catchments (less than 1 km^2; Bahat et al., 2009), in catchments of few dozen km^2 (Zoccatelli et al., 2011) and larger catchments (> 100 km^2; Arnaud et al., 2011). It is therefore essential to use rainfall data with high spatial and temporal resolution for hydrological modeling purpose. Such resolution can be provided for small and medium sized catchments by weather radar data (e.g., 1.5 km^2 and 3 min, Peleg et al., 2013) and for large catchments by satellite data (e.g., Nikolopoulos et al., 2010; Shamir et al., 2013).

General circulation models (GCMs) used for climate studies and climate projections have coarser spatial and temporal resolution than usually required for hydrological simulations. For example, most GCM simulations that are available from the Coupled Model Intercomparison Project Phase 5 (CMIP5) have a spatial scale larger than 10 000 km^2 and monthly to 6 h reporting intervals (an extensive overview of CMIP5 is given by Moss et al., 2010 and Taylor et al., 2012), which is inadequate to serve as input for catchment scale hydrologic modeling.

To generate GCM-forced rainfall input with spatial and temporal scales that are appropriate for catchment scale hydrologic modeling, dynamical and statistical downscaling methods were developed (e.g., Fowler et al., 2007; Hewitson and Crane, 1996; Wilby and Wigley, 1997). Dynamical downscaling uses output from the GCM as boundary conditions for a nested regional circulation model (RCM) with higher spatial and temporal resolution that accounts for local and regional climate processes and orographic influences. Due to the significant computational time involved, dynamical downscaling efforts often yield a single downscaled realization of physically consistent input vectors for hydrologic modeling (e.g., downscaling 30 years of GCM data produces a single 30 year simulation). This approach often requires bias adjustment for the downscaled high resolution fields and/or for the resultant streamflow (Georgakakos et al., 2012a).

A less computationally intensive approach uses statistical downscaling of the GCM fields. In statistical downscaling, statistical relationships are formed between local observations (e.g., temperature or rainfall intensity at ground gauges) and large scale atmospheric variables simulated by the GCMs (e.g., sea level pressure or specific humidity). This downscaling method permits the incorporation of changes in the mean and variability of climate in a statistically consistent and computationally inexpensive way (Semenov and Barrow, 1997). It does, however, simplify the large-to-small scale process dynamics and arrives at relationships for such dynamics that are calibrated based on the historical climate. Statistical downscaling is often integrated with a weather generator (WG), which is a stochastic model that simulates a likely weather time series realizations that are based on statistical analysis of observed local data (Wilks and Wilby, 1999). In contrast to dynamical downscaling, an ensemble of many realizations can be generated to represent the statistical properties of the weather. Statistical downscaling is used in this work as a first step to understand the sensitivity of the hydrologic regime to climate change.

WG can be used to generate rainfall that represents different statistical characteristics of the regional synoptic sys-

tems (e.g., Robertson et al., 2004; Kioutsioukis et al., 2008; Samuels et al., 2009; Shamir et al., 2014b). In these cases, however, the rainfall was generated with a daily temporal resolution for point locations. Thober et al. (2014) presented a daily but spatially high-resolution rainfall generator (4 km^2) that satisfies the requirements for assessments of hydrological response at regional and continental scales. Two WG models that were recently presented by Paschalis et al. (2013) and Peleg and Morin (2014), simulate rainfall in high spatial and temporal resolution (4 km^2 and 5 min and 0.25 km^2 and 5 min, respectively). This is a resolution that is adequate for hydrological modeling of small size (less than 50 km^2) catchments assuming reliable rainfall estimates (Borga et al., 2014). The latter rainfall generator model (high resolution synoptically conditioned weather generator (HiReS-WG), Peleg and Morin, 2014) which is used in this study, is unique in that it was designed to capture convective characteristics in a climate regime that is dominated by convective rainfall.

In this study we present a new modeling framework to assess projected climate impact on the hydrological regime by generating high spatiotemporal resolution rain fields with statistics that are dependent on the regional meteorological synoptic systems (Fig. 1). The suggested modeling framework integrates methods and models from various disciplines such as a classification of synoptic systems, remote sensing of rainfall, stochastic convective rainfall generator, and hydrological modeling. This integration enables, for the first time, an assessment of the impact of climate change on the hydrological regime in an environment that is strongly influenced by convective rainfall. To exemplify the presented modeling framework, two medium size catchments in Israel were selected for a case study. Section 2 describes the suggested conceptual modeling framework, and Sect. 3 presents the case study. The projected rainfall and runoff changes are presented in Sect. 4, and discussion and conclusions are in Sect. 5.

2   A framework for assessing the sensitivity of the hydrologic regime to climate change

A conceptual framework for assessing the sensitivity of the hydrologic regime to climate change is presented on the left side of Fig. 1. This framework integrates several tools and models that were previously developed by the authors. The steps of the conceptual modeling framework and their relation with the previously developed tools and models are presented in the following section. First, using reanalysis data, a synoptic classification is performed to determine the regional synoptic system prevalent during rainfall events and its occurrence frequency for the historical climate record. Second, a historical record of remotely sensed rainfall estimate is used to derive the relevant rainfall spatiotemporal statistical properties for each synoptic system. Third, based on the first and second steps, the WG is used to generate an

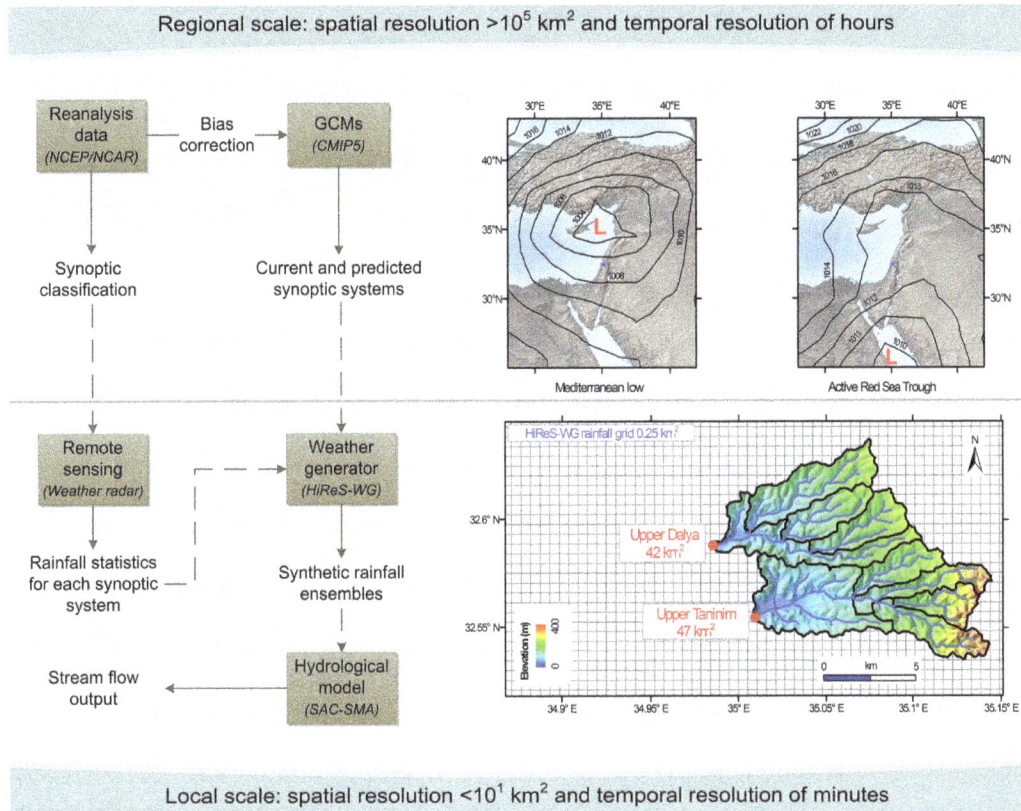

Figure 1. A framework flow chart (left) for assessing the sensitivity of the hydrologic regime to climate change. Models mentioned in parenthesis (italic caption) relate to the models or methods used for the case study of upper Dalya and upper Taninim catchments. The catchments are presented in the lower right side (red dots represent the hydrometric stations) and the prevailing two wet synoptic systems affecting the region are presented in the upper right side (contours represent sea level pressure in hPa, blue star indicate the catchments location).

ensemble of likely rainfall realizations that represents the historical climate. Fourth, the GCM synoptic variables are bias corrected to match the synoptic variables derived from the reanalysis data during the analyzed historic period, and projected changes in the occurrence frequency of future synoptic systems are then estimated from the bias corrected GCM. Fifth, the WG is used to generate an ensemble of likely rainfall scenarios that represent the occurrence frequency of the GCM's projected synoptic systems. Sixth, the historic and projected synthetic rainfall ensembles are used as input to a hydrologic model to assess the sensitivity of the hydrologic regime to modeled climate change.

This framework is adequate for regions that experience rainfall from distinctively different synoptic systems that can be represented by uniquely identified statistical indices. In such regions, not only the effect of changes in total amounts of rainfall, but also the effect of changes in occurrence frequency of the synoptic systems that cause rainfall should be examined. A high correspondence between annual amounts of rainfall and the occurrence frequency of wet synoptic systems was found for the study region (Saaroni et al., 2010). Therefore, the proposed framework focuses on the change in

projected future rainfall that resulted from changes in the occurrence frequency of the regional wet synoptic systems. The effect of projected changes in rainfall spatiotemporal characteristics was not examined in this study because future projections with the adequate skill and resolution that are required for such assessment are not yet available for the application area.

3   The case study of the Dalya and Taninim catchments

The framework discussed in the previous section was applied to the upper Dalya and upper Taninim catchments. We focused on the projected change in rainfall and streamflow volumes by comparing the beginning of the 21st century (historical period; 1996–2005) to mid-21st century (future period; 2046–2055). To achieve this goal the following methods and models (Fig. 1, left side in italic) were used: (1) a synoptic classification was carried out using the NCEP/NCAR reanalysis data (Peleg and Morin, 2012); (2) convective rainfall space–time characterization and the associated empirical distributions per synoptic system were computed using data

Table 1. Summary of data sources and resolutions used in this study.

Data type	Source	Resolution
Weather radar	Shacham (EMS) Mekorot	5 min and $1.4° \times 1$ km
Rain gauges (national network)	Israel Meteorological Service	Daily
Rain gauges (dense network)	Hydrometeorology lab at the Hebrew University	1 min
Hydrometric stations	Israel Water Authority	Changes within flow event

from the Shacham–Mekorot weather radar (Peleg and Morin, 2012; Peleg et al., 2013); (3) selected GCM simulations from CMIP5 were bias corrected and changes in the occurrence frequency of the synoptic systems between the historical and future periods were estimated (Peleg et al., 2014); (4) rainfall ensembles with convective features for current and projected climate were generated by the HiReS-WG (Peleg and Morin, 2014); and (5) the streamflow in the catchments' outlets was simulated using the Sacramento Soil Moisture Accounting Model (SAC-SMA) (Shamir et al., 2014a).

3.1 Study area

The upper Dalya (Bat Shlomo hydrometric station, 42 km^2) and upper Taninim (Amiqam hydrometric station, 47 km^2) catchments are located in the Ramot Menashe region, north-western Israel (Fig. 1, right side). The terrain near the catchments outlets is mostly flat, ascending moderately eastward up to 380 m a.s.l. Both upper Dalya and upper Taninim are ephemeral streams, similar in size but slightly different in their geologic formation and soil cover. The Adulam formation, dated to the lower Eocene and composed mainly of limestone and chalk, cover most of the drainage area of the upper Dalya catchment. This formation covers also large parts of the upper Taninim catchment, while the highest parts of the catchment consist of older (Paleocene) Taqiye formation, composed mainly of chalk and marl. The area is mostly cultivated agriculture that has experienced a severe soil erosion in the last several decades. The soil thickness varies between 240 and 340 mm for the upper Dalya catchment and between 240 and 500 mm for the upper Taninim catchment. In both catchments the soil is classified as Rendzina with clay texture in the top soil. The lithology of the upper parts of Dalya and Taninim catchments is concordant with other areas of the Ramot Menashe region (Grodek et al., 2012).

The region has a Mediterranean climate with wet winters (October–May) and dry and hot summers (June–September). The annual rainfall is highly variable with rainfall exceeded 1000 mm in the wettest observed year (1991/1992), while less than 400 mm was recorded in driest observed year (1981/1982). A considerable number of rainfall events in this area are caused by convective processes (Peleg and Morin, 2014). Records of rainfall and streamflow exist for these catchments from a weather radar, rain gauges, and hydrometric stations (a summary of available data is in Table 1). Following a rigorous quality control assessment

applied to the weather radar information, data for twelve hydrological years (1 October–30 September) were compiled for this study (i.e., 1991/1992–1997/1998, 1999/2000–2002/2003 and 2004/2005).

3.2 Rainfall patterns and synoptic classification

The wet synoptic systems (i.e., synoptic systems that might lead to rainfall over the catchments) and their rainfall statistics were studied for the northern coastal region of Israel by Peleg and Morin (2012, 2014). A summary of their findings that is relevant to this study follows.

The wet synoptic systems were classified using the NCEP/NCAR meteorological reanalysis data (Kalnay et al., 1996), extracted for the nearest location available to the catchments (35° E, 32.55° N, Fig. 1). The grid cell is located offshore; approximately extending from the northern Israel coastline to Cyprus. The above-mentioned 12 years of data were examined and 882 6 h intervals with rainfall over the catchments observed by the weather radar were identified. The synoptic classification during these periods was performed using a cluster analysis technique that considers four variables from the reanalysis data: (1) sea level pressure, (2) specific humidity at 700 hPa, (3) geopotential height at 500 hPa; and (4) zonal wind at 850 hPa. Initially, three wet synoptic systems classes were identified (Peleg and Morin, 2012). However, the first two were merged into one synoptic class – known as the Mediterranean low (ML), an extratropical cyclone, see Fig. 1). The ML is the prevailing synoptic system that generates rainfall over the study region (Alpert et al., 2004; Peleg et al., 2014; Peleg and Morin, 2012; Saaroni et al., 2010). The second synoptic system (Fig. 1) was linked to the active Red Sea trough (ARST), which occurs mainly during the transition seasons (Tsvieli and Zangvil, 2005) and mainly affects the southern and eastern parts of Israel (Kahana et al., 2002). The ARST is defined as a sea-level trough that extends from eastern Africa along the Red Sea towards the Mediterranean (Ashbel, 1938). The cluster analysis classified 94 % of the wet synoptic systems in the study area as ML, and (6 %) as ARST (Peleg and Morin, 2012).

Empirical distributions that describe the rainfall statistics during ML and ARST events were derived from analysis of 191 586 radar volume scans of the C-band Shacham (EMS) Mekorot weather radar system, located ∼ 60 km south of the study area (Peleg and Morin, 2012). The convective features were spatially determined using a segmentation method and

Table 2. Summary of the control (NCEP/NCAR reanalysis) and the climate models (CMIP5) used in the study.

Modeling center	Model name	Grid location	Spatial resolution
National Centers for Environmental Prediction	NCEP/NCAR reanalysis	35° E, 32.5° N	2.5° × 2.5°
Beijing Climate Center, China Meteorological Administration	BCC-CSM1.1	33.75° E, 32.1° N	2.8° × 2.8°
NOAA Geophysical Fluid Dynamics Laboratory	GFDL-ESM2G	33.75° E, 33.4° N	2° × 2.5°
Met Office Hadley Centre (Realization contributed by Instituto Nacional de Pesquisas Espaciais)	HadGEM2-ES	33.75° E, 33.1° N	1.25° × 1.87°

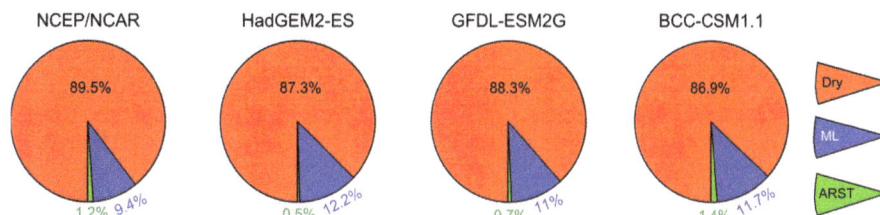

Figure 2. Annual average of 6 h occurrence frequency of dry, ML, and ARST synoptic systems derived from the NCEP/NCAR reanalysis data and the three GCMs used in this study for the historical period (after Peleg et al., 2014). The GCMs slightly overestimate the wet synoptic systems frequency by 2.1 % (HadGEM2-ES), 1.2 % (GFDL-ESM2G) and 2.5 % (BCC-CSM1.1).

temporally analyzed using a rain cell tracking algorithm (Peleg and Morin, 2012). The empirical distributions include for example: the number of convective rain cells, their areal extent, maximal and areal mean of rain intensity, orientation and advection properties, the low-intensity rainfall area, and mean rainfall intensity (Peleg and Morin, 2012, 2014). In addition to the radar data, a dense rain-gauge network in a 4 km^2 plot located in the upper Taninim catchment was used to assess the small scale rainfall spatial correlation at the sub-grid resolution (Peleg et al., 2013).

3.3 Climate change projection using CMIP5 models

The Coupled Model Intercomparison Project Phase 5 (CMIP5, see Moss et al., 2010 and Taylor et al., 2012 for more details) simulations were used to compare the wet synoptic systems' occurrence between the historical (beginning of the 21st century) and future (mid-21st-century) periods. The analyzed simulations were obtained from models with 6 h temporal resolution that correspond with the NCEP/NCAR reanalysis, and for two of the IPCC Assessment Report 5 emission scenarios: (1) the high-emission scenario (RCP8.5); and (2) a midrange-mitigation-emission scenario (RCP4.5). Peleg et al. (2014) present synoptic system classification and occurrence frequency related to each system for the eastern Mediterranean based on NCEP/NCAR reanalysis and GCMs for the two selected scenarios. A quantile-quantile bias correction was applied for the meteorological variables of the CMIP5 models, using the NCEP/NCAR reanalysis data as a reference for the corrections. For the current analysis three GCM CMIP5 models were selected: HadGEM2-ES, GFDL-ESM2G and BCC-CSM1.1. These GCMs were selected because of their 6 h temporal resolution, the close proximity of the pixels from which the models' meteorological variables were derived to the study region (Table 2), and the relatively good representation of synoptic systems occurrence frequency by these models (Fig. 2). It was found that the following are likely: rain in the region will become less frequent because of a reduction in the occurrence frequency of wet synoptic systems, the wet season period (i.e., the period from the first to last rainy day in the year) will be shortened mainly from the ending of the wet season and the occurrence frequency of ARST will increase at the expense of the ML; the magnitudes of the above changes for each GCM and emission scenario are summarized in Table 3 (see a detailed analysis and discussion by Peleg et al., 2014).

Based on the above analyses two scenarios of projected synoptic system frequency change were examined here. RCP4.5 scenario: the wet season period was projected to be shortened by 6.5 % from its ending, the occurrence of wet synoptic systems was projected to be reduced by 16 %, and the occurrence of ARST was projected to increase by 4.25 % at the expense of the ML. RCP8.5 scenario: the wet season period was projected to be shortened by 9.5 % from its ending, the occurrence of wet synoptic systems was projected to be reduced by 16 %, and the occurrence of ARST was projected to increase by 8.5 % at the expense of the ML. These two projections were combined subjectively from the changes found for each GCM (Table 3) giving some higher weights to higher values of change found.

Table 3. Projected changes in regional synoptic system frequency derived from three GCMs comparing the beginning and the middle of the 21st century for the northern Israeli coastline (after Peleg et al., 2014).

	HadGEM2-ES		GFDL-ESM2G		BCC-CSM1.1	
	RCP4.5	RCP8.5	RCP4.5	RCP8.5	RCP4.5	RCP8.5
Change in occurrence frequency of wet synoptic systems	−10%	−10%	−12%	−12%	−17%	−17%
Change in duration of wet season period	−2.6%	−6%	+6.6%	+1.9%	−12.6%	−12.7%
Change in occurrence frequency of ARSTs (on the expense of occurrence frequency of MLs)	+5.8%	+10.8%	0%	+2.2%	+1%	+1.2%

3.4 HiReS-WG

The high resolution synoptically conditioned weather generator (HiReS-WG) is a stochastic model that generates rain fields with a substantial proportion of convective features (Peleg and Morin, 2014). The HiReS-WG generates rain fields based on the empirical distributions of the rainfall characteristics subjected to the classified wet synoptic system. The rain fields were generated for the catchments in a spatial resolution of $0.25\,\mathrm{km}^2$ (see Fig. 1 lower right side) and a temporal resolution of 5 min. This is a sufficiently high spatiotemporal resolution that can be adequately used to simulate the hydrological response of the studied catchments. An ensemble of 300 years of likely rainfall realizations that represents the historical period was generated. The ensemble was further divided into 10 series of 30 years each, in order to assess the inter- and intra-annual variability of the rainfall. In addition, ensembles were generated for the future period for each scenario and were also similarly divided into 10 data series, representing the projected changes in the occurrence of the wet synoptic systems, as discussed above.

3.5 SCA-SMA hydrological model

The Sacramento Soil Moisture Accounting Model (SAC-SMA) is a conceptual, continuous, and aerial-lumped model that describes the wetting and drying processes in the soil. A detailed description of the continuous-time form formulation of the model that was implemented in this study as described by Georgakakos (1986). The SAC-SMA robustness was demonstrated by the results of the Distributed Modeling Intercomparison Project (Reed et al., 2004; Smith et al., 2004) and this model is used in many operational setups for various water resources management and flood warning practices. In recent years the SAC-SMA was also used in various climate change impact studies (e.g., Georgakakos et al., 2012a, b; Carpenter, 2011; Kerkhoven and Gan, 2011; Koutroulis et al., 2013; Kwon et al., 2011; Najafi et al., 2011).

The development of the SAC-SMA model for the study area is detailed in Shamir et al. (2014a) and a short summary is provided below. The model was calibrated for the hydrometric stations of Bat Shlomo (upper Dalya catchment) and Amiqam (upper Taninim catchment; Fig. 1). The rainfall data are from the Shacham (EMS) Mekorot weather radar system for a period of 12 years (mentioned above). Daily rainfall data from 26 rain gauges within 100 km distance from the radar were used for the radar-gauge adjustment and another 13 rain gauges in the surroundings of the Dalya and Taninim catchments were used for validation (Peleg and Morin, 2012). Rainfall data were initially calculated using the weather radar reflectivity data by applying a fixed reflectivity-rainfall power law relationship and then readjusted for each year using the weighted regression method (Gabella et al., 2001; Morin and Gabella, 2007). The model was constructed for four and five sub-basins for the upper Dalya and upper Taninim catchments, respectively. The rainfall input to the model was spatially aggregated to reflect mean areal values for these sub-basins. Initial SAC-SMA parameter values were estimated using field survey soil data and GIS layers of terrain, soil, and lithology. The SAC-SMA model was implemented to run in 5 min intervals, which is the native resolution of the radar data and the output of the HiReS-WG and is adequate because of the small sub-basins size and their rapid response time. The model was calibrated to capture the general hydrologic regime rather than to simulate the peak and the timing of a specific flow event. The SAC-SMA model was set to run continuously for each year from 1 October to 31 May, where for each year the initial conditions of the soil components were set to dry conditions ($\sim 1\%$ of saturation).

Comparing the observed average mean areal rainfall over the hydrometric stations drainage area with the streamflow indicate that initial cumulative rainfall is required in the beginning of the rainy season before measurable streamflow is recorded in the stations. This phenomenon was previously reported in other Mediterranean karst catchments (e.g., Ben-Zvi, 1998; Hartmann et al., 2014; Rimmer et al., 2006; Samuels et al., 2009). Note that even very intense and long rainfall events in the beginning of the wet season (e.g., 1991/1992 and 1994/1995) did not generate measurable streamflow at the stations. The SAC-SMA model structure was modified to include an initial loss reservoir that starts empty every year (1 October) and has to be satisfied first before the rainfall is applied as input to the rest of the SAC-SMA model components.

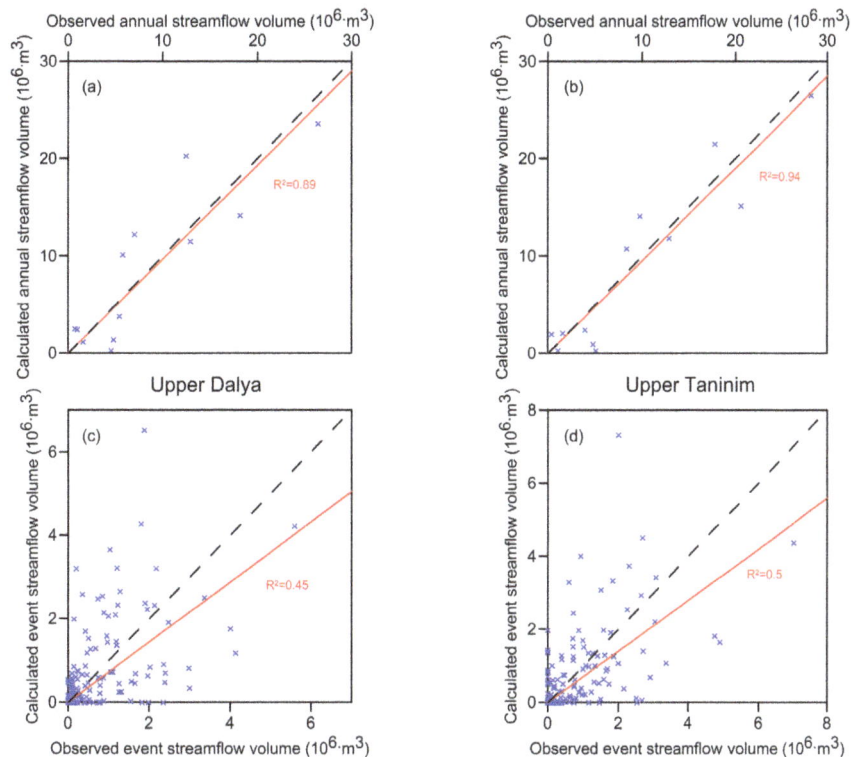

Figure 3. Panels **(a)** and **(b)** present calculated (from SAC-SMA) vs. observed (from hydrometric stations) annual streamflow volumes for the upper Dalya and upper Taninim catchments, respectively, for 12 years of data. The R^2 values relate to the linear fit that runs through the origin (red line). Dashed line represents perfect match between observed and calculated volumes (slope of 1). Panels **(c)** and **(d)** are the same but for event streamflow volumes.

The SAC-SMA simulations match the observed annual volumes at the hydrometric stations of the upper Dalya and upper Taninim catchments (Fig. 3a and b), with high R^2 values (0.89 and 0.94, respectively) for the linear fit (slope of 0.96 and 0.95, respectively). To examine the fit at the event level we defined rain and streamflow events as follows: a rain event begins when rain first appears over the catchment and ends when there is a dry-spell intermission that exceeds at least 3 h before the next pulse of rain occurs; a rain event is being accounted for only if its mean areal rainfall exceeds 10 mm. A streamflow event begins when a corresponding rain event begins. The end of the streamflow event was determined by fitting an exponential decay to the discharge recession limb (using a decay constant of $0.01\,h^{-1}$ for both catchments), cut at a threshold closer to zero ($\sim 0.03\,m^3s^{-1}$). An example is shown in Fig. 4 for a 2-week period in January 1991 for the upper Dalya catchment. The SAC-SMA simulates the events streamflow volumes reasonably well (Fig. 3c and d), with an acceptable slope of the linear fit between observed and calculated data of 0.72 for the upper Dalya catchment and 0.7 for the upper Taninim catchment (R^2 of 0.45 and 0.5, respectively).

A rainfall ensemble of 300 years was generated by the HiReS-WG for historical climate conditions and was used as forcing to the SAC-SMA model. Figure 5 presents three

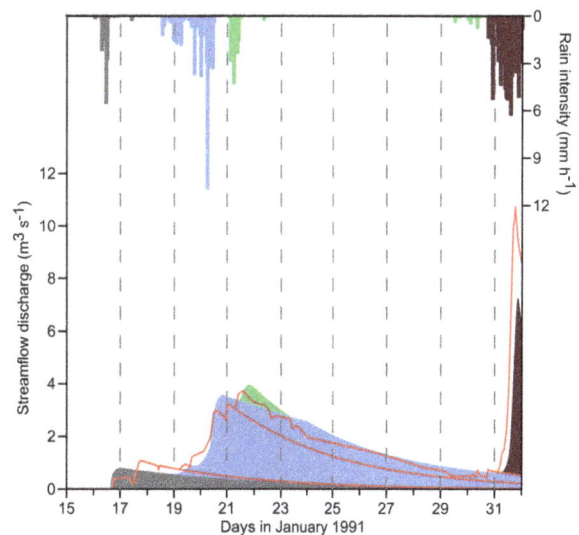

Figure 4. An example of rainfall and streamflow discharge from the upper Dalya catchment separated into different events of rainfall and streamflow. Red line represents the observed hydrograph obtained from the hydrometric station and filled areas represent the hydrograph simulated by the SAC-SMA. See text for details about the separation procedure for rain and streamflow events.

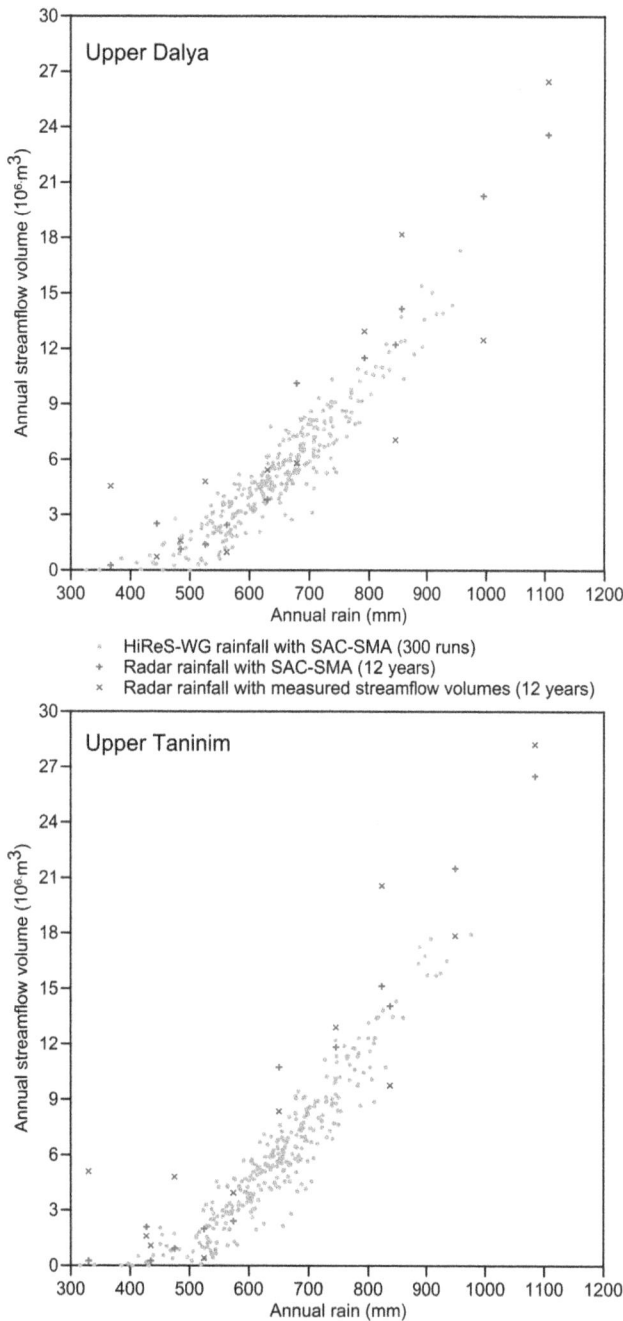

Figure 5. The annual streamflow volumes vs. the annual rainfall depth over the catchments. Three data samples are presented: observed rainfall from weather radar and observed volumes from hydrometric stations (blue), observed rainfall from weather radar and simulated volumes from SAC-SMA (red) and rainfall realizations generated by the HiReS-WG and simulated volumes from SAC-SMA (grey).

combinations of annual rainfall and streamflow data: (1) observed rainfall (from weather radar) and observed streamflow, (2) observed rainfall and its associated SAC-SMA simulated streamflow, and (3) HiReS-WG simulated rainfall and

its associated SAC-SMA simulated streamflow. It seems that the HiReS-WG and SAC-SMA integration simulates well the observed period with the exception of the wettest annual rainfall (1181 mm observed in 1991/1992). This is attributed to the HiReS-WG limitation in generating uncharacteristically long lasting rain events as in 1991/1992 (discussed in details by Peleg and Morin, 2014). For example, the average duration of the longest 10 % rain events during the observed 12 years was 13 h, while the average duration of the longest 10 % rain events during 1991/1992 was 17 h.

4 Projected change in the hydrological regime induced by climate change

4.1 Changes in rainfall amounts

Three ensembles, each with 10 data series of 30 years of rainfall, were generated by the HiReS-WG. The first ensemble represents the historical period (beginning of the 21st century) and the second and third ensembles represent the projected future (mid 21st century) for the RCP4.5 and RCP8.5 emission scenarios, respectively. As seen in Fig. 6, the mean annual rainfall is projected to decrease from 646 to 547 mm (15 % reduction) and 531 mm (18 %) for the RCP4.5 and RCP8.5 emission scenarios, respectively. The decrease in mean annual rainfall was found to be statistically significant for both emission scenarios using a two-sided z test (with a significance level $[\alpha]$ of 0.05).

A range of reduction values can be estimated by examining the change between the minimum (maximum) rainfall annual mean that was derived from the 10 data series of the three ensembles. For the RCP4.5 and RCP8.5 emission scenarios, the projected range of the mean annual rainfall reduction over the catchments is 2–23 % and 7–25 %, respectively. In addition, the standard deviation of the annual rainfall derived from the ensembles is projected to decrease from 104.5 mm for the historical period to 99.9 mm and 95.8 mm for the RCP4.5 and RCP8.5 scenarios, respectively. The changes in standard deviation however, were found insignificant ($\alpha = 0.05$) using F test for equality of two standard deviations; samples were normally distributed as tested by Shapiro–Wilk W test ($\alpha = 0.05$). In contrast to the projected decrease in mean annual rainfall and standard deviations, the coefficient of variation (CV) of the annual rainfall is projected to increase from an average of 16.5 % for the historical period to 17.9 and 18.2 % for the RCP4.5 and RCP8.5 emission scenarios, respectively, indicating an increase of the inter-annual changes relative to the mean rainfall.

Local sensitivity tests were conducted for the parameters that were modified in the ensembles of the future projection. Rainfall ensembles of 300 years each were generated for the following cases: a shortening of the wet season period, an increase in the occurrence frequency of ARST on the expense of occurrence frequency of ML, and a reduction of

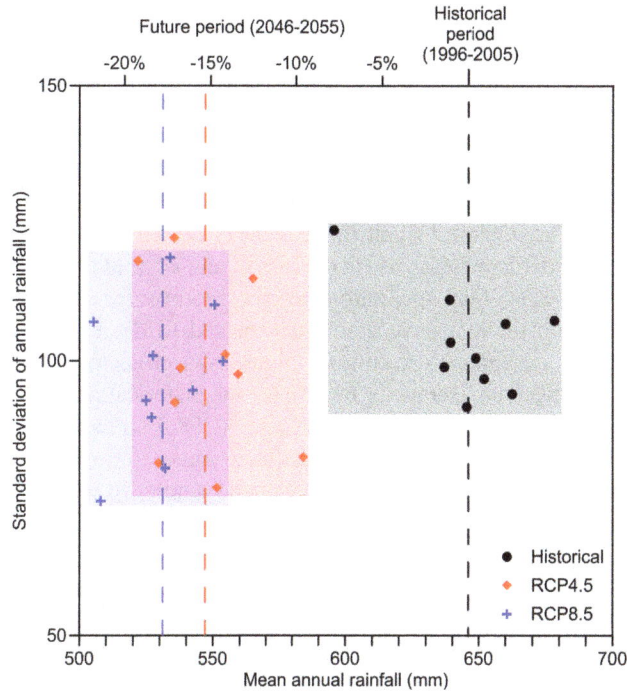

Figure 6. The standard deviation of the annual rainfall of each 30-year ensemble selected from the 300-year simulation (black for historical, red for RCP4.5 scenario and blue for RCP8.5 scenario) is presented vs. the mean annual rainfall for each ensemble. Dashed lines represent the average annual rainfall of each ensemble, and colored boxes show the extent of annual rainfall mean and standard deviation for the different periods and scenarios. The rainfall ensembles represent upper Dalya and upper Taninim catchments combined.

Figure 7. Same as Fig. 6 but for annual streamflow volume for upper Dalya catchment.

Table 4. Summary of the local sensitivity tests. The average annual rainfall of each ensemble and its projected change comparing to the historical ensemble (in parenthesis) are presented.

Ensemble	Mean annual rainfall (mm)
Historical period (1996–2005)	646
Shortening of wet season period by 10 %	640 (−0.9 %)
Increasing occurrence frequency of ARSTs on the expense of occurrence frequency of MLs by 10 %	643 (−0.5 %)
Reducing occurrence frequency of wet synoptic systems by 10 %	547 (−15.3 %) *

* Significant changes at the 0.05 level

occurrence frequency of wet synoptic systems. For all cases, a 10 % change in the parameter values that were estimated from the historic data was examined. The average annual rainfall of each ensemble and its change compared to the historical period are summarized in Table 4.

The reduction in the mean annual rainfall (by 15 %; statistically significant using a two-sided z test, $\alpha = 0.05$) was most sensitive to the reduction in occurrence frequency of the wet synoptic systems (by 10 %). Increased frequency of ARST at the expense of ML (by 10 %) had a small (a reduction of 0.5 %) and insignificant effect on rainfall in this region. This is a surprising result because of the marked differences in rainfall characteristics between the ML and ARST synoptic conditions. For example, convective features caused by ARST compared to ML are larger in area but with weaker rain intensities (Peleg and Morin, 2012, 2014); it is possible that some of these differences affect the total rainfall in an opposite manner such that the integrated effect is small. Shortening of the wet season period (by 10 %) had no detectible effect on the mean annual rainfall (a reduction of 0.9 %, statistically insignificant). This is because the shortening of the wet season period by 10 % implies a reduction

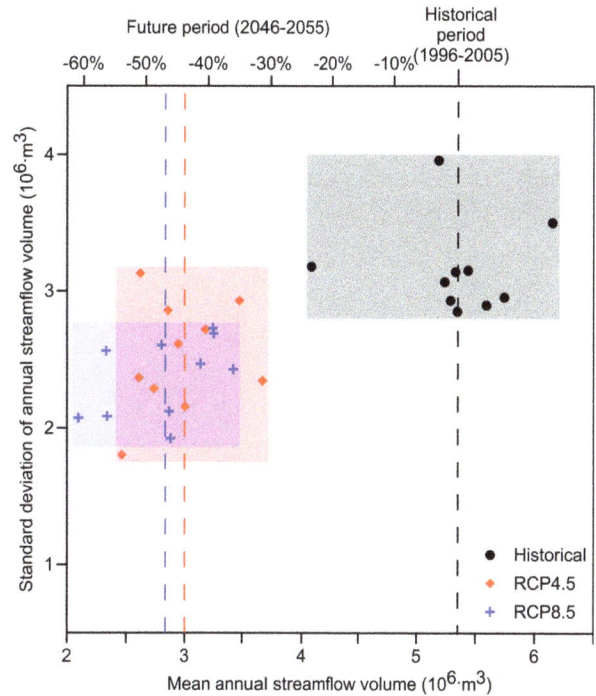

of about 1–2 rain events only when the occurrence frequency of rain events for either the beginning or end of the wet season is low.

4.2 Changes in streamflow volumes

The mean annual streamflow volumes for the 10 data series of the historical period and of the future period (RCP4.5 and RCP8.5 emission scenarios) are presented in Fig. 7 for the upper Dalya catchment. Streamflow was calculated by the SAC-SMA using input from the above mentioned rainfall HiReS-WG ensembles. The mean annual streamflow volume calculated from the ensembles is projected to decrease from 5.34×10^6 m^3 to 2.96×10^6 m^3 (45 % reduction) for the RCP4.5 scenario and to 2.84×10^6 m^3 (47 %) for the RCP8.5 scenario. A range of reduction values were computed (Fig. 7) using the maximum and minimum streamflow

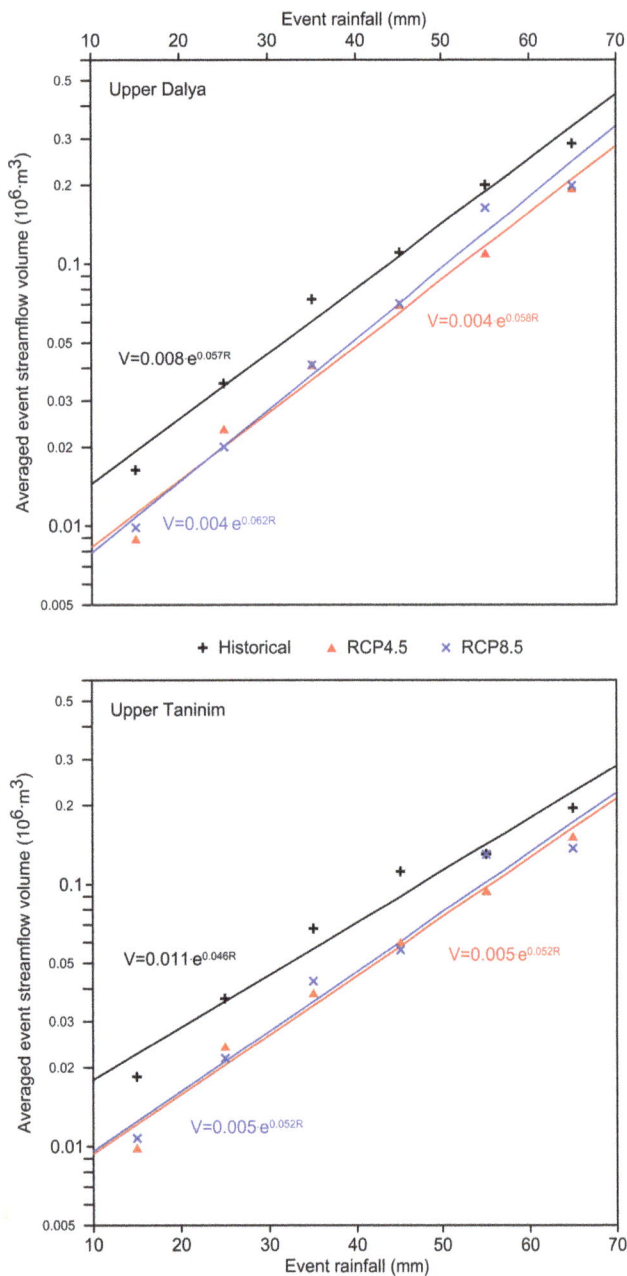

Figure 8. The average event streamflow volume vs. event rainfall depth for the historical period (black), RCP4.5 simulations (red) and RCP8.5 simulations (blue) with exponential fits (lines) for upper Dalya (upper graph) and upper Taninim (lower graph) catchments. Note that rainfall is quantized in 10 mm intervals and the points in the graph represent the average of all events within each 10 mm interval of rainfall.

volumes of the data series for each period, as explained in the previous section. For the RCP4.5 and RCP8.5 emission scenarios, the projected reduction range of the mean annual streamflow volume is 10–60 % and 16–66 %, respectively. As seen in the analysis of the rainfall ensembles, the standard deviation of the annual streamflow volume was pro-

jected to decrease from 3.17×10^6 m^3 for the historical climate to 2.53×10^6 m^3 and 2.39×10^6 m^3 for the RCP4.5 and RCP8.5 emission scenarios, respectively. The decrease in the mean annual streamflow volume was found to be statistically significant (using a two-sided z test, $\alpha = 0.05$). Because the samples of the standard deviation are not normally distributed (tested by Shapiro–Wilk W, $\alpha = 0.05$), we did not assess the statistical significance of the projected reduction in the standard deviation. As in the case of annual rainfall, while both mean annual streamflow volume and average standard deviation of annual volume are expected to decrease, the CV of the annual streamflow volume was projected to increase from an average of 61.9 % for the historical period to 87.5 and 86.2 % for the RCP4.5 and RCP8.5 scenarios, respectively. This paragraph discuss the changes in streamflow volumes for the upper Dalya catchment; similar trends were found for the upper Taninim catchment.

In addition to the annual scale, changes in events' streamflow volumes were examined. In general, a comparison between streamflow volume and rain depth at the event scale presents a large scatter because of factors such as antecedent moisture conditions and spatial and temporal distribution of rainfall. Therefore, we have examined the average streamflow volume for all events with rainfall depth binned in 10 mm intervals (Fig. 8); an exponential fit was applied for each period for both catchments (lowest R^2 calculated was 0.96). The average events' streamflow volumes for the historical period were found to be higher than for the future period for upper Dalya by a factor of 1.9–2 and 1.4–1.9 for the RCP4.5 and RCP8.5, respectively, and for upper Taninim by a factor of 1.5–2.1 for both emission scenarios. The decrease in events' streamflow volumes for a given event rainfall depth is a result of the reduction in occurrence frequency of wet synoptic systems which implies longer dry periods between events. Thus, the upper Dalya and upper Taninim catchments were projected to produce less runoff because of drier soil and drier antecedent moisture conditions that precede rainfall events.

Last, the effect of climate change on the duration of the streamflow season (i.e., from the first to the last appearance of streamflow in the catchment's outlet) was examined. Figure 9 presents the wet season periods and streamflow seasons for both catchments. It can be seen that streamflow season is projected to be shortened mainly because of delayed onset. This shorter streamflow season is attributed to the increase in dry duration between rainfall events, which cause drier soil conditions and delays the first streamflow event. Although to a lesser extent, the projected shorter streamflow season was also due to an earlier ending caused by projected earlier ending of wet synoptic systems, as discussed in Sect. 3.3. The shorter streamflow season was projected to be more severe for the RCP8.5 scenario (26 and 28 % for the upper Dalya and upper Taninim catchments, respectively) compared to the RCP4.5 scenario (22 % for both catchments).

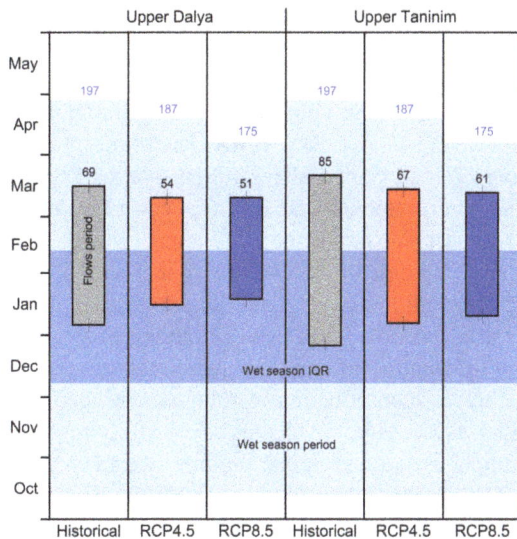

Figure 9. The period of the wet season (light blue, numbers represent duration in days), its interquartile range (IQR, dark blue) and the streamflow season (central boxes, numbers represent duration in days) for upper Dalya and Taninim catchments for historical and future periods.

5 Discussion and conclusions

In this paper we present a case study in which we assessed the sensitivity of the hydrological regime of two medium-size catchments (upper Dalya and upper Taninim, $\sim 50\,\text{km}^2$ each) to projected climate changes by generating high spatiotemporal resolution rain fields using the HiReS-WG (adequate for convective features) and applying them as input to the SAC-SMA hydrological model. The climate projections were done by modifying the occurrence frequency of the regional meteorological wet synoptic systems based on analysis of three GCMs CMIP5 simulations for the RCP4.5 and RCP8.5 emission scenarios.

Previous studies used daily rainfall resolution, for example: Chiew and McMahon (2002) examined the projected effects of climate change on runoff, evapotranspiration and soil moisture for eight Australian catchments, ranging from a medium scale ($27\,\text{km}^2$) to large scale ($544\,\text{km}^2$); Jones et al. (2006) estimated the hydrological sensitivity to changes in rainfall and potential evaporation in 22 catchments in Australia, ranging from medium ($52\,\text{km}^2$) to large ($1300\,\text{km}^2$) scale; Chiew et al. (2010) conducted a comprehensive study on eight medium-to-large catchments ($161–1540\,\text{km}^2$) in Western Australia by comparing the predicted changes in rainfall and runoff characteristics; Vaze and Teng (2011) conducted another study in Australia to predict changes in runoff for catchments of medium-to-large scales ($50–2000\,\text{km}^2$). Most of the above studies used rainfall data for point locations.

The use of the HiReS-WG model to generate high spatiotemporal resolution rain fields for hydrological projection

for small-to-medium sized catchments enables the simulation of streamflow not only for the annual or seasonal scale but also for the examination of the characteristics of flow events that last less than a few hours. In this study we demonstrated this capability by examining the projected change in the volumes of streamflow events for the upper Dalya and upper Taninim catchments. It was found that a major reduction in the projected average volumes of the streamflow events is expected.

The synthetic rainfall ensembles generated by the HiReS-WG indicated that the mean annual rainfall is projected to be reduced by 15 % (in a range of 2–23 %) and 18 % (7–25 %) for the RCP4.5 and RCP8.5 emission scenarios, respectively, comparing 1996–2005 to 2046–2055 periods. This reduction is larger than the regional reduction projected by other studies conducted for northern Israel. For example, Evans (2009) projected an annual decrease in rainfall of 20 mm (3 %) for 2050 compared to 2005; future rainfall reductions for the end of the 21st-century of 30–75 mm (4–12 %) were reported by Alpert et al. (2008) while Oenol and Semazzi (2009) reported a decrease in winter rainfall of 24 %. These studies used the A2 emission scenario (presented in the IPCC AR4 report) and were averaged for a much larger domain than the one examined in this study. In addition, their projections were derived from RCMs. In this study we focused on changes in the characteristics of the wet synoptic systems. We found that the characteristics of the projected wet synoptic systems are expected to have a shorter season, fewer occurrences of wet synoptic systems, and an increase in the occurrences of ARST at the expense of ML. Other factors related to rainfall characteristics might also be subject to change, for example, Black (2009) and Krichak et al. (2011) discussed that the proportion of convective rainfall out of the total rainfall might increase as a result of projected increase in surface temperature. Such factors weren't considered in this study but could be included in future studies. To the best of our knowledge, currently there are no future projections that consider changes in rainfall spatiotemporal statistics for this region.

Along with the projected decrease in mean annual rainfall, a decrease in the annual rainfall standard deviations and an increase in the annual rainfall CV were found. This implies an increase of the inter-annual changes relative to the mean rainfall for the future period. This is a similar trend to that found by Samuels et al. (2013) for Tel Aviv, located on the central coastline of Israel, for the A1B emission scenario. They reported a median decrease of 9.8 % in annual rainfall for 2035–2060 with a minor annual rainfall decrease of 1.7 % for the upper 5 percentiles. This implies that the extreme wet years will be affected by climate change in a different way than the mean annual rainfall. Similar results were found for the projected change in the mean annual streamflow volume, as it was also found that the mean and standard deviation are projected to decrease and the CV is projected to increase.

The mean annual streamflow volumes were projected to be reduced by 10–66 % (RCP4.5 and RCP8.5 emission scenarios). Only three studies that assessed climate change projected impact on the hydrological regime for this region (using A1B emission scenario) were found. Samuels et al. (2009) evaluated the impact of projected increase of multi-year droughts and extreme rainfall events on the streamflow of the Upper Jordan River and its tributaries. They found that while the projected increase in extreme rainfall events increases the streamflow intensity and change the extreme events recurrence distribution; no substantial changes were found for the low flow regime. Samuels et al. (2010) reported a projected reduction in mean daily base flow and surface flow of 10–11 % and 17 %, respectively, for the large-sized catchment ($> 1000\,\mathrm{km}^2$) of the upper Jordan River, comparing 1980–2005 and 2036–2060 periods. Rimmer et al. (2011) calculated trends of the annual incoming water volumes for the Sea of Galilee (northern Israel), comparing the periods 1979–2009 and 2015–2060 and using four climate models. The trends were found to vary between $-9.05 \times 10^6\,\mathrm{m}^3\,\mathrm{year}^{-1}$ and $1.94 \times 10^6\,\mathrm{m}^3\,\mathrm{year}^{-1}$; the maximum projected annual inflow reduction trend was equal to $\sim 1.3\,\%$. All studies projected a reduction in streamflow volumes, but differences exist between projected reduction amounts as there are differences between the studies: the domain size, the emission scenarios, the historical and future periods that were considered, and the catchments properties (e.g., Dalya and Taninim are ephemeral streams and the Upper Jordan River is a perennial river).

Changes in rainfall amounts are usually amplified in streamflow discharge and volume (Chiew and McMahon, 2002; Chiew et al., 2010; Li et al., 2013; Peleg and Gvirtzman, 2010). This effect was well demonstrated in the results of this study when examining the non-linearity of the synoptic-rainfall-streamflow interactions, which resolved in percentage change in runoff of about 3 times the percentage change in rainfall (the rainfall is projected to reduce by 2–25 % and the annual streamflow volume by 10–66 %). Future drier soil conditions resulting from the shortening of the wet season and reducing the number of rainfall events (which cause an increase in the length of dry periods between rainfall events) are the reasons for this amplification. Chiew et al. (2006) defined this amplification, the change in annual runoff in comparison to the change in annual rainfall, as the precipitation elasticity of streamflow (ϵ_p). Chiew et al. (2010) reported a change in annual runoff that is up to 2 times the change in annual rainfall, but in an earlier study Chiew and McMahon (2002) stated that "in ephemeral catchment with low runoff coefficients the percentage change in runoff can be more than 4 times the percentage change in rainfall". ϵ_p values similar in magnitude to those of this study ($\epsilon_p > 3$) were found in other catchments worldwide, representing a variety of climates Chiew et al. (2006).

Lastly, limitation of the HiReS-WG and SAC-SMA should be taken into consideration when discussing the effect of climate change on rainfall and streamflow. First, the HiReS-WG is unable to generate the observed extreme rainfall year (1991/1992); this is attributed to the HiReS-WG inability of generating rain events with extreme long-durations (Peleg and Morin, 2014). The SAC-SMA was often implemented to a coarser spatiotemporal resolution than an area of tens km^2 and 5 min intervals that was applied in this study. The River Forecast Centers of the US National Weather Service commonly implement the SAC-SMA in basins larger than $50\,\mathrm{km}^2$ and with hourly or longer temporal resolution (e.g., Shamir et al., 2006). The coarse resolution allows for averaging in time and space, and the hydrologic response is less abrupt therefore contributing to robust model performance. Nevertheless, the SAC-SMA was successfully implemented in the upper Dalya and upper Taninim catchments (Shamir et al., 2014a). When considering the results presented here of the effect of climate change on the hydrological regime it should be noted that besides the rainfall changes discussed above, no other changes were made. SAC-SMA parameters (land use, vegetation type and coverage, evaporation, etc.) were regarded as stationary; thus enabling the examination of how the projected changes in the regional synoptic systems would affect the runoff without taking into consideration other non-synoptic parameters (some are expected to change, e.g., evaporation is projected to increase along with an increase in temperature).

We conclude that for small-to-medium sized catchments influenced by convective rainfall, the HiReS-WG, integrated in the suggested modeling framework, is a good tool for projecting changes in the hydrological regime resulting from climate change. The HiReS-WG can be applied to catchments in other climate environments affected by convective rainfall. Further research assessing projected changes in rainfall spatiotemporal statistics is still required to better represent climate change impact on rainfall; once such information is available, its integration into the HiReS-WG is straightforward.

Acknowledgements. The study was funded by the Israel Water Authority, the Israel Ministry of Environmental Protection, the Israel Ministry of Agriculture and Rural Development, the KKL-JNF, the Israel–USA Bi-National Science Foundation (BSF-2008046), and the Israel Science Foundation's Recanati and IDB Group Foundation (grant no. 332/11). We acknowledge the World Climate Research Programmer's Working Group on Coupled Modelling, which is responsible for CMIP, and we thank the climate modeling groups (listed in Table 2 of this paper) for making the model's output available. For CMIP, the US Department of Energy's Program for Climate Model Diagnosis and Intercomparison provides coordinating support and led the development of software infrastructure in partnership with the Global Organization for Earth System Science Portals. NCEP reanalysis data was provided by the NOAA/OAR/ESRL PSD, Boulder, Colorado, USA, from their website at http://www.esrl.noaa.gov/psd/.

Edited by: L. Samaniego

References

Alpert, P., Osetinsky, I., Ziv, B., and Shafir, H.: Semi-objective classification for daily synoptic systems: application to the Eastern Mediterranean climate change, Int. J. Climatol., 24, 1001–1011, doi:10.1002/joc.1036, 2004.

Alpert, P., Krichak, S. O., Shafir, H., Haim, D., and Osetinsky, I.: Climatic trends to extremes employing regional modeling and statistical interpretation over the E. Mediterranean, Global Planet. Change, 63, 163–170, doi:10.1016/j.gloplacha.2008.03.003, 2008.

Arnaud, P., Lavabre, J., Fouchier, C., Diss, S., and Javelle, P.: Sensitivity of hydrological models to uncertainty in rainfall input, Hydrolog. Sci. J., 56, 397–410, doi:10.1080/02626667.2011.563742, 2011.

Ashbel, D.: Great floods in Sinai Peninsula, Palestine, Syria and the Syrian Desert, and the influence of the red sea on their formation, Q. J. Roy. Meteorol. Soc., 64, 635–639, doi:10.1002/qj.49706427716, 1938.

Bahat, Y., Grodek, T., Lekach, J., and Morin, E.: Rainfall-runoff modeling in a small hyper-arid catchment, J. Hydrol., 373, 204–217, doi:10.1016/j.jhydrol.2009.04.026, 2009.

Ben-Zvi, A.: Enhancement of runoff from a small watershed by cloud seeding, J. Hydrol., 101, 291–303, doi:10.1016/0022-1694(88)90041-8, 1998.

Black, E.: The impact of climate change on daily precipitation statistics in Jordan and Israel, Atmos. Sci. Lett., 10, 192–200, doi:10.1002/asl.233, 2009.

Borga, M., Stoffel, M., Marchi, L., Marra, F., and Jakob, M.: Hydrogeomorphic response to extreme rainfall in headwater systems: flash floods and debris flows, J. Hydrol., 518, 194–205, doi:10.1016/j.jhydrol.2014.05.022, 2014.

Carpenter, T. M. M.: An interdisciplinary approach to characterize flash flood occurrence frequency for mountainous Southern California, Ph.D. thesis, UC San Diego, California, 2011.

Chiew, F. H. S. and McMahon, T. A.: Modelling the impacts of climate change on Australian streamflow, Hydrol. Process., 16, 1235–1245, doi:10.1002/hyp.1059, 2002.

Chiew, F. H. S., Peel, M. C., McMahon, T. A., and Siriwardena, LW.: Precipitation elasticity of streamflow in catchments across the world, IAHS publication, 308, 256–262, 2006.

Chiew, F. H. S., Kirono, D. G. C., Kent, D. M., Frost, A. J., Charles, S. P., Timbal, B., Nguyen, K. C., and Fu, G.: Comparison of runoff modelled using rainfall from different downscaling methods for historical and future climates, J. Hydrol., 387, 10–23, doi:10.1016/j.jhydrol.2010.03.025, 2010.

Evans, J. P.: 21st century climate change in the Middle East, Climatic Change, 92, 417–432, doi:10.1007/s10584-008-9438-5, 2009.

Fowler, H. J., Blenkinsop, S., and Tebaldi, C.: Linking climate change modelling to impacts studies: recent advances in downscaling techniques for hydrological modelling, Int. J. Climatol., 27, 1547–1578, doi:10.1002/joc.1556, 2007.

Gabella, M., Joss, J., Perona, G., and Galli, G.: Accuracy of rainfall estimates by two radars in the same Alpine environment using gage adjustment, J. Geophys. Res.-Atmos., 106, 5139–5150, doi:10.1029/2000jd900487, 2001.

Georgakakos, K. P.: A generalized stochastic hydrometeorological model for flood and flash-flood forecasting. 1. Formulation, Water Resour. Res., 22, 2083–2095, doi:10.1029/WR022i013p02083, 1986.

Georgakakos, A. P., Yao, H., Kistenmacher, M., Georgakakos, K. P., Graham, N. E., Cheng, F. Y., Spencer, C., and Shamir, E.: Value of adaptive water resources management in Northern California under climatic variability and change: reservoir management, J. Hydrol., 412, 34–46, doi:10.1016/j.jhydrol.2011.04.038, 2012a.

Georgakakos, K. P., Graham, N. E., Cheng, F. Y., Spencer, C., Shamir, E., Georgakakos, A. P., Yao, H., and Kistenmacher, M.: Value of adaptive water resources management in northern California under climatic variability and change: dynamic hydroclimatology, J. Hydrol., 412, 47–65, doi:10.1016/j.jhydrol.2011.04.032, 2012b.

Grodek, T., Jacoby, Y., Morin, E., and Katz, O.: Effectiveness of exceptional rainstorms on a small Mediterranean basin, Geomorphology, 159, 156–168, doi:10.1016/j.geomorph.2012.03.016, 2012.

Hartmann, A., Mudarra, M., Andero, B., Marin, A., Wagener, T., and Lange, J.: Modeling spatiotemporal impacts of hydroclimatic extremes on groundwater recharge at a Mediterranean karst aquifer, Water Resour. Res., 50, 6507–6521, doi:10.1002/2014WR015685, 2014.

Hewitson, B. C. and Crane, R. G.: Climate downscaling: techniques and application, Clim. Res., 7, 85–95, doi:10.3354/cr007085, 1996.

Jones, R. N., Chiew, F. H. S., Boughton, W. C., and Zhang, L.: Estimating the sensitivity of mean annual runoff to climate change using selected hydrological models, Adv. Water Resour., 29, 1419–1429, doi:10.1016/j.advwatres.2005.11.001, 2006.

Kahana, R., Ziv, B., Enzel, Y., and Dayan, U.: Synoptic climatology of major floods in the Negev Desert, Israel, Int. J. Climatol., 22, 867–882, doi:10.1002/joc.766, 2002.

Kalnay, E., Kanamitsu, M., Kistler, R., Collins, W., Deaven, D., Gandin, L., Iredell, M., Saha, S., White, G., Woollen, J., Zhu, Y., Chelliah, M., Ebisuzaki, W., Higgins, W., Janowiak, J., Mo, K. C., Ropelewski, C., Wang, J., Leetmaa, A., Reynolds, R., Jenne, R., and Joseph, D.: The NCEP/NCAR 40 year reanalysis project, B. Am. Meteorol. Soc., 77, 437–471, doi:10.1175/1520-0477(1996)077<0437:tnyrp>2.0.co;2, 1996.

Kerkhoven, E. and Gan, T. Y.: Unconditional uncertainties of historical and simulated river flows subjected to climate change, J. Hydrol., 396, 113–127, doi:10.1016/j.jhydrol.2010.10.042, 2011.

Kioutsioukis, I., Melas, D., and Zanis, P.: Statistical downscaling of daily precipitation over Greece, Int. J. Climatol., 28, 679–691, doi:10.1002/joc.1557, 2008.

Koutroulis, A. G., Tsanis, I. K., Daliakopoulos, I. N., and Jacob, D.: Impact of climate change on water resources status: a case study for Crete Island, Greece, J. Hydrol., 479, 146–158, doi:10.1016/j.jhydrol.2012.11.055, 2013.

Krichak, S. O., Breitgand, J. S., Samuels, R., and Alpert, P.: A double-resolution transient RCM climate change simulation experiment for near-coastal eastern zone of the Eastern Mediterranean region, Theor. Appl. Climatol., 103, 167–195, doi:10.1007/s00704-010-0279-6, 2011.

Kwon, H.-H., Sivakumar, B., Moon, Y.-I., and Kim, B.-S.: Assessment of change in design flood frequency under climate change using a multivariate downscaling model and a precipitation-runoff model, Stoch. Env. Res. Risk A., 25, 567–581, doi:10.1007/s00477-010-0422-z, 2011.

Li, F., Zhang, Y., Xu, Z., Teng, J., Liu, C., Liu, W., and Mpelasoka, F.: The impact of climate change on runoff in the southeastern Tibetan Plateau, J. Hydrol., 505, 188–201, doi:10.1016/j.jhydrol.2013.09.052, 2013.

Morin, E. and Gabella, M.: Radar-based quantitative precipitation estimation over Mediterranean and dry climate regimes, J. Geophys. Res., 112, D20108, doi:10.1029/2006jd008206, 2007.

Morin, E. and Yakir, H.: Hydrological impact and potential flooding of convective rain cells in a semi-arid environment, Hydrolog. Sci. J., 59, 1353–1362, doi:10.1080/02626667.2013.841315, 2014.

Morin, E., Goodrich, D. C., Maddox, R. A., Gao, X. G., Gupta, H. V., and Sorooshian, S.: Spatial patterns in thunderstorm rainfall events and their coupling with watershed hydrological response, Adv. Water Resour., 29, 843–860, doi:10.1016/j.advwatres.2005.07.014, 2006.

Moss, R. H., Edmonds, J. A., Hibbard, K. A., Manning, M. R., Rose, S. K., van Vuuren, D. P., Carter, T. R., Emori, S., Kainuma, M., Kram, T., Meehl, G. A., Mitchell, J. F. B., Nakicenovic, N., Riahi, K., Smith, S. J., Stouffer, R. J., Thomson, A. M., Weyant, J. P., and Wilbanks, T. J.: The next generation of scenarios for climate change research and assessment, Nature, 463, 747–756, doi:10.1038/nature08823, 2010.

Najafi, M. R., Moradkhani, H., and Jung, I. W.: Assessing the uncertainties of hydrologic model selection in climate change impact studies, Hydrol. Process., 25, 2814–2826, doi:10.1002/hyp.8043, 2011.

Nikolopoulos, E. I., Anagnostou, E. N., Hossain, F., Gebremichael, M., and Borga, M.: Understanding the scale relationships of uncertainty propagation of satellite rainfall through a distributed hydrologic model, J. Hydrometeorol., 11, 520–532, doi:10.1175/2009jhm1169.1, 2010.

Oenol, B. and Semazzi, F. H. M.: Regionalization of climate change simulations over the eastern Mediterranean, J. Climate, 22, 1944–1961, doi:10.1175/2008jcli1807.1, 2009.

Paschalis, A., Molnar, P., Fatichi, S., and Burlando, P.: A stochastic model for high-resolution space-time precipitation simulation, Water Resour. Res., 49, 8400–8417, doi:10.1002/2013wr014437, 2013.

Paschalis, A., Fatichi, S., Molnar, P., Rimkus, S., and Burlando, P.: On the effects of small scale space–time variability of rainfall on basin flood response, J. Hydrol., 514, 313–327, doi:10.1016/j.jhydrol.2014.04.014, 2014.

Peleg, N. and Gvirtzman, H.: Groundwater flow modeling of two-levels perched karstic leaking aquifers as a tool for estimating recharge and hydraulic parameters, J. Hydrol., 388, 13–27, doi:10.1016/j.jhydrol.2010.04.015, 2010.

Peleg, N. and Morin, E.: Convective rain cells: radar-derived spatiotemporal characteristics and synoptic patterns over the eastern Mediterranean, J. Geophys. Res.-Atmos., 117, D15116, doi:10.1029/2011jd017353, 2012.

Peleg, N. and Morin, E.: Stochastic convective rain-field simulation using a high-resolution synoptically conditioned weather generator (HiReS-WG), Water Resour. Res., 50, 2124–2139, doi:10.1002/2013wr014836, 2014.

Peleg, N., Ben-Asher, M., and Morin, E.: Radar subpixel-scale rainfall variability and uncertainty: lessons learned from observations of a dense rain-gauge network, Hydrol. Earth Syst. Sci., 17, 2195–2208, doi:10.5194/hess-17-2195-2013, 2013.

Peleg, N., Bartov, M., and Morin, E.: CMIP5-predicted climate shifts over the east mediterranean: implications for the transition region between mediterranean and semi-arid climates, Int. J. Climatol., doi:10.1002/joc.4114, online first, 2014.

Reed, S., Koren, V., Smith, M., Zhang, Z., Moreda, F., Seo, D. J., and Participants, D.: Overall distributed model intercomparison project results, J. Hydrol., 298, 27–60, doi:10.1016/j.jhydrol.2004.03.031, 2004.

Rimmer, A. and Salingar, Y.: Modelling precipitation-streamflow processes in karst basin: The case of the Jordan River sources, Israel, J. Hydrol., 331, 524–542, doi:10.1016/j.jhydrol.2006.06.003, 2006.

Rimmer, A., Givati, A., Samuels, R., and Alpert, P.: Using ensemble of climate models to evaluate future water and solutes budgets in Lake Kinneret, Israel, J. Hydrol., 410, 248–259, doi:10.1016/j.jhydrol.2011.09.025, 2011.

Robertson, A. W., Kirshner, S., and Smyth, P.: Downscaling of daily rainfall occurrence over northeast Brazil using a hidden Markov model, J. Climate, 17, 4407–4424, doi:10.1175/jcli-3216.1, 2004.

Rozalis, S., Morin, E., Yair, Y., and Price, C.: Flash flood prediction using an uncalibrated hydrological model and radar rainfall data in a Mediterranean watershed under changing hydrological conditions, J. Hydrol., 394, 245–255, doi:10.1016/j.jhydrol.2010.03.021, 2010.

Saaroni, H., Halfon, N., Ziv, B., Alpert, P., and Kutiel, H.: Links between the rainfall regime in Israel and location and intensity of Cyprus lows, Int. J. Climatol., 30, 1014–1025, doi:10.1002/joc.1912, 2010.

Samuels, R., Rimmer, A. and Alpert, P.: Effect of extreme rainfall events on the water resources of the Jordan River, J. Hydrol., 375, 512–523, doi:10.1016/j.jhydrol.2009.07.001, 2009.

Samuels, R., Rimmer, A., Hartmann, A., Krichak, S., and Alpert, P.: Climate change impacts on Jordan river flow: downscaling application from a regional climate model, J. Hydrometeorol., 11, 860–879, doi:10.1175/2010jhm1177.1, 2010.

Samuels, R., Harel, M., and Alpert, P.: A new methodology for weighting high-resolution model simulations to project future rainfall in the Middle East, Clim. Res., 57, 51–60, doi:10.3354/cr01147, 2013.

Semenov, M. A. and Barrow, E. M.: Use of a stochastic weather generator in the development of climate change scenarios, Climatic Change, 35, 397–414, doi:10.1023/a:1005342632279, 1997.

Shamir, E., Carpenter, T. M., Fickenscher, P., and Georgakakos, K. P.: Evaluation of the National Weather Service operational hydrologic model and forecasts for the American River basin, J. Hydrol. Eng., 11, 392–407, doi:10.1061/(asce)1084-0699(2006)11:5(392), 2006.

Shamir, E., Georgakakos, K., Spencer, C., Modrick, T. M., Murphy Jr., M., and R., J.: Evaluation of real-time flash flood forecasts for Haiti during the passage of Hurricane Tomas, November 4–6, 2010, Nat. Hazards, 67, 459–482, doi:10.1007/s11069-013-0573-6, 2013.

Shamir, E., Georgakakos, K. P., Peleg, N., and Morin, E.: Hydrologic Model Development for the Dalia-Taninim watersheds in Israel, Tech. rep., HRC, San Diego, CA, available at: www.hrc-lab.org/projects/projectpdfs/HRCTN66_20140210.pdf (last access: 15 September 2014), 2014a.

Shamir, E., Megdal, S., Carrillo, C., Castro, C., Chang, H.-I., Chief, K., Corkhill, F., Eden, S., Georgakakos, K., Nelson, K., and Prietto, J.: Climate change and water resources management in the Upper Santa Cruz River, Arizona, J. Hydrol., doi:10.1016/j.jhydrol.2014.11.062, 2014b.

Singh, V. P.: Effect of spatial and temporal variability in rainfall and watershed characteristics on stream flow hydrograph, Hydrol. Process., 11, 1649–1669, doi:10.1002/(sici)1099-1085(19971015)11:12<1649::aid-hyp495>3.0.co;2-1, 1997.

Smith, J. A., Baeck, M. L., Morrison, J. E., and Sturdevant-Rees, P.: Catastrophic rainfall and flooding in Texas, J. Hydrometeorol., 1, 5–25, doi:10.1175/1525-7541(2000)001<0005:crafit>2.0.co;2, 2000.

Smith, M. B., Seo, D. J., Koren, V. I., Reed, S. M., Zhang, Z., Duan, Q., Moreda, F., and Cong, S.: The distributed model intercomparison project (DMIP): motivation and experiment design, J. Hydrol., 298, 4–26, doi:10.1016/j.jhydrol.2004.03.040, 2004.

Taylor, K. E., Stouffer, R. J., and Meehl, G. A.: An overview of CMIP5 and the experiment design, B. Am. Meteorol. Soc., 93, 485–498, doi:10.1175/bams-d-11-00094.1, 2012.

Thober, S., Mai, J., Zink, M., and Samaniego, L.: Stochastic temporal disaggregation of monthly precipitation for regional gridded data sets, Water Resour. Res., 50, 8714–8735, doi:10.1002/2014WR015930, 2014.

Tsvieli, Y. and Zangvil, A.: Synoptic climatological analysis of "wet" and "dry" Red Sea Troughs over Israel, Int. J. Climatol., 25, 1997–2015, doi:10.1002/joc.1232, 2005.

Vaze, J. and Teng, J.: Future climate and runoff projections across New South Wales, Australia: results and practical applications, Hydrol. Process., 25, 18–35, doi:10.1002/hyp.7812, 2011.

Wilby, R. L. and Wigley, T. M. L.: Downscaling general circulation model output: a review of methods and limitations, Prog. Phys. Geog., 21, 530–548, doi:10.1177/030913339702100403, 1997.

Wilks, D. S. and Wilby, R. L.: The weather generation game: a review of stochastic weather models, Prog. Phys. Geog., 23, 329–357, doi:10.1177/030913339902300302, 1999.

Yakir, H. and Morin, E.: Hydrologic response of a semi-arid watershed to spatial and temporal characteristics of convective rain cells, Hydrol. Earth Syst. Sci., 15, 393–404, doi:10.5194/hess-15-393-2011, 2011.

Zoccatelli, D., Borga, M., Viglione, A., Chirico, G. B., and Blöschl, G.: Spatial moments of catchment rainfall: rainfall spatial organisation, basin morphology, and flood response, Hydrol. Earth Syst. Sci., 15, 3767–3783, doi:10.5194/hess-15-3767-2011, 2011.

8

Identifying the origin and geochemical evolution of groundwater using hydrochemistry and stable isotopes in the Subei Lake basin, Ordos energy base, Northwestern China

F. Liu[1,2], X. Song[1], L. Yang[1], Y. Zhang[1], D. Han[1], Y. Ma[1], and H. Bu[1]

[1]Key Laboratory of Water Cycle and Related Land Surface Processes, Institute of Geographic Sciences and Natural Resources Research, Chinese Academy of Sciences, 11 A, Datun Road, Chaoyang District, Beijing, 100101, China
[2]University of Chinese Academy of Sciences, Beijing, 100049, China

Correspondence to: X. Song (songxf@igsnrr.ac.cn)

Abstract. A series of changes in groundwater systems caused by groundwater exploitation in energy base have been of great concern to hydrogeologists. The research aims to identify the origin and geochemical evolution of groundwater in the Subei Lake basin under the influence of human activities. Water samples were collected, and major ions and stable isotopes ($\delta^{18}O$, δD) were analyzed. In terms of hydrogeological conditions and the analytical results of hydrochemical data, groundwater can be classified into three types: the Quaternary groundwater, the shallow Cretaceous groundwater and the deep Cretaceous groundwater. Piper diagram and correlation analysis were used to reveal the hydrochemical characteristics of water resources. The dominant water type of the lake water was Cl-Na type, which was in accordance with hydrochemical characteristics of inland salt lakes; the predominant hydrochemical types for groundwater were HCO_3–Ca, HCO_3–Na and mixed HCO_3–Ca–Na–Mg types. The groundwater chemistry is mainly controlled by dissolution/precipitation of anhydrite, gypsum, halite and calcite. The dedolomitization and cation exchange are also important factors. Rock weathering is confirmed to play a leading role in the mechanisms responsible for the chemical composition of groundwater. The stable isotopic values of oxygen and hydrogen in groundwater are close to the local meteoric water line, indicating that groundwater is of modern local meteoric origin. Unlike significant differences in isotopic values between shallow groundwater and deep groundwater in the Habor Lake basin, shallow Cretaceous groundwater and deep Cretaceous groundwater have similar isotopic characteristics in the Subei Lake basin. Due to the evaporation effect and dry climatic conditions, heavy isotopes are more enriched in lake water than in groundwater. The low slope of the regression line of $\delta^{18}O$ and δD in lake water could be ascribed to a combination of mixing and evaporation under conditions of low humidity. Comparison of the regression line for $\delta^{18}O$ and δD showed that lake water in the Subei Lake basin contains more heavily isotopic composition than that in the Habor Lake basin, indicating that lake water in the discharge area has undergone stronger evaporation than lake water in the recharge area. Hydrochemical and isotopic information of utmost importance has been provided to decision makers by the present study so that a sustainable groundwater management strategy can be designed for the Ordos energy base.

1 Introduction

The Ordos Basin is located in Northwestern China, which covers an area of $28.2 \times 10^4\,km^2$ in total and comprises the second largest coal reserves in China (Dai et al., 2006). It was authorized as a national energy base in 1998 by the former State Planning Commission (Hou et al., 2006). More than 400 lake basins with diverse sizes are distributed in the Ordos Basin. The Dongsheng–Shenfu coalfield, situated in the Inner Mongolia Autonomous Region, is an important component of the Ordos energy base. It is the largest explored coalfield with enormous potential for future development. The proven reserves of coal are 230 billion tons. The coal is extracted from Jurassic strata and subsurface mining is common. Local residents there mostly depend on groundwater on account of the serious shortage of surface water. Water resources support the exploitation of coal and development of related industries. In China, since 2011, all new construction projects must carry out an environment evaluation of groundwater consistent with the technical guidelines of the PRC Ministry of Environmental Protection (2011). It is of greatest significance in mining areas, because water resources are an essential component of the mining process (Agartan and Yazicigil, 2012). Over the past several decades, the quantity and quality of groundwater resources have been affected by the rapid development of coal mining. Haolebaoji well field of Subei Lake basin is a typical, large well field and acts as an important water source for this coalfield. However, large-scale and intensive groundwater exploitation could remarkably influence the hydrochemical field of groundwater systems in the study area. In recent years, with the fast development of Ordos energy base, more and more well fields have been built in some lake basins (including Haolebaoji well field newly built in the Subei Lake basin) in order to meet the increasing demand on water resources. However, due to a lack of adequate hydrogeological knowledge about these specific lake basins and reasonable groundwater management strategies, water resources in these specific lake basins are currently subject to increasing pressure from altered hydrology associated with water extraction for regional development and groundwater over-exploitation has taken place. If it continues, it may cause a series of negative impacts on the groundwater-dependent ecosystem around these lakes. Thus, studies about the lake basins are urgently needed so as to obtain comprehensive knowledge of the hydrochemical and isotopic characteristics, and geochemical evolution of groundwater under the background of intensive groundwater exploitation.

Research of groundwater and hydrogeology in the Ordos Basin has been conducted by numerous Chinese scholars and institutes because the Ordos Basin plays a vital role in natural resources exploitation and national economic development. Most importantly, China Geological Survey Bureau has conducted some regional-scale research on groundwater resources of Ordos Basin beginning in 1980s (Zhang et al., 1986; Hou et al., 2008). The previous research has clarified geology and hydrogeology and has provided a comprehensive overview of quantity and quality of groundwater in this region, laying a solid foundation for the present study. However, regional-scale groundwater investigations may not provide much accurate information on the groundwater flow characteristics in small basins (Toth, 1963). Hence, it is also significant to implement local groundwater resource investigations. As Winter (1999) concluded that lakes in different part of groundwater flow systems have different flow characteristics. Data on hydrochemistry and stable isotopes of water were used to study the origin and geochemical evolution of groundwater in the Habor Lake basin (Yin et al.,

2009), which is located in the recharge zone. But other lakes in the runoff and discharge area still have not been studied so far. Due to the particularity of the discharge area, a variety of hydrochemical effects such as evaporation, decarbonation, strong mixing action, etc., take place and result in extremely complicated hydrochemical and isotopic characteristics. In addition, intensive groundwater withdrawal has dramatically changed the local hydrologic cycle in these specific lake basins, groundwater flow field and hydrochemistry have been changed significantly, and a series of ecological environment problems have taken place. Therefore, given that these potential problems originate from human activity, it is essential to conduct hydrochemical and isotopic study of Subei Lake basin located in the discharge area.

Isotopic and geochemical indicators often serve as effective methods for solving multiple problems in hydrology and hydrogeology, especially in semi-arid and arid regions (Clark and Fritz, 1997; Cook and Herczeg, 1999). These techniques have been widely used to obtain groundwater information such as its source, recharge and the interaction between groundwater and surface water (De Vries and Simmers, 2002; Yuko et al., 2002; Yang et al., 2012a). The technique of stable isotopes as excellent tracers has been widely used by many scholars in the study of hydrological cycle (Chen et al., 2011; Cervi et al., 2012; Garvelmann et al., 2012; Yang et al., 2012a; Hamed and Dhahri, 2013; Kamdee et al., 2013). Greater knowledge on the origin and behavior of major ions in groundwater can enhance the understanding of the geochemical evolution of groundwater. Measurement of the relative concentration of major ions in groundwater from different aquifers can provide information on the geochemical reactions within the aquifer and the possible evolutionary pathways of groundwater (Cook and Herczeg, 1999).

The aim of the research is to recognize the origin and geochemical evolution of groundwater in the Subei Lake basin under the influence of human activities. The main objectives are to (1) ascertain the origin of groundwater and (2) determine the geochemical factors and mechanisms controlling the chemical composition of groundwater. In the context of a large number of well fields built in some lake basins in order to meet the increasing demand of water resources, the results of the present study will be valuable in obtaining a deeper insight into hydrogeochemical changes caused by human activity, and providing significant information on, for example, the water quality situation and geochemical evolution of groundwater to decision makers so that they can make sustainable groundwater management strategies for other similar small lake basins and even the Ordos energy base.

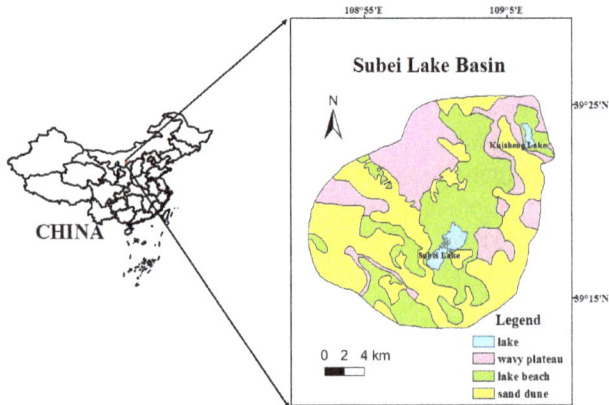

Figure 1. Location of the study area and geomorphic map.

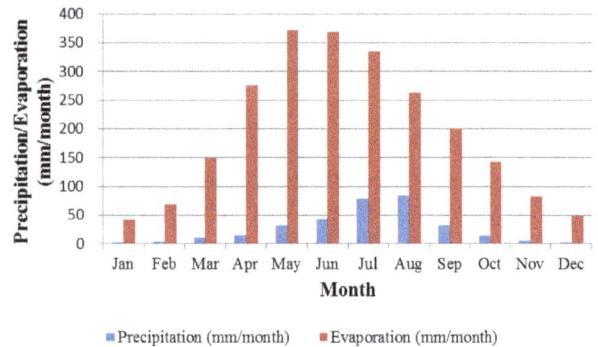

Figure 2. Average monthly precipitation and evaporation in the study area.

2 Study area

2.1 Physiography

The study area is situated in the northern part of the Ordos Basin, which is located at the junction of Uxin Banner, Hanggin Banner and Ejin Horo Banner in Ordos City and is mainly administratively governed by Uxin Banner of Ordos City. It covers an area of almost $400 \, \text{km}^2$ within $39°13'30''$–$39°25'40''$ N and $108°51'24''$–$109°08'40''$ E. Its length is 23 km from east to west and its width is 22 km from north to south (Fig. 1).

The continental semi-arid to arid climate controls the whole study area, which is characterized by long, cold winters and short, hot summers (Li et al., 2010, 2011). According to the data of the Wushenzhao meteorological station, the average monthly temperature ranges from $-11.5 \, °C$ in January to $21.9 \, °C$ in July. The mean annual precipitation in the study area was $324.3 \, \text{mm yr}^{-1}$ from 1985 to 2008. The total annual precipitation varied greatly from year to year with a minimum of 150.2 mm in 2000 and a maximum of 432.3 mm in 1985. The majority of the precipitation falls in the form of rain during the 3-month period from June to August, with more than 63.6 % of annual precipitation (Fig. 2). The mean annual evaporation is $2349.1 \, \text{mm yr}^{-1}$ (from 1985 to 2008) at Wushenzhao station (Fig. 2), which far exceeds rainfall for the area. The average value of monthly evaporation is lowest in January ($42.4 \, \text{mm month}^{-1}$) and highest from May to July, with maximum evaporation in May ($377.4 \, \text{mm month}^{-1}$).

As a small-scale lake basin, the general geomorphic types of Subei Lake basin are wavy plateau, lake beach and sand dunes (Fig. 1). The terrain of Subei Lake's west, east and north sides is relatively higher with altitudes between 1370 and 1415 m; the terrain of its south side is slightly lower with elevations between 1290 and 1300 m. The topography of the center area of Subei Lake basin is flat and low-lying. There are no perennial or ephemeral rivers within the study area; the main surface water bodies are Subei and Kuisheng lakes, and they are situated in the same watershed considering actual hydrogeological conditions and groundwater flow field. In response to precipitation, diffuse overland flow and groundwater recharge the Subei and Kuisheng lakes (Hou et al., 2006; Wang et al., 2010). Subei Lake is located in the low-lying center of the study area (Fig. 1), which is an inland lake characterized by high alkalinity; Kuisheng Lake is also a perennial water body and it is located in northeastern corner of the study area, only covering $2 \, \text{km}^2$ (Fig. 1).

2.2 Geologic and hydrogeologic setting

Subei Lake basin is a relatively closed hydrogeological unit given that a small quantity of lateral outflow occurs in a small part of southern boundary (Wang et al., 2010). The Quaternary sediments and Cretaceous formation can be observed in the study area. The Quaternary sediments are mainly distributed around the Subei Lake with relatively smaller thickness. Generally the thickness of Quaternary sediments is below 20 m. The Quaternary layer is chiefly composed of the interlaced layers of sand and mud. The Cretaceous formations mainly consist of sedimentary sandstones and generally outcrop in the regions with relatively higher elevation. The maximum thickness of Cretaceous rocks could be nearly 1000 m in the Ordos Plateau (Yin et al., 2009), so the Cretaceous formation composed of mainly sandstone is the major water-supplying aquifer of the investigated area. Calcite, dolomite, anhydrite, aragonite, gypsum, halite and feldspar are major minerals in the Quaternary and Cretaceous strata (Hou et al., 2006).

Groundwater resources are very abundant in the investigated area, and phreatic aquifer and confined aquifer can be observed in this region. According to Wang et al. (2010) and the data from Inner Mongolia Second Hydrogeology Engineering Geological Prospecting Institute, the phreatic aquifer is composed of Quaternary and Cretaceous sandstones, with its thickness ranging from 10.52 to 63.54 m. In terms of borehole data, the similar groundwater levels in the Quaternary and Cretaceous phreatic aquifers indicate a very close hy-

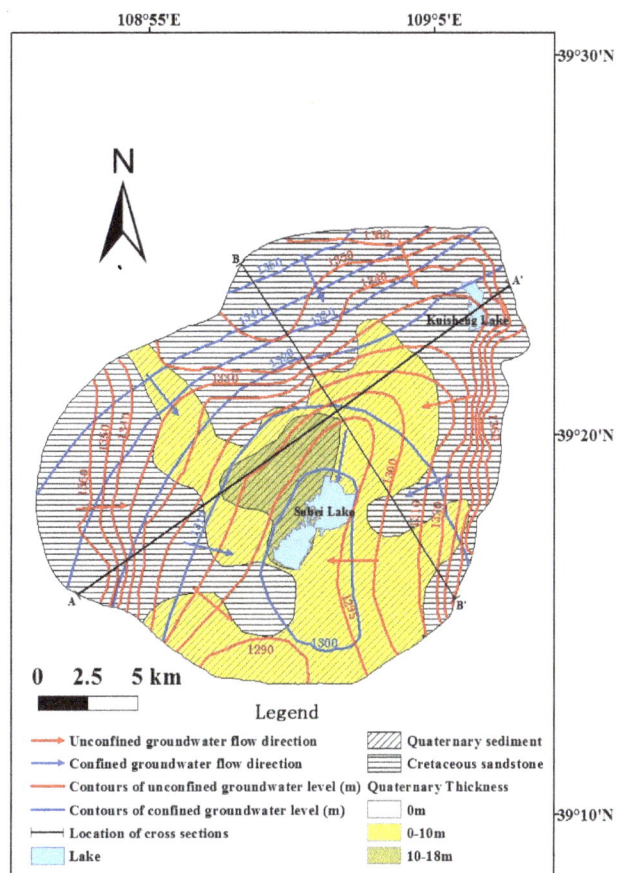

Figure 3. Hydrogeological map of the study area. Data were revised from original source (Inner Mongolia Second Hydrogeology Engineering Geological Prospecting Institute, 2010).

draulic connection between the Quaternary layer and Cretaceous phreatic aquifer, which could be viewed as an integrated unconfined aquifer in the area. The depth to water table in unconfined aquifer is influenced by the terrain change, of which the minimum value is below 1 m in the low-lying region, and the maximum value could be up to 13.24 m. The hydraulic conductivity of the aquifer changes between 0.16 and 17.86 m day^{-1}. The specific yield of unconfined aquifer varies from 0.058 to 0.155. The recharge source of groundwater in the unconfined aquifer is mainly the infiltration recharge of precipitation; it can be also recharged by lateral inflow from groundwater outside the study area. Besides the above recharge terms, leakage recharge from the underlying confined aquifer and infiltration recharge of irrigation water can also provide a small percentage of groundwater recharge. Evaporation is the main discharge way of the unconfined groundwater. In addition, lateral outflow, artificial exploitation and leakage discharge are also included in the main discharge patterns. Unconfined groundwater levels were contoured to illustrate the general flow field in the area (Fig. 3). Groundwater levels were monitored during September 2003.

As is shown in Fig. 3, lateral outflow occurs in a small part of southern boundary determined by analyzing the contours and flow direction of groundwater. The groundwater flows predominantly from surrounding uplands to low lands, which is under the control of topography. On the whole, groundwater in phreatic aquifer flows toward Subei Lake and recharges lake water (Fig. 3).

The unconfined and confined aquifers are separated by an uncontinuous aquitard. Generally speaking, permeable layers and aquitards intervein in the vertical profile of the aquifer system. Nevertheless, aquitards may pinch out in many places, so the aquifer system acts as a single hydrogeologic unit. In the present study, the covering aquitard is composed of the mudstone layer, which is mainly distributed in the second sand layer, and discontinued mudstone lens could also be observed in Cretaceous strata (Fig. 4). The phreatic aquifer is underlain by a confined aquifer composed of Cretaceous rocks. Due to huge thickness and high permeability of confined aquifer, it is regarded as the most promising water-supplying aquifer for domestic and industrial uses. The hydraulic conductivity of confined aquifer changes between 0.14 and 27.04 m day^{-1}. The hydraulic gradient varies from 0.0010 to 0.0045 and the storage coefficient changes between 2.17×10^{-5} and 1.98×10^{-3}. The confined aquifer primarily receives leakage recharge from the unconfined groundwater. The flow direction of confined groundwater is similar to that of unconfined groundwater (Fig. 3). Artificial exploitation is the major way in which confined groundwater is drained.

In the present study, the depth of sampling wells, in combination with hydrogeological map of the study area, is used to classify the groundwater as Quaternary groundwater, shallow Cretaceous groundwater and deep Cretaceous groundwater. As a research on an adjacent, specific, shallow groundwater system of Ordos Basin shows that the circulation depth is 120 m (Yin et al., 2009). It is difficult to determine the circulation depth of shallow groundwater in fact because the circulation depth of local flow systems changes depending on the topography and the permeability of local systems (Yin et al., 2009). In this study, Quaternary groundwater was defined on the basis of the distribution of Quaternary sediments thickness and depth of sampling wells. According to Hou et al. (2006), the maximum circulation depth of local groundwater flow system in the study area is also 120 m, determined by using a large amount of hydrochemical and isotopic data; 120 m is chosen as the maximum circulation depth of the local groundwater system and is used to divide the Cretaceous groundwater samples into two groups: samples taken in wells shallower than 120 m were classified as shallow Cretaceous groundwater, while samples taken in wells deeper than 120 m were deep Cretaceous groundwater.

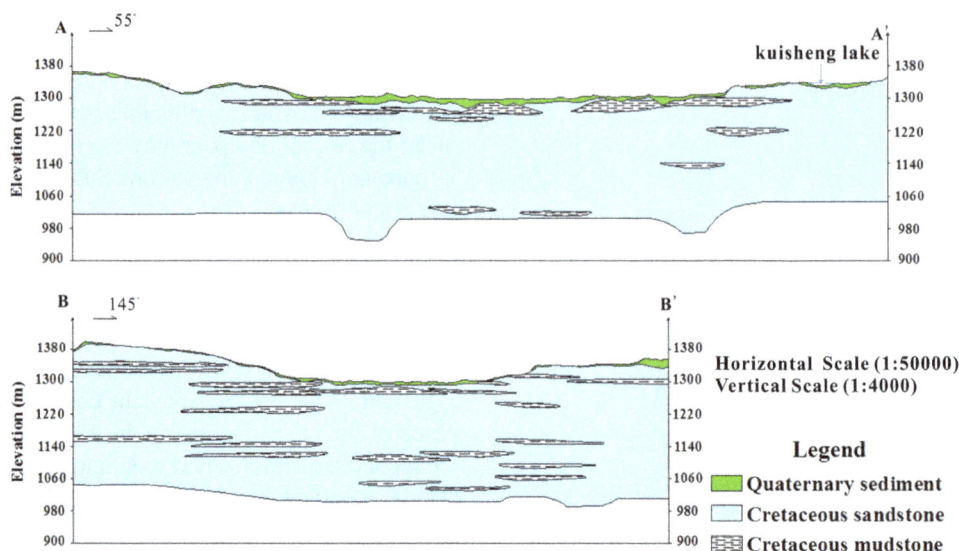

Figure 4. Geologic sections of the study area. Data were revised from original source (Inner Mongolia Second Hydrogeology Engineering Geological Prospecting Institute, 2010).

3   Methods

3.1   Water sampling

Two important sampling actions were conducted in the study area during August and December 2013, respectively. A total of 95 groundwater samples and seven lake water samples were collected. The first sampling action was during the rainy season and the other was during the dry season. The sampling locations are shown in Fig. 5. The water samples were taken from wells for domestic and agricultural purposes, ranging in depth from 2 to 300 m. The length of screen pipes in all sampling wells ranges from 1 to 10 m and every sampling well has only one screen pipe rather than multiple screens. The distance between the bottom of the screen pipe and the total well depth ranges from 0 to 3 m in the study area, and the bottom depth of screen pipe was assigned to the water samples. The samples from the wells were mostly taken using pumps installed in these wells and after removing several well volumes prior to sampling. The 100 and 50 mL polyethylene bottles were pre-rinsed with water sample three times before the final water sample was collected. Lake water samples were collected at Subei Lake, Kuisheng Lake and Shahaizi Lake. Cellulose membrane filters (0.45 μm) were used to filter samples for cation and anion analysis. All samples were sealed with adhesive tape so as to prevent evaporation. GPS was applied to locate the sampling locations.

3.2   Analytical techniques

Electrical conductivity (EC), pH value and water temperature of each sample were measured in situ using an EC/pH meter (WM-22EP, DKK-TOA, Japan), which was previously calibrated. Dissolved oxygen concentration and oxidation–reduction potential were also determined using a HACH HQ30d Single-Input Multi-Parameter Digital Meter. In situ hydrochemical parameters were monitored until these values reached a steady state.

The hydrochemical parameters were analyzed at the Center for Physical and Chemical Analysis of Institute of Geographic Sciences and Natural Resources Research, Chinese Academy of Sciences (IGSNRR, CAS). Major ion compositions were measured for each sample including K^+, Na^+, Ca^{2+}, Mg^{2+}, Cl^-, SO_4^{2-} and NO_3^-. An inductively coupled plasma optical emission spectrometer (ICP-OES) (Perkin-Elmer Optima 5300DV, USA) was applied to analyze major cations. Major anions were measured by ion chromatography (ICS-2100, Dionex, USA). HCO_3^- concentrations in all groundwater samples were determined by the titration method using 0.0048 M H_2SO_4 on the day of sampling; methyl orange endpoint titration was adopted with the final pH of 4.2–4.4. Due to the extremely high alkalinity of lake water samples, HCO_3^- concentrations in all lake water samples were analyzed by titration using 0.1667 M H_2SO_4. CO_3^{2-} concentrations were also analyzed by titration; phenolphthalein was used as an indicator of endpoint titration.

Hydrogen (δD) and oxygen (δ^{18}O) composition in the water samples were analyzed using a liquid water isotope analyzer (LGR, USA) at the Institute of Geographic Sciences and Natural Resources Research, Chinese Academy of Sciences (IGSNRR, CAS). Results were expressed in the standard δ notation as per mil (‰) difference from Vienna standard mean ocean water (VSMOW, 0‰) with analytical precisions of $\pm$1‰ (δD) and $\pm$0.1‰ (δ^{18}O).

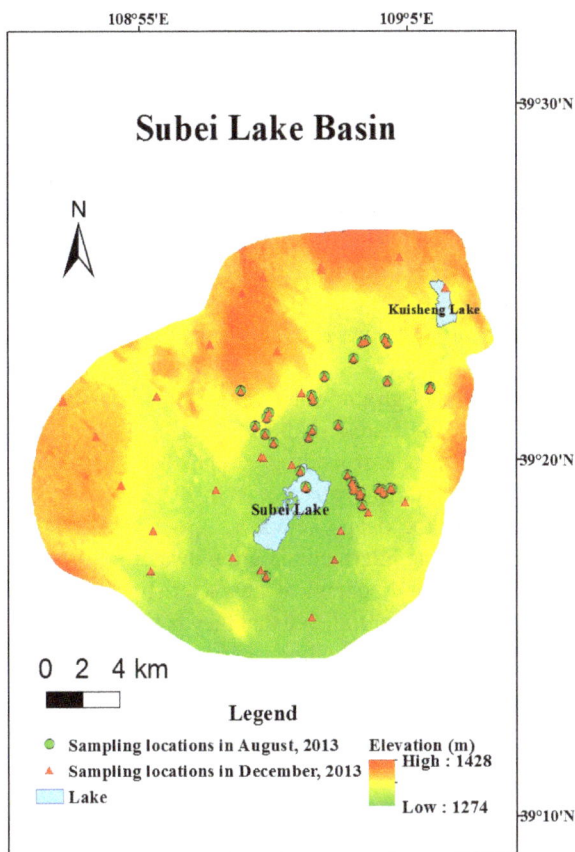

Figure 5. Sampling locations in August and December 2013.

Table 1. The chemical composition and isotopic data of lake water in August and December 2013.

ID	Date	EC (μS cm^{-1})	T (°C)	pH	DO (mg L^{-1})	ORP (mV)	K$^+$ (mg L^{-1})	Na$^+$ (mg L^{-1})	Ca^{2+} (mg L^{-1})	Mg^{2+} (mg L^{-1})	Cl$^-$ (mg L^{-1})	SO$_4^{2-}$ (mg L^{-1})	HCO$_3^-$ (mg L^{-1})	CO$_3^{2-}$ (mg L^{-1})	NO$_3^-$ (mg L^{-1})	TDS (mg L^{-1})	δD (‰)	δ^{18}O (‰)
EEDS08	29 Aug 2013	130 400	22.5	10.11	11.06	−1.8	1956	42 020	2.28	3.01	37 440.28	22 066.83	6000.3	19 356.45	98.93	125 943.93	−1	19.4
EEDS09	29 Aug 2013	190 100	24.3	10.25	15.8	−14.8	6475	96 530	0.00	2.4	108 517.4	37 581.86	12 661.65	46 565.53	511.48	302 514.49	15	29.4
EEDS38	30 Aug 2013	1017	23.7	8.86	7.65	61.8	10.63	97.59	17.21	70.03	32.71	92.85	480.68	0	1.07	562.43	−45	−5.8
EEDS08	6 Dec 2013	120 400	1.8	8.9		17.6	1997.73	36 617.7	11.52	3.7	30 787.74	7513.4	5186.7	19 406.47	87.48	99 019.09	−18	5.8
EEDS09	6 Dec 2013	229 000	2.3	8.49		39.5	7567	77 840	36.34	11.39	113 003.44	5276.76	11 593.8	13 754.59	207.99	223 494.4	−9	16.2
EEDS38	4 Dec 2013	4200	1.1	10.47	17.96	26.3	602	38.88	9.06	352.2	164.54	448.23	1423.8	900.3	9.53	3236.64	−28	−2.6
EEDS60	4 Dec 2013	14 080	2.7	9.04	10.58	23.6	56.154	3393.74	4.27	41.49	1418.76	386.04	1067.85	2600.87	11.77	8447.02	−28	−1.9

4   Results

4.1   Hydrochemical characteristics

In situ water quality parameters such as pH, electrical conductivity (EC), temperature, dissolved oxygen concentration (DO), oxidation–reduction potential (ORP) and total dissolved solids (TDS) as well as analytical data of the major ions composition in groundwater and lake water samples are shown in Table 1 and Table S1 in the Supplement. Based on the chemical data, hydrochemical characteristics of groundwater and lake water are discussed.

The chemical composition for lake water showed that Na$^+$ accounted for, on average, 93 % of total cations and Cl$^-$ accounted for, on average, 58 % of total anions. Thus, Na$^+$ and Cl$^-$ were the dominant elements (Fig. 6), which was in accordance with hydrochemical characteristics of inland salt lakes. This was also observed in lake water of Habor Lake basin located in the recharge area (Yin et al., 2009). The pH of lake water varied from 8.86 to 10.25 with an average of 9.74 in August and from 8.49 to 10.47 with an average of 9.23 in December; it can be seen that the pH was relatively stable and was always more than 8.4 without obvious seasonal variation, which indicated that the dissolved carbon-

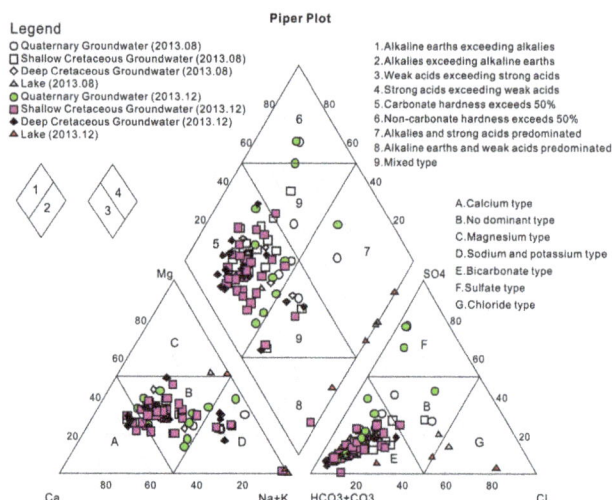

Figure 6. Piper diagram of groundwater and lake water in August and December 2013.

ates were in the HCO_3^- and CO_3^{2-} forms simultaneously. The temperatures of lake water ranged from 1.1 to 24.3 °C with large seasonal variations, implying that the surface water body was mainly influenced by hydrometeorological factors. The dissolved oxygen concentration of lake water showed an upward tendency from August (mean value: $11.50 \, mg \, L^{-1}$) to December (mean value: $14.27 \, mg \, L^{-1}$) because the relationship between water temperature and DO is inverse when oxygen content in the air stays relatively stable. With the decreasing water temperature, the dissolved oxygen value rises. The average value of ORP ranged from 15.1 mV in August to 26.8 mV in December, which was in accordance with the upward tendency of DO. It showed that lake water had stronger oxidation in December than that in August and there is a close relationship between DO and ORP. The average values of major ions concentrations showed a downward trend except for Ca^{2+}, Mg^{2+} from August to December. Specifically, the average values of Ca^{2+} and Mg^{2+} increased from 6.50 to $15.30 \, mg \, L^{-1}$ and 25.15 to $102.20 \, mg \, L^{-1}$, respectively; other ions concentrations were reduced to different degrees. The same variation trend of major ions from August to December could be found in the Habor Lake basin (Yin et al., 2009) as well. Before August, the strong evaporation capacity of lake water exceeded the finite recharge amount, which caused lake water to be enriched. After August, lake water was recharged and diluted by groundwater and a large amount of fresh overland flow from precipitation. The EC values varied between 1017 and $229\,000 \, \mu S \, cm^{-1}$. This relatively large range of variation was closely related to the oscillation of the TDS values, which ranged from 0.56 to $302.5 \, g \, L^{-1}$. The results showed that lake water chemistry was controlled by strong evaporation and recharge from overland flow and groundwater.

The hydrochemical data of groundwater were plotted on a Piper triangular diagram (Piper, 1953), which is perhaps the most commonly used method for identifying hydrochemical patterns of major ion composition (Fig. 6). With respect to cations, most of samples are scattered in zones A, B and D of the lower-left triangle, indicating that some are calcium-type, some are sodium-type water but most are of a mixed type; regarding anions, most groundwater samples are plotted in zone E of the lower-right triangle (Fig. 6), showing that bicarbonate-type water is predominant. The predominant hydrochemical types are HCO_3–Ca, HCO_3–Na and mixed HCO_3–Ca–Na–Mg types. Figure 6 also indicates that there are three groups of groundwater in the Subei Lake basin: the Quaternary groundwater, shallow Cretaceous groundwater and deep Cretaceous groundwater. The shallow Cretaceous groundwater refers to groundwater in the local groundwater system, and the deep Cretaceous groundwater refers to groundwater in the intermediate groundwater system of Ordos Basin. The hydrochemical characteristics of the three groups of groundwater indicate that they have undergone different degrees of mineralization.

With respect to the Quaternary groundwater, the pH varied from 7.64 to 9.04 with an average of 8.09 in August and changed from 7.49 to 9.26 with an average of 8.08 in December, indicating an alkaline nature. The TDS varied from 396 to $1202 \, mg \, L^{-1}$, 314 to $1108 \, mg \, L^{-1}$ with averages of 677 and $625 \, mg \, L^{-1}$, respectively, in August and December. The major cations were Na^+, Ca^{2+} and Mg^{2+}, while the major anions were HCO_3^- and SO_4^{2-}.

As for the shallow Cretaceous groundwater (< 120 m), the pH varied from 7.37 to 8.3 with an average of 7.77 in August and oscillated from 7.49 to 9.37 with an average of 8.14 in December. The TDS varied from 249 to $1383 \, mg \, L^{-1}$ and from 217 to $1239 \, mg \, L^{-1}$, with averages of 506 and $400 \, mg \, L^{-1}$, respectively, in August and December.

For the deep Cretaceous groundwater (> 120 m), the pH varied from 7.75 to 8.09 with an average of 7.85 in August and fluctuated from 7.99 to 8.82 with an average of 8.23 in December. The TDS varied from 266 to $727 \, mg \, L^{-1}$, 215 to $464 \, mg \, L^{-1}$ with averages of 377 and $296 \, mg \, L^{-1}$, respectively, in August and December.

4.2 Stable isotopic composition in groundwater and surface water

In the present study, the results of the stable isotope analysis for groundwater and lake water are plotted in Fig. 7. In a previous study, the local meteoric water line (LMWL) in the northern Ordos Basin had been developed by Yin et al. (2010). The LMWL is $\delta D = 6.45\delta^{18}O - 6.51$ ($r^2 = 0.87$), which is similar to that developed by Hou et al. (2007) ($\delta D = 6.35 \, \delta^{18}O - 4.69$). In addition, it is very clear in the plot that the LMWL is located below the global meteoric water line (GMWL) defined by Craig (1961) ($\delta D = 8 \, \delta^{18}O + 10$), which suggests the occurrence of secondary

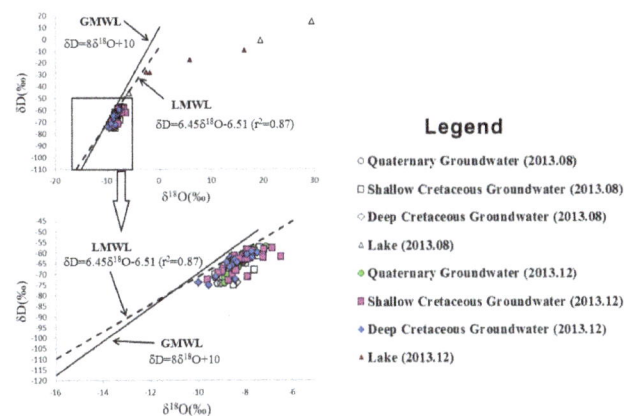

Figure 7. Relationship between hydrogen and oxygen isotopes in groundwater and lake water in August and December 2013.

evaporation during rainfall. The LMWL is controlled by local hydrometeorological factors, including the origin of the vapor mass, re-evaporation during rainfall and the seasonality of precipitation (Clark and Fritz, 1997).

The linear regression curve equation of $\delta^{18}O$ and δD in groundwater can be defined as $\delta D = 6.3\ \delta^{18}O - 13.0$ ($r^2 = 0.62$). Groundwater follows the LMWL in the study area, indicating that it is of modern local meteoric origin rather than the recharge from precipitation in paleoclimate conditions. In August, the stable isotope values in the Quaternary groundwater were found to range from -9.2 to -8.0‰ in ^{18}O with an average of -8.8‰ and from -74 to -62‰ in ^2H with an average of -71‰; the shallow Cretaceous groundwater had $\delta^{18}O$ ranging from -9.3 to -7.5‰ and δD varying from -75 to -57‰. The average values of $\delta^{18}O$ and δD of the shallow Cretaceous groundwater were -8.3 and -66‰, respectively. $\delta^{18}O$ and δD of the deep Cretaceous groundwater ranged from -9.3 to -7.8‰ and from -74 to -61‰, respectively. The average values of $\delta^{18}O$ and δD were -8.4 and -67‰, respectively. In December, the stable isotope values in the Quaternary groundwater ranged from -8.9 to -7.2‰ in ^{18}O with an average of -8.2‰ and from -74 to -57‰ in ^2H with an average of -65‰; $\delta^{18}O$ and δD of the shallow Cretaceous groundwater ranged from -9.7 to -6.5‰ and from -73 to -58‰, respectively. The average values of $\delta^{18}O$ and δD were -8.2 and -64‰, respectively. $\delta^{18}O$ of the deep Cretaceous groundwater varied from -10.0 to -7.5‰ and δD ranged from -75 to -60‰. The average values of $\delta^{18}O$ and δD of the deep Cretaceous groundwater were -8.5 and -66‰, respectively.

The regression curve equation of $\delta^{18}O$ and δD in lake water could be defined as $\delta D = 1.47\ \delta^{18}O - 29.09$ ($r^2 = 0.95$), where $\delta^{18}O$ ranged from -5.8 to 29.4‰ and δD ranged from -46 to 15‰ with averages of 14.3 and -10‰ in August, while in December, $\delta^{18}O$ and δD of lake water ranged from -2.6 to 16.2‰ and from -28 to -9‰, respectively. The average values of $\delta^{18}O$ and δD were 4.4 and -21‰, respec-

tively, in December. The low slope of the regression line of $\delta^{18}O$ and δD in lake water could be ascribed to a combination of mixing and evaporation under conditions of low humidity.

4.3 Linkage among geochemical parameters of groundwater

Correlations among groundwater-quality parameters are shown in Table 2 and Fig. 8. All of the major cations and anions are significantly correlated with TDS (Table 2), which shows that these ions have been dissolved into groundwater continuously and resulted in the rise of TDS.

As is shown in Table 2, the concentration of Mg^{2+} is correlated with HCO_3^- and SO_4^{2-}, with correlation coefficients of 0.582 and 0.819, respectively. The concentration of Ca^{2+} is well correlated with SO_4^{2-} with a correlation coefficient of 0.665. Cl^- has good correlation with Na^+ with a large correlation coefficient of 0.824.

The results of linear regression analysis of some pairs of ions are displayed in Fig. 8. There is good correlation between Cl^- and Na^+ in Quaternary groundwater and shallow Cretaceous groundwater; Ca^{2+} and SO_4^{2-} have a good positive correlation in Quaternary groundwater and shallow Cretaceous groundwater. Mg^{2+} is well correlated with HCO_3^- in shallow Cretaceous groundwater.

5 Discussion

Generally speaking, water–rock interactions are the most important factors influencing the observed geochemical composition of groundwater (Appelo and Willemsen, 1987); the geochemical and isotopic results of this work are no exception. In terms of dissolved minerals and the correlation of geochemical parameters, the dominant geochemical processes and formation mechanisms could be found (Su et al., 2009). The weathering and dissolution of minerals in the host rocks and ion exchange are generally the main source of ions in groundwater based on available research. The stable isotopes signatures in lake water can reveal the predominant mechanism controlling the chemical composition of lake water.

5.1 Geochemical processes of groundwater

As displayed in the correlation analysis of geochemical parameters, a good correlation between Mg^{2+} and HCO_3^- indicates that the weathering of dolomite releases ions to the groundwater, as expressed in Reaction (R1). The fact that Mg^{2+} is well correlated with HCO_3^- could be found in the Habor Lake basin of Ordos Plateau (Yin et al., 2009). Ca^{2+} has good correlation with SO_4^{2-}, implying that the dissolution of gypsum and anhydrite may be the key processes controlling the chemical composition of groundwater in the discharge area, which can be explained by Reaction (R2). Just as with the achievements obtained by Hou et al. (2006), gyp-

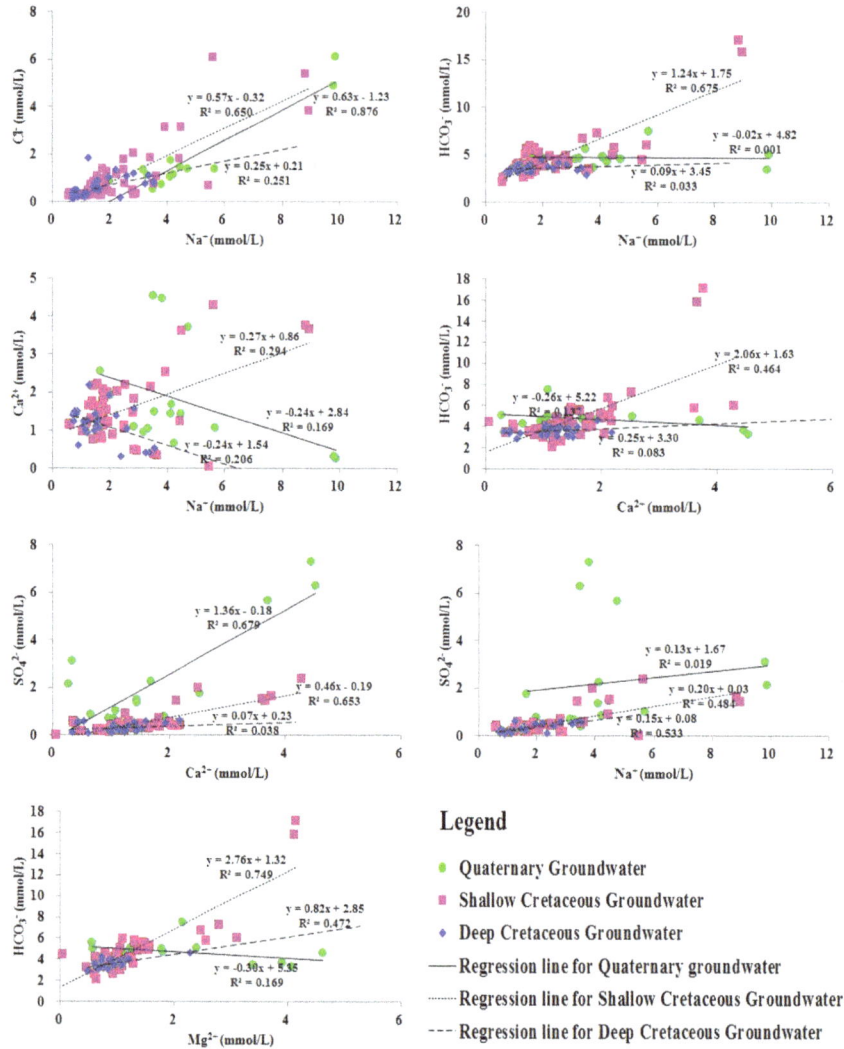

Figure 8. Relationship between some pairs of ions in groundwater.

sum and anhydrite are present in these strata, so it is reasonable to consider that gypsum and anhydrite are the source of the SO_4^{2-}. However, in Yin's study, there is poor correlation between Ca^{2+} and SO_4^{2-} in the Habor Lake basin (Yin et al., 2009). It can be explained by geochemical evolution of groundwater along flow path from the recharge area to the discharge area. There is poor correlation between Na^+ and SO_4^{2-}, suggesting that the weathering of Glauber's salt ($Na_2SO_4 \cdot 10H_2O$) may not be the major sources of such ions in groundwater. On the contrary, a good correlation between Na^+ and SO_4^{2-} can be found in the Habor Lake basin (Yin et al., 2009). It indicates that Glauber's salt may be more abundant in the recharge area (Habor Lake basin) than in the discharge area (Subei Lake basin). Although there is no obvious correlation between Ca^{2+} and HCO_3^-, it is reasonable to regard the dissolution of carbonate minerals as a source of Ca^{2+} and HCO_3^- due to the widespread occurrence of carbonate rocks in the study area, as conveyed in Reaction (R3).

The concentration of Mg^{2+} is well correlated with SO_4^{2-}, suggesting the possible dissolution of gypsum, followed by cation exchange. The pH is negatively correlated with Ca^{2+}; it is likely that the dissolution of carbonate minerals is constrained due to the reduction of the hydrogen ion concentration in water at higher pH. It can be judged from the above analysis that during groundwater flow, the following reactions are very likely to take place in the study area:

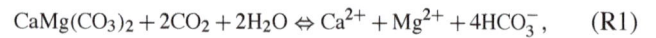

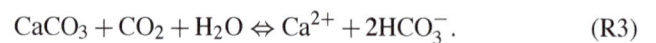

$$CaMg(CO_3)_2 + 2CO_2 + 2H_2O \Leftrightarrow Ca^{2+} + Mg^{2+} + 4HCO_3^-, \quad \text{(R1)}$$

$$CaSO_4 \Leftrightarrow Ca^{2+} + SO_4^{2-}, \quad \text{(R2)}$$

$$CaCO_3 + CO_2 + H_2O \Leftrightarrow Ca^{2+} + 2HCO_3^-. \quad \text{(R3)}$$

In order to explore the mechanism of salinity in semi-arid regions, the plot of Na^+ versus Cl^- is widely used (Magaritz et

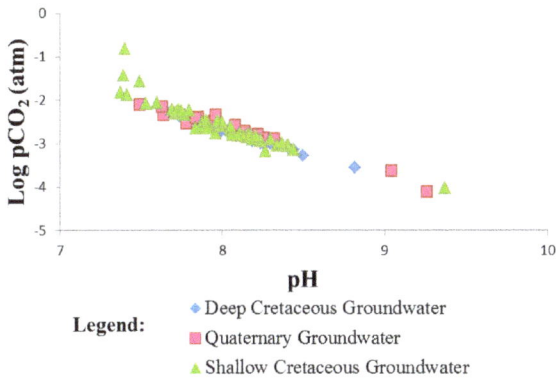

Figure 9. Geochemical relationship of pH vs. log (pCO_2) in groundwater.

al., 1981; Dixon and Chiswell, 1992; Sami, 1992). The concentration of Cl^- is well correlated with Na^+, suggesting that the dissolution of halite may be the major source of Na^+ and Cl^-. Theoretically, the dissolution of halite will release equal amounts of Na^+ and Cl^- into the solution. Nevertheless, the results deviate from the anticipated 1 : 1 relationship. Almost all samples have more Na^+ than Cl^-. The molar Na / Cl ratio varies from 0.68 to 16.00 with an average value of 3.48. A greater Na / Cl ratio may be ascribed to the feldspar weathering and the dissolution of other Na-containing minerals. The relatively high Na^+ concentration in the groundwater could also be illustrated by cation exchange between Ca^{2+} or Mg^{2+} and Na^+, as is discussed later.

The partial pressure of carbon dioxide (pCO_2) values were calculated by the geochemical computer code PHREEQC (Parkhurst and Appelo, 2004). The pCO_2 values of groundwater range from $10^{-0.82}$ to $10^{-4.1}$ atm. The vast majority of groundwater samples (about 96 %) have higher pCO_2 values than the atmospheric pCO_2, which is equal to $10^{-3.5}$ atm (Van der Weijden and Pacheco, 2003), indicating that groundwater has received CO_2 from root respiration and the decomposition of soil organic matter. Figure 9 indicates that the pCO_2 values are negatively correlated with pH values; the partial pressure values of CO_2 decrease as pH values increase (Rightmire, 1978; Adams et al., 2001). It likely has a connection with relatively longer aquifer residence time, more physical, chemical reactions with aquifer minerals and biological reactions of microorganism that produce CO_2 taking place. According to Hou et al. (2008), feldspars can be observed in the Cretaceous formations and it is possible that the following reaction occurs:

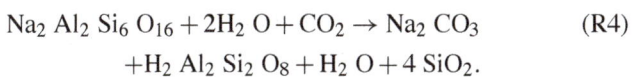

$$Na_2\ Al_2\ Si_6\ O_{16} + 2H_2\ O + CO_2 \rightarrow Na_2\ CO_3 \qquad (R4)$$
$$+H_2\ Al_2\ Si_2\ O_8 + H_2\ O + 4\ SiO_2.$$

This reaction will consume CO_2 and give rise to the increase of the concentration of Na^+ and HCO_3^-. As a result, the partial pressure of CO_2 will decrease and the pH will increase. In terms of a statistical analysis, the average pH values of the

Quaternary groundwater and the shallow Cretaceous groundwater are 8.08 and 7.99, respectively, lower than that of the deep Cretaceous groundwater (8.11). However, the average pCO_2 values of the Quaternary groundwater and the shallow Cretaceous groundwater are $10^{-2.67}$ and $10^{-2.58}$ atm, respectively, higher than that of the deep Cretaceous groundwater (about $10^{-2.79}$ atm). The negative correlation characteristic between pCO_2 and pH shows that the dissolution of feldspar takes place along groundwater flow path. The phenomenon also occurs in the Habor Lake basin according to Yin et al. (2009).

Cation exchange is an important process of water–rock interactions that obviously influences the major ion composition of groundwater (Xiao et al., 2012). Although the cation exchange is widespread in the geochemical evolution of all groundwater, it is essential to know and identify the various changes undergone by water during their traveling processes in the groundwater system under the influence of anthropogenic activities. In the present study, the molar Na / Ca ratio changes between 0.5 and 106.09 with an average of 3.80, suggesting the presence of Na^+ and Ca^{2+} exchange. It can be conveyed in the following reaction:

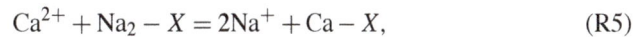

$$Ca^{2+} + Na_2 - X = 2Na^+ + Ca - X, \qquad (R5)$$

where X is sites of cation exchange.

Schoeller proposed that chloro-alkaline indices could be used to study the cation exchange between the groundwater and its host environment during residence or travel (Schoeller, 1965; Marghade et al., 2012; Li et al., 2013). The Schoeller indices, such as CAI-I and CAI-II, are calculated by the following equations, where all ions are expressed in $mEq\ L^{-1}$:

$$CAI-I = \frac{Cl^- - (Na^+ + K^+)}{Cl^-}, \qquad (1)$$

$$CAI-II = \frac{Cl^- - (Na^+ + K^+)}{HCO_3^- + SO_4^{2-} + CO_3^{2-} + NO_3^-}. \qquad (2)$$

When the Schoeller indices are negative, an exchange of Ca^{2+} or Mg^{2+} in groundwater with Na^+ or K^+ in aquifer materials takes place, Ca^{2+} or Mg^{2+} will be removed from solution and Na^+ or K^+ will be released into the groundwater. At the same time, negative value indicates chloro-alkaline disequilibrium and the reaction is known as cation–anion exchange reaction. During this process, the host rocks are the primary sources of dissolved solids in the water. In another case, if the positive values are obtained, then the inverse reaction possibly occurs and it is known as base exchange reaction. In the present study, almost all groundwater samples had negative Schoeller index values (Table S1), which indicates cation–anion exchange (chloro-alkaline disequilibrium). The results indeed clearly show that Na^+ and K^+ are released by the Ca^{2+} and Mg^{2+} exchange, which is

Table 2. Correlation coefficient of major parameters in groundwater.

	K^+	Na^+	Ca^{2+}	Mg^{2+}	Cl^-	SO_4^{2-}	HCO_3^-	TDS	pH
K^+	1.000	0.538	0.309	0.560	0.553	0.300	0.572	0.534	−0.063
Na^+		1.000	0.217	0.651	0.824	0.485	0.602	0.728	−0.072
Ca^{2+}			1.000	0.754	0.375	0.665	0.478	0.796	−0.600
Mg^{2+}				1.000	0.655	0.819	0.582	0.939	−0.382
Cl^-					1.000	0.375	0.576	0.776	−0.144
SO_4^{2-}						1.000	0.160	0.770	−0.226
HCO_3^-							1.000	0.625	−0.398
TDS								1.000	−0.378
pH									1.000

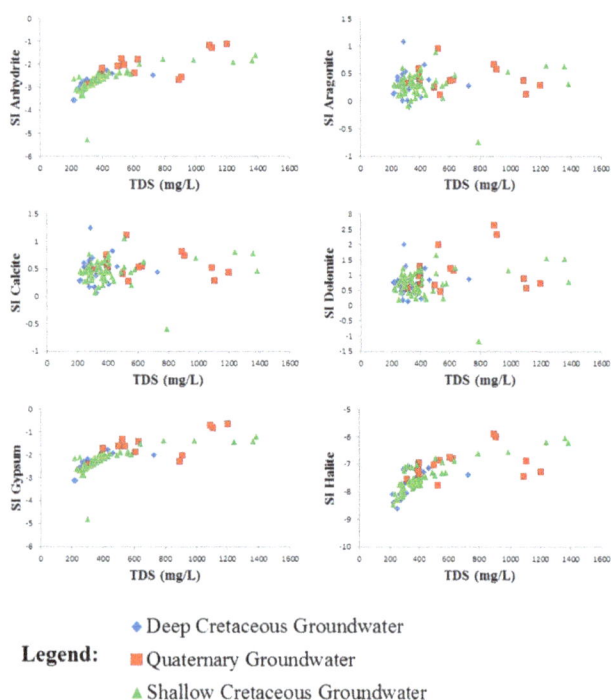

Legend:
- Deep Cretaceous Groundwater
- Quaternary Groundwater
- Shallow Cretaceous Groundwater

Figure 10. Plots of saturation indices with respect to some minerals in groundwater.

a common form of cation exchange in the study area. This also further confirms that the cation exchange is one of the major contributors to higher concentrations of Na^+ in the groundwater, and it is still an important geochemical process of groundwater in the Subei Lake basin under the influence of human activities.

5.2 The formation mechanisms of groundwater and surface water

The saturation index is a vital geochemical parameter in the fields of hydrogeology and geochemistry, often useful for identifying the existence of some common minerals in the groundwater system (Deutsch, 1997). In this present study,

saturation indices (SIs) with respect to gypsum, anhydrite, calcite, dolomite, aragonite and halite were calculated in terms of the following equation (Lloyd and Heathcote, 1985):

$$SI = \log\left(\frac{IAP}{k_s(T)}\right) \tag{3}$$

where IAP is the relevant ion activity product, which can be calculated by multiplying the ion activity coefficient γ_i and the composition concentration m_i, and $k_s(T)$ is the equilibrium constant of the reaction considered at the sample temperature. The geochemical computer model PHREEQC (Parkhurst and Appelo, 2004) was used to calculate the saturation indices. When the groundwater is saturated with some minerals, SI equals zero; positive values of SI represent oversaturation, and negative values show undersaturation (Appelo and Postma, 1994; Drever, 1997).

Figure 10 indicates the plots of SI versus the total dissolved solids (TDS) for all the groundwater samples. The calculated values of SI for anhydrite, gypsum and halite oscillate between −5.27 and −1.11, between −4.8 and −0.65 and between −8.61 and −5.9, with averages of −2.62, −2.16 and −7.49, respectively. It shows that the groundwater in the study area was below the equilibrium with anhydrite, gypsum and halite, indicating that these minerals are anticipated to dissolve. However, the SIs of aragonite, calcite, and dolomite range from −0.74 to 1.09, −0.59 to 1.25 and −1.16 to 2.64, with averages of 0.32, 0.48 and 0.81, respectively. On the whole, the groundwater samples were dynamically saturated to oversaturated with aragonite, calcite and dolomite, implying that the three major carbonate minerals may have affected the chemical composition of groundwater in the Subei Lake basin. The results show that the groundwater may well produce the precipitation of aragonite, calcite and dolomite. Saturation of aragonite, calcite and dolomite could be attained quickly due to the existence of abundant carbonate minerals in the groundwater system.

The soluble ions in natural waters mainly derive from rock and soil weathering (Lasaga et al., 1994), anthropogenic input and partly from the precipitation input. In order to make an analysis of the formation mechanisms of hydrochemistry,

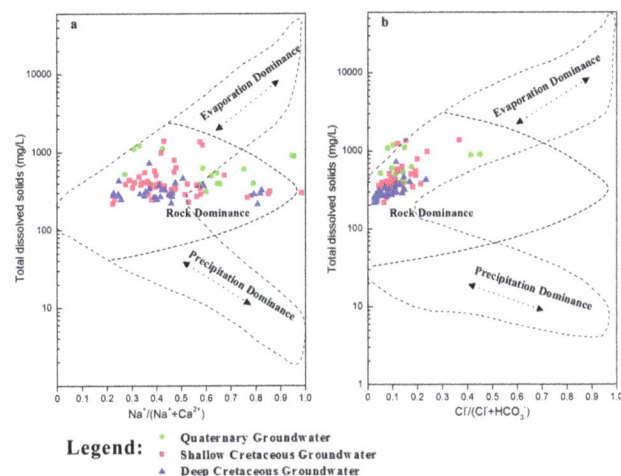

Figure 11. Gibbs diagram of groundwater samples in the Subei Lake basin: **(a)** TDS vs. $Na^+ / (Na^+ + Ca^{2+})$, **(b)** TDS vs. $Cl^- / (Cl^- + HCO_3^-)$.

Gibbs diagrams have been widely used in hydrogeochemical studies (Feth and Gibbs, 1971; Naseem et al., 2010; Marghade et al., 2012; Yang et al., 2012b; Xing et al., 2013). Gibbs (1970) recommended two diagrams to assess the dominant effects of precipitation, rock weathering and evaporation on geochemical evolution of groundwater in semi-arid and arid regions. The diagrams show the weight ratios of $Na^+ / (Na^+ + Ca^{2+})$ and $Cl^- / (Cl^- + HCO_3^-)$ against TDS, and precipitation dominance, rock dominance, and evaporation dominance are included in the controlling mechanisms (Gibbs, 1970). The distributed characteristic of samples in Fig. 11 shows that rock weathering is the dominant mechanism in the geochemical evolution of the groundwater in the study area. The ratio of $Na^+ / (Na^+ + Ca^{2+})$ was mostly less than 0.5 in shallow and deep Cretaceous groundwater, with low TDS values (Fig. 11). It shows that rock weathering was the main mechanism controlling the chemical compositions of shallow and deep Cretaceous groundwater. In the Quaternary groundwater, about two-thirds of samples had a ratio of $Na^+ / (Na^+ + Ca^{2+})$ greater than 0.5 and higher TDS between 314 and 1202 mg/L, which indicated that the Quaternary groundwater was not only controlled by rock weathering, but also by the process of evaporation–crystallization. It is obvious that the weight ratio of $Na^+ / (Na^+ + Ca^{2+})$ spreads from low to high without a great variation of TDS, which indicated that cation exchange also played a role by increasing Na^+ and decreasing Ca^{2+} under the background of rock dominance. During the cation exchange process, the TDS values do not change obviously because $2 \, mmol \, L^{-1}$ of Na^+ is released by $1 \, mmol \, L^{-1} \, Ca^{2+}$ exchange, and the weight of $1 \, mmol \, L^{-1}$ of Ca^{2+} ($40 \, mg \, L^{-1}$) is nearly equal to that of $2 \, mmol \, L^{-1}$ of Na^+ ($46 \, mg \, L^{-1}$).

In August, the average isotopic values of deep Cretaceous groundwater ($\delta^{18}O$: $-8.4‰$, δD: $-67‰$) were enriched

compared with the Quaternary groundwater ($\delta^{18}O$: $-8.8‰$, δD: $-71‰$), but in December, the average isotopic values of deep Cretaceous groundwater ($\delta^{18}O$: $-8.5‰$, δD: $-66‰$) were depleted compared with the Quaternary groundwater ($\delta^{18}O$: $-8.2‰$, δD: $-65‰$); the stable isotopic values of Quaternary groundwater had a wider range from August to December than those of deep Cretaceous groundwater. This may be explained by heavy isotope enrichment in the Quaternary groundwater caused by evaporation given that there was effectively no precipitation in the study area during the period from August to December; meanwhile, the deep Cretaceous groundwater may have been mainly recharged by lateral inflow from groundwater outside the study area, resulting in smaller seasonal fluctuations in the isotopic values.

Furthermore, the average values of $\delta^{18}O$ and δD of the shallow Cretaceous groundwater are -8.3 and $-66‰$ and -8.2 and $-64‰$, respectively, in August and December; meanwhile, the average values of $\delta^{18}O$ and δD of the deep Cretaceous groundwater are -8.4 and $-67‰$ and -8.5 and $-66‰$, respectively, in August and December. Thus, given the precision of the analysis, shallow Cretaceous groundwater and deep Cretaceous groundwater have similar isotopic characteristics in the Subei Lake basin, which indicates that they may be replenished by the similar water source due to the similar geological setting. This also validates the existence of leakage. The similar isotopic characteristic of groundwater from the Cretaceous aquifer may be ascribed to the increasingly close relationship between shallow Cretaceous groundwater and deep Cretaceous groundwater due to changes in the hydrodynamic field caused by intensive groundwater exploitation. Conversely, the phenomenon of deep groundwater depleted in heavy isotopes compared with shallow groundwater was found in the Habor Lake basin located in the recharge area (Yin et al., 2009).

The hydrogen and oxygen isotopes signatures in lake water show that it contains abnormally high levels of heavy isotopic composition. Compared with the stable isotopic values in groundwater, it is evident that lake water has undergone a greater degree of enrichment in heavy isotopes, which further illustrates that fractionation by strong evaporation is occurring predominantly in the lake water. This also proves to be in accordance with the unique hydrochemical characteristics of the lake water. In addition, the slope and intercept of the regression line for $\delta^{18}O$ and δD in lake water were 1.47 and -29.09, lower than the slope and intercept (7.51, -7.12) observed for lake water in the Habor Lake basin (Yin et al., 2009). By comparison, it is clearly confirmed that lake water in the discharge area has undergone stronger evaporation than lake water in the recharge area. As a result, lake water in the Subei Lake basin contains more heavily isotopic composition than that in the Habor Lake basin.

6 Conclusions

The present study examines the hydrochemical and isotopic composition of the groundwater and surface water in the Subei Lake basin with various methods such as correlation analysis, saturation index, Piper diagram and Gibbs diagrams. The combination of major elements geochemistry and stable isotopes (δ^{18}O, δD) has provided a comprehensive understanding of the hydrodynamic functioning and the processes of mineralization that underpin the geochemical evolution of the whole water system. The hydrochemical data show that three groups of groundwater are present in the Subei Lake basin: the Quaternary groundwater, shallow Cretaceous groundwater and deep Cretaceous groundwater. The analysis of groundwater chemistry clarifies that the chemistry of lake water was controlled by strong evaporation and recharge from overland flow and groundwater; meanwhile the major geochemical processes responsible for the observed chemical composition in groundwater are the dissolution/precipitation of anhydrite, gypsum, halite and calcite and the weathering of feldspar and dolomite. Furthermore, the cation exchange has also played an extremely vital role in the groundwater evolution. The absolute predominance of rock weathering in the geochemical evolution of groundwater in the study area is confirmed by the analytical results of Gibbs diagrams. The stable isotopic data indicate that groundwater is of modern local meteoric origin rather than the recharge from precipitation in paleoclimate conditions. Unlike significant differences in isotopic values between shallow groundwater and deep groundwater in the Habor Lake basin, shallow Cretaceous groundwater and deep Cretaceous groundwater have similar isotopic characteristics in the Subei Lake basin. Due to the evaporation effect and dry climatic conditions, heavy isotopes are more enriched in lake water than groundwater. The low slope of the regression line of δ^{18}O and δD in lake water could be ascribed to a combination of mixing and evaporation under conditions of low humidity. A comparison of the regression line for δ^{18}O and δD shows that lake water in the Subei Lake basin contains more heavily isotopic composition than that in the Habor Lake basin, indicating that lake water in the discharge area has undergone stronger evaporation than lake water in the recharge area.

Much more accurate groundwater information has been obtained by conducting this study on Subei Lake basin, which will further enhance the knowledge of geochemical evolution of the groundwater system in the whole Ordos Basin and provide comprehensive understanding of Subei Lake basin, typical of lake basins in the discharge area where significant changes in the groundwater system have taken place under the influence of human activity. More importantly, it could provide valuable groundwater information for decision makers and researchers to formulate scientifically reasonable groundwater resource management strategies in these lake basins of Ordos Basin so as to minimize the nega-

tive impacts of anthropogenic activities on the water system. In addition, given that there have been a series of ecological and environmental problems, more ecohydrological studies in these lake basins are urgently needed from the perspective of the future sustainable development of natural resources.

Acknowledgements. This research was supported by the State Basic Research Development Program (973 Program) of China (grant no. 2010CB428805) and Greenpeace International. The authors are grateful for our colleagues for their assistance in sample collection and analysis. Special thanks go to the editor and the two anonymous reviewers for their critical reviews and valuable suggestions.

Edited by: S. Uhlenbrook

References

Adams, S., Titus, R., Pietersen, K., Tredoux, G., and Harris, C.: Hydrochemical characteristics of aquifers near Sutherland in the Western Karoo, South Africa, J. Hydrol., 241, 91–103, doi:10.1016/S0022-1694(00)00370-X, 2001.

Agartan, E. and Yazicigil, H.: Assessment of water supply impacts for a mine site in western Turkey, Mine Water Environ., 31, 112–128, doi:10.1007/s10230-011-0167-z, 2012.

Appelo, C. A. J. and Postma, D.: Geochemistry, groundwater and pollution, AA Balkema, Rotterdam, 1994.

Appelo, C. A. J. and Willemsen, A.: Geochemical calculations and observations on salt water intrusions, I. A combined geochemical/mixing cell model, J. Hydrol., 94, 313–330, 1987.

Cervi, F., Ronchetti, F., Martinelli, G., Bogaard, T. A., and Corsini, A.: Origin and assessment of deep groundwater inflow in the Ca' Lita landslide using hydrochemistry and in situ monitoring, Hydrol. Earth Syst. Sci., 16, 4205–4221, doi:10.5194/hess-16-4205-2012, 2012.

Chen, J., Liu, X., Wang, C., Rao, W., Tan, H., Dong, H., Sun, X., Wang, Y., and Su, Z.: Isotopic constraints on the origin of groundwater in the Ordos Basin of northern China, Environ. Earth Sci., 66, 505–517, doi:10.1007/s12665-011-1259-6, 2011.

Clark, I. D. and Fritz, P.: Environmental isotopes in hydrogeology, CRC press, Boca Raton, Florida, 1997.

Cook, P. G. and Herczeg, A. L.: Environmental tracers in subsurface hydrology, Kluwer, Dordrecht, 1999.

Craig, H.: Isotopic variations in meteoric waters, Science, 133, 1702–1703, doi:10.1126/science.133.3465.1702, 1961.

Dai, S. F., Ren, D. Y., Chou, C. L., Li, S. S., and Jiang, Y. F.: Mineralogy and geochemistry of the No. 6 coal (Pennsylvanian) in the Junger Coalfield, Ordos Basin, China, Int. J. Coal Geol., 66, 253–270, doi:10.1016/j.coal.2005.08.003, 2006.

Deutsch, W. J.: Groundwater geochemistry: fundamentals and applications to contamination, CRC press, Boca Raton, Florida, 1997.

De Vries, J. J. and Simmers, I.: Groundwater recharge: an overview of processes and challenges, Hydrogeol. J., 10, 5–17, doi:10.1007/s10040-001-0171-7, 2002.

Dixon, W. and Chiswell, B.: The use of hydrochemical sections to identify recharge areas and saline intrusions in alluvial aquifers, southeast Queensland, Australia, J. Hydrol., 135, 259–274, doi:10.1016/0022-1694(92)90091-9, 1992.

Drever, J. I.: The geochemistry of natural waters: surface and groundwater environments, Prentice Hall, New Jersey, 1997.

Feth, J. H. and Gibbs, R. J.: Mechanisms controlling world water chemistry: evaporation-crystallization process, Science, 172, 871–872, 1971.

Garvelmann, J., Külls, C., and Weiler, M.: A porewater-based stable isotope approach for the investigation of subsurface hydrological processes, Hydrol. Earth Syst. Sci., 16, 631–640, doi:10.5194/hess-16-631-2012, 2012.

Gibbs, R. J.: Mechanisms controlling world water chemistry, Science, 170, 1088–1090, doi:10.1126/science.170.3962.1088, 1970.

Hamed, Y. and Dhahri, F.: Hydro-geochemical and isotopic composition of groundwater, with emphasis on sources of salinity, in the aquifer system in Northwestern Tunisia, J. Afr. Earth Sci., 83, 10–24, doi:10.1016/j.jafrearsci.2013.02.004, 2013.

Hou, G., Zhao, M., and Wang, Y.: Groundwater investigation in the Ordos Basin, China Geological Survey, Beijing, 2006 (in Chinese).

Hou, G., Su, X., Lin, X., Liu, F., Yi, S., Dong, W., Yu, F., Yang, Y., and Wang, D.: Environmental isotopic composition of natural water in Ordos Cretaceous Groundwater Basin and its significance for hydrological cycle, J. Jilin Univ. (Earth Science Edition), 37, 255–260, 2007.

Hou, G., Liang, Y., Su, X., Zhao, Z., Tao, Z., Yin, L., Yang, Y., and Wang, X.: Groundwater systems and resources in the Ordos Basin, China, Acta Geol. Sin., 82, 1061–1069, 2008.

Kamdee, K., Srisuk, K., Lorphensri, O., Chitradon, R., Noipow, N., Laoharojanaphand, S., and Chantarachot, W.: Use of isotope hydrology for groundwater resources study in Upper Chi river basin, J. Radioanal. Nuclear Chem., 297, 405–418, doi:10.1007/s10967-012-2401-y, 2013.

Lasaga, A. C., Soler, J. M., Ganor, J., Burch, T. E., and Nagy, K. L.: Chemical-weathering rate laws and global geochemical cycles, Geochim. Cosmochim. Ac., 58, 2361–2386, doi:10.1016/0016-7037(94)90016-7, 1994.

Li, P., Wu, J., and Qian, H.: Groundwater quality assessment and the forming mechanism of the hydrochemistry in Dongsheng Coalfield of Inner Mongolia, J. Water Resour. Water Eng., 21, 38–41, 2010 (in Chinese).

Li, P., Qian, H., and Wu, J. H.: Application of set pair analysis method based on entropy weight in groundwater quality assessment – a case study in Dongsheng City, Northwest China, E-J. Chem., 8, 851–858, 2011.

Li, P., Qian, H., Wu, J., Zhang, Y., and Zhang, H.: Major ion chemistry of shallow groundwater in the Dongsheng Coalfield, Ordos Basin, China, Mine Water Environ., 32, 195–206, doi:10.1007/s10230-013-0234-8, 2013.

Lloyd, J. W. and Heathcote, J.: Natural inorganic hydrochemistry in relation to groundwater, Oxford University Press, New York, 1985.

Magaritz, M., Nadler, A., Koyumdjisky, H., and Dan, J.: The use of Na-Cl ratios to trace solute sources in a semi-arid zone, Water Resour. Res., 17, 602–608, doi:10.1029/Wr017i003p00602, 1981.

Marghade, D., Malpe, D. B., and Zade, A. B.: Major ion chemistry of shallow groundwater of a fast growing city of Central India, Environ. Monit. Assess., 184, 2405–2418, doi:10.1007/s10661-011-2126-3, 2012.

Naseem, S., Rafique, T., Bashir, E., Bhanger, M. I., Laghari, A., and Usmani, T. H.: Lithological influences on occurrence of high-fluoride groundwater in Nagar Parkar area, Thar Desert, Pakistan, Chemosphere, 78, 1313–1321, doi:10.1016/j.chemosphere.2010.01.010, 2010.

Parkhurst, D. L. and Appelo, C.: PHREEQC2 user's manual and program, Water-Resources Investigations Report, US Geological Survey, Denver, Colorado, 2004.

Piper, A. M.: A graphic procedure in the geochemical interpretation of water analysis, US Department of the Interior, Geological Survey, Water Resources Division, Ground Water Branch, Washington, 1953.

PRC Ministry of Environmental Protection: Technical guidelines for environment impact assessment-groundwater environment, China Environmental Science Press,Beijing, 2011.

Rightmire, C. T.: Seasonal-variation in pCO_2 and ^{13}C content of soil atmosphere, Water Resour. Res., 14, 691–692, doi:10.1029/Wr014i004p00691, 1978.

Sami, K.: Recharge mechanisms and geochemical processes in a semiarid sedimentary basin, Eastern Cape, South-Africa, J. Hydrol., 139, 27–48, doi:10.1016/0022-1694(92)90193-Y, 1992.

Schoeller, H.: Qualitative evaluation of groundwater resources, In: Methods and techniques of groundwater investigations and development, UNESCO, Paris, 54–83, 1965.

Su, Y., Zhu, G., Feng, Q., Li, Z., and Zhang, F.: Environmental isotopic and hydrochemical study of groundwater in the Ejina Basin, northwest China, Environ. Geol., 58, 601–614, doi:10.1007/s00254-008-1534-3, 2009.

Toth, J.: A theoretical analysis of groundwater flow in small drainage basins, J. Geophys. Res., 68, 4795–4812, doi:10.1029/Jz068i008p02354, 1963.

Van der Weijden, C. H., and Pacheco, F. A. L.: Hydrochemistry, weathering and weathering rates on Madeira island, J. Hydrol., 283, 122–145, doi:10.1016/S0022-1694(03)00245-2, 2003.

Wang, W., Yang, G., and Wang, G.: Groundwater numerical model of Haolebaoji well field and evaluation of the environmental problems caused by exploitation, South-to-North Water Trans. Water Sci. Technol., 8, 36–41, 2010 (in Chinese).

Winter, T. C.: Relation of streams, lakes, and wetlands to groundwater flow systems, Hydrogeol. J., 7, 28–45, doi:10.1007/s100400050178, 1999.

Xiao, J., Jin, Z., Zhang, F., and Wang, J.: Solute geochemistry and its sources of the groundwaters in the Qinghai Lake catchment, NW China, J. Asian Earth Sci., 52, 21–30, doi:10.1016/j.jseaes.2012.02.006, 2012.

Xing, L., Guo, H., and Zhan, Y.: Groundwater hydrochemical characteristics and processes along flow paths in the North China Plain, J. Asian Earth Sci., 70–71, 250–264, doi:10.1016/j.jseaes.2013.03.017, 2013.

Yang, L., Song, X., Zhang, Y., Han, D., Zhang, B., and Long, D.: Characterizing interactions between surface water and groundwater in the Jialu River basin using major ion chemistry and stable isotopes, Hydrol. Earth Syst. Sci., 16, 4265–4277, doi:10.5194/hess-16-4265-2012, 2012a.

Yang, L., Song, X., Zhang, Y., Yuan, R., Ma, Y., Han, D., and Bu, H.: A hydrochemical framework and water quality assessment of river water in the upper reaches of the Huai River Basin, China, Environ. Earth Sci., 67, 2141–2153, doi:10.1007/s12665-012-1654-7, 2012b.

Yin, L., Hou, G., Dou, Y., Tao, Z., and Li, Y.: Hydrogeochemical and isotopic study of groundwater in the Habor Lake Basin of the Ordos Plateau, NW China, Environ. Earth Sci., 64, 1575–1584, doi:10.1007/s12665-009-0383-z, 2009.

Yin, L., Hou, G., Su, X., Wang, D., Dong, J., Hao, Y., and Wang, X.: Isotopes (δD and δ^{18}O) in precipitation, groundwater and surface water in the Ordos Plateau, China: implications with respect to groundwater recharge and circulation, Hydrogeol. J., 19, 429–443, doi:10.1007/s10040-010-0671-4, 2010.

Yuko, A., Uchida, T., and Ohte, N.: Residence times and flow paths of water in steep unchannelled catchments, Tanakami, Japan, J. Hydrol., 261, 173–192, 2002.

Zhang, J., Fang, H., and Ran, G.: Groundwater resources assessment in the Ordos Cretaceous Artesian Basin, Inner Mongolia Bureau of Geology and Mineral Resources, Hohhot, 1986.

9

Reducing the ambiguity of karst aquifer models by pattern matching of flow and transport on catchment scale

S. Oehlmann[1], T. Geyer[1,2], T. Licha[1], and M. Sauter[1]

[1]Geoscience Center, University of Göttingen, Göttingen, Germany
[2]Landesamt für Geologie, Rohstoffe und Bergbau, Regierungspräsidium Freiburg, Freiburg, Germany

Correspondence to: S. Oehlmann (sandra.oehlmann@geo.uni-goettingen.de)

Abstract. Assessing the hydraulic parameters of karst aquifers is a challenge due to their high degree of heterogeneity. The unknown parameter field generally leads to a high ambiguity for flow and transport calibration in numerical models of karst aquifers. In this study, a distributed numerical model was built for the simulation of groundwater flow and solute transport in a highly heterogeneous karst aquifer in south-western Germany. Therefore, an interface for the simulation of solute transport in one-dimensional pipes was implemented into the software COMSOL Multiphysics® and coupled to the three-dimensional solute transport interface for continuum domains. For reducing model ambiguity, the simulation was matched for steady-state conditions to the hydraulic head distribution in the model area, the spring discharge of several springs and the transport velocities of two tracer tests. Furthermore, other measured parameters such as the hydraulic conductivity of the fissured matrix and the maximal karst conduit volume were available for model calibration. Parameter studies were performed for several karst conduit geometries to analyse the influence of the respective geometric and hydraulic parameters and develop a calibration approach in a large-scale heterogeneous karst system.

Results show that it is possible not only to derive a consistent flow and transport model for a $150\,km^2$ karst area but also to combine the use of groundwater flow and transport parameters thereby greatly reducing model ambiguity. The approach provides basic information about the conduit network not accessible for direct geometric measurements. The conduit network volume for the main karst spring in the study area could be narrowed down to approximately $100\,000\,m^3$.

1 Introduction

Karst systems play an important role in water supply worldwide (Ford and Williams, 2007). They are characterized as dual-flow systems where flow occurs in the relatively lowly conductive fissured matrix and in highly conductive karst conduits (Reimann et al., 2011). There are a number of process-based modelling approaches available for simulating karst aquifer behaviour. Overviews on the various types of distributed process and lumped-parameter models are provided by several authors (Teutsch and Sauter, 1991; Jeannin and Sauter, 1998; Kovács and Sauter, 2007; Hartmann et al., 2014). In most cases, lumped-parameter models are applied, since they are less demanding on input data (Geyer et al., 2008; Perrin et al., 2008; Hartmann et al., 2013; Schmidt et al., 2013). These models consider neither the actual flow process nor the heterogeneous spatial distribution of aquifer parameters, but are able to simulate the integral aquifer behaviour, e.g. karst spring responses. The spatial distribution of model parameters and state variables, e.g. the hydraulic head distribution, need to be addressed with distributed numerical models should the necessary field data be available (e.g. Oehlmann et al., 2013; Saller et al., 2013). A distributed modelling approach suited for the simulation of strongly heterogeneous and anisotropic aquifers with limited data availability is the hybrid modelling approach. The approach simulates the fast flow component in the highly conductive karst conduit system in discrete one-dimensional elements and couples it to a two- or three-dimensional continuum representing the fissured matrix of the aquifer (Oehlmann et al., 2013). Hybrid models are rarely applied to real karst systems because they have a high demand of input data (Reimann et al., 2011). They are, however, regularly applied in long-term karst genetic simulation scenarios (e.g. Clemens et al., 1996; Bauer et al., 2003; Hubinger and Birk, 2011). In these models not only groundwater flow but also solute transport is coupled in the fissured matrix and in the karst conduits. Aside from karst evolution such coupling enables models to simulate tracer or contaminant transport in the karst conduit system (e.g. Birk et al., 2005). In addition to serving for predictive purposes, such models can be used for deriving information about the groundwater catchment itself (Rehrl and Birk, 2010).

A major problem for characterizing the groundwater system with numerical models is generally model ambiguity. The large number of calibration parameters is usually in conflict with a relatively low number of field observations, e.g. different hydraulic parameter fields and process variables may give a similar fit to the observed data but sometimes very different results for prognostic simulations (Li et al., 2009). Especially the geometric and hydraulic properties of the karst conduit system are usually unknown and difficult to characterize with field experiments for a whole spring catchment (Worthington, 2009). With artificial tracer test data the maximum conduit volume can be estimated but an unknown contribution of fissured matrix water prevents further conclusions on conduit geometry (Birk et al., 2005; Geyer et al., 2008). It is well known that the use of several objective functions, i.e. several independent field observations, can significantly reduce the number of plausible parameter combinations (Ophori, 1999). Especially in hydrology (e.g. Khu et al., 2008; Hunter et al., 2005) and also for groundwater systems (e.g. Ophori, 1999; Hu, 2011; Hartmann et al., 2013), this approach has been successfully applied with a wide range of observation types, e.g. groundwater recharge, hydraulic heads, remote sensing and solute transport. Particularly, the simulation of flow and transport is known to reduce model ambiguity and yield information on karst conduit geometry (e.g. Birk et al., 2005; Covington et al., 2012; Luhmann et al., 2012; Hartmann et al., 2013). Usually, automatic calibration schemes performing a multi-objective calibration for several parameters are used for this purpose (Khu et al., 2008). However, for complex modelling studies calculation times might be large due to the high number of model runs needed (Khu et al., 2008) and a precise conceptual model is essential as basis for the automatic calibration (Madsen, 2003). In general, numerical models of karst aquifers are difficult to build because of their highly developed heterogeneity (Rehrl and Birk, 2010). Thus, automatic calibration procedures are better suited for conceptual and lumped-parameter models, where calibration parameters include effective geometric properties and no spatial representation of the hydraulic parameter field and conduit geometry is necessary. Complex distributed numerical approaches generally require longer simulation times due to the necessary spatial resolution. Long simulation times limit the number of model runs that can reasonably be performed and man-

ual calibration based on hydrogeological knowledge is necessary (e.g. Saller et al., 2013). Therefore, applied distributed numerical models in karst systems usually focus on a smaller number of objective functions. They generally cannot simulate the hydraulic head distribution in the area, spring discharge and tracer breakthrough curves simultaneously on catchment scale. Some studies combine groundwater flow with particle tracking for tracer directions (e.g. Worthington, 2009; Saller et al., 2013) without simulating tracer transport. On the other hand there are studies simulating breakthrough curves without calibrating for measured hydraulic heads (e.g. Birk et al., 2005). For developing process-based models which can be used as prognostic tools, e.g. for the delineation of protection zones, the simulation should be able to reproduce groundwater flow and transport within a groundwater catchment. Especially in complex hydrogeological systems, this approach would reduce model ambiguity, which is a prerequisite in predicting groundwater resources and pollution risks.

This study shows how the combination of groundwater flow and transport simulation can be used not only to develop a basis for further prognostic simulations in a heterogeneous karst aquifer with a distributed modelling approach on catchment scale, but also to reduce model ambiguity and draw conclusions on the spatially distributed karst network geometries and the actual karst conduit volume. The approach shows the kind and minimum number of field observations needed for this aim. Furthermore, a systematic calibration strategy is presented to reduce the number of necessary model runs and the simulation time compared to standard multi-objective calibrations. For this purpose a hybrid model was built and a pattern matching procedure was applied for a well-studied karst aquifer system in south-western Germany. The model was calibrated for three major observed parameters: the hydraulic head distribution derived from measurements in 20 boreholes, the spring discharge of six springs and the tracer breakthrough curves of two tracer tests.

2 Modelling approach

The simulation is based on the mathematical flow model discussed in detail by Oehlmann et al. (2013). The authors set up a three-dimensional hybrid model for groundwater flow with the software COMSOL Multiphysics®. As described by Oehlmann et al. (2013) the simulation was conducted simultaneously in the three-dimensional fissured matrix, in an individual two-dimensional fault zone and in one-dimensional karst conduit elements to account for the heterogeneity of the system. Results showed that the karst conduits widen towards the springs and therefore, a linear relationship between the conduit radius and the conduit length s [L] was established. Values for s start with zero at the point farthest away from the spring and increase towards the respective karst spring.

In agreement with these results and karst genesis simulations by Liedl et al. (2003), the conduit radius is calculated as

$$r_c = ms + b, \tag{1}$$

where r_c [L] is the radius of a conduit branch and m and b are the two parameters defining the conduit size. b [L] is the initial radius of the conduit at the point farthest away from the spring and m [–] is the slope with which the conduit radius increases along the length of the conduit s.

In the following the equations used for groundwater flow and transport are described. The subscript "m" denotes the fissured matrix, "f" the fault zone and "c" the conduits hereby allowing a clear distinction between the respective parameters. Parameters without a subscript are the same for all karst features in the model.

2.1 Groundwater flow

Groundwater flow was simulated for steady-state conditions. This approach seems appropriate since this work focuses on the simulation of tracer transport in the conduit system during tracer tests, which are ideally conducted under quasi-steady-state flow conditions. Therefore, the simulations refer to periods with a small change of spring discharge, e.g. base flow recession, and are not designed to predict conditions during intensive recharge/discharge events. The groundwater flow in the three-dimensional fissured matrix was simulated with the continuity equation and the Darcy equation (Eq. 2a und b).

$$Q_m = \nabla(\rho u_m), \tag{2a}$$
$$u_m = -K_m \nabla H_m, \tag{2b}$$

where Q_m is the mass source term [$M L^{-3} T^{-1}$], ρ the density of water [$M L^{-3}$] and u_m the Darcy velocity [$L T^{-1}$]. In Eq. (2b) K_m is the hydraulic conductivity of the fissured matrix [$L T^{-1}$] and H_m the hydraulic head [L].

Two-dimensional fracture flow in the fault zone was simulated with the COMSOL® fracture flow interface. The interface only allows for the application of the Darcy equation inside of fractures, so laminar flow in the fault zone was assumed. In order to obtain a process-based conceptualization of flow, the hydraulic fault conductivity K_f was calculated by the cubic law (Eq. 3):

$$K_f = \frac{d_f^2 \rho g}{12\mu}, \tag{3}$$

where d_f is the fault aperture [L], ρ the density of water [$M L^{-3}$], g the gravity acceleration [$L T^{-2}$] and μ the dynamic viscosity of water [$M T^{-1} L^{-1}$].

For groundwater flow in the karst conduits, the Manning equation was used (Eq. 4).

$$u_c = \frac{1}{n}\left(\frac{r_c}{2}\right)^{\frac{2}{3}} \sqrt{\frac{dH_c}{dx}}, \tag{4}$$

where u_c is the specific discharge in this case equalling the conduit flow velocity [$L T^{-1}$], n the Manning coefficient [$T L^{-1/3}$], $r_c/2$ the hydraulic radius [L] and dH_c/dx the hydraulic gradient [–]. The Manning coefficient is an empirical value for the roughness of a pipe with no physical nor measurable meaning. The hydraulic radius is calculated by dividing the cross section by the wetted perimeter, which in this case corresponds to the total perimeter of the pipe (Reimann et al., 2011).

The whole conduit network was simulated for turbulent flow conditions. Due to the large conduit diameters (0.01–6 m, Sect. 5) this assumption is a good enough approximation. Hereby, strong changes in flow velocities due to the change from laminar to turbulent flow can be avoided. At the same time, the model does not require an estimation of the critical Reynolds number, which is difficult to assess accurately.

The three-dimensional flow in the fissured matrix and the one-dimensional conduit flow were coupled through a linear exchange term that was defined according to Barenblatt et al. (1960) as

$$q_{ex} = \frac{\alpha}{L}(H_c - H_m), \tag{5}$$

where q_{ex} is the water exchange between conduit and fissured matrix [$L^2 T^{-1}$] per unit conduit length L [L], H_m the hydraulic head in the fissured matrix [L], H_c the hydraulic head in the conduit [L] and α the leakage coefficient [$L^2 T^{-1}$]. The leakage coefficient was defined as

$$\alpha = 2\pi r_c K_m, \tag{6}$$

where $2\pi r_c$ is the conduit perimeter [L]. Other possible influences, e.g. the lower hydraulic conductivity at the solid–liquid interface of the pipe and the fact that water is not exchanged along the whole perimeter but only through the fissures are not considered. The exact value of these influences is unknown and the exchange parameter mainly controls the reaction of the karst conduits and the fissured matrix to hydraulic impulses. Since the flow simulation is performed for steady-state conditions this simplification is not expected to exhibit significant influence on the flow field.

2.2 Solute transport

Transient solute transport was simulated based on the steady-state groundwater flow field. COMSOL Multiphysics® offers a general transport equation with its solute transport interface. This interface was applied for the three-dimensional fissured matrix. In this work saturated, conservative transport was simulated, with an advection–dispersion equation (Eq. 7)

$$\frac{\partial}{\partial t}(\theta_m c_m) + \nabla(u_m c_m) = \nabla[(D_{Dm} + D_e)\nabla c_m] + S_m, \quad (7)$$

where θ_m is the matrix porosity [–], c_m the solute concentration [$M L^{-3}$], D_{Dm} the mechanical dispersion [$L^2 T^{-1}$] and D_e the molecular diffusion [$L^2 T^{-1}$]. S_m is the source term [$L^3 T^{-1}$].

The solute transport interface cannot be applied to one-dimensional elements within a three-dimensional model. COMSOL® offers a so-called coefficient form edge PDE interface to define one-dimensional mathematical equations. There, a partial differential equation is provided (COMSOL AB, 2012) which can be adapted as needed and leads to Eq. (8) in its application for solute transport in karst conduits:

$$\theta_c \frac{\partial c_c}{\partial t} + \nabla(-D_c \nabla c_c + u_c c_c) = f, \quad (8)$$

where θ_c is the conduit porosity which is set equal to 1, D_c [$L^2 T^{-1}$] the diffusive/dispersive term $D_c = (D_{Dc} + D_e)$, f the source term and u_c [$L T^{-1}$] the flow velocity inside the conduits, which corresponds to the advective transport component. Flow divergence cannot be neglected, as is often the case in other studies (e.g. Hauns et al., 2001; Birk et al., 2006; Coronado et al., 2007). Different conduit sizes and in- and outflow along the conduits lead to significant velocity divergence in the conduit system. This needs to be considered for mass conservation during the simulation. The mechanical conduit dispersion D_{Dc} was calculated with Eq. (9) (Hauns et al., 2001).

$$D_{Dc} = \varepsilon u_c, \quad (9)$$

where ε is the dispersivity in the karst conduits [L].

The source term f [$M T^{-1} L^{-1}$] in Eq. (8) equals in this case the mass flux of solute per unit length L [L] due to matrix–conduit exchange of solute c_{ex}:

$$f = c_{ex} = -D_e \frac{2\pi r_c}{L}(c_m - c_c) - q_{ex} c_i. \quad (10)$$

The first term of the right-hand side of Eq. (10) defines the diffusive exchange due to the concentration difference between conduit and fissured matrix. The second term is a conditional term adding the advective exchange of solute due to water exchange. The concentration of the advective exchange c_i is defined as

$$c_i = \begin{cases} c_c & \text{if } q_{ex} > 0 \\ c_m & \text{if } q_{ex} \le 0 \end{cases}. \quad (11)$$

When q_{ex} is negative, the hydraulic head in the fissured matrix is higher than in the conduit (Eq. 5) and water with the

solute concentration of the fissured matrix c_m enters the conduit. When it is positive, water with the solute concentration c_c of the conduit leaves the conduit and enters the fissured matrix. Since one-dimensional transport is simulated in a three-dimensional environment, the left-hand side of Eq. (8) is multiplied with the conduit cross section πr_c^2 [L^2]. These considerations lead to the following transport equation for the karst conduits:

$$\pi r_c^2 \frac{\partial c_c}{\partial t} + \pi r_c^2 \nabla(-D_c \nabla c_c + u_c c_c)$$
$$= -D_e \frac{2\pi r_c}{L}(c_m - c_c) - q_{ex} c_i. \quad (12)$$

3 Field site and model design

The field site is the Gallusquelle spring area on the Swabian Alb in south-western Germany. The size of the model area is approximately 150 km^2, including the catchment area of the Gallusquelle spring and surrounding smaller spring catchments (Oehlmann et al., 2013). The Gallusquelle spring is the main point outlet with a long-term average annual discharge of 0.5 m^3 s^{-1}. The model area is constrained by three rivers and no-flow boundaries derived from tracer test information and the dip of the aquifer base (Oehlmann et al., 2013) (Fig. 1).

The aquifer consists of massive and bedded limestone of the stratigraphic units Kimmeridgian 2 and 3 (ki 2/3) (Golwer, 1978; Gwinner, 1993). The marly limestones of the underlying Kimmeridgian 1 (ki 1) mainly act as an aquitard. In the west of the area where they get close to the surface, they are partly karstified and contribute to the aquifer (Sauter, 1992; Villinger, 1993). The Oxfordian 2 (ox 2) that lies beneath the ki 1 consists of layered limestones. It is more soluble than the ki 1 but only slightly karstified because of the protective effect of the overlying geological units. In the catchment areas of the Fehla-Ursprung and the Balinger springs close to the western border (Fig. 1a) the ox 2 partly contributes to the aquifer. For simplicity, only two vertical layers were differentiated in the model: the aquifer and the underlying aquitard.

The geometry of the conduit system was transferred from the COMSOL® model calibrated for flow by Oehlmann et al. (2013). It is based on the occurrence of dry valleys in the investigation area and artificial tracer test information (Gwinner, 1993). The conduit geometry for the Gallusquelle spring was also employed for distributed flow simulations by Doummar et al. (2012) and Mohrlok and Sauter (1997) (Fig. 1). In this work, all highly conductive connections identified by tracer tests in the field were simulated as discrete one-dimensional karst conduit elements. The only exception is a connection in the west of the area that runs perpendicular to the dominant fault direction and reaches the Fehla-Ursprung spring at the northern boundary (Fig. 1). While the element was regarded as a karst conduit by Oehlmann et

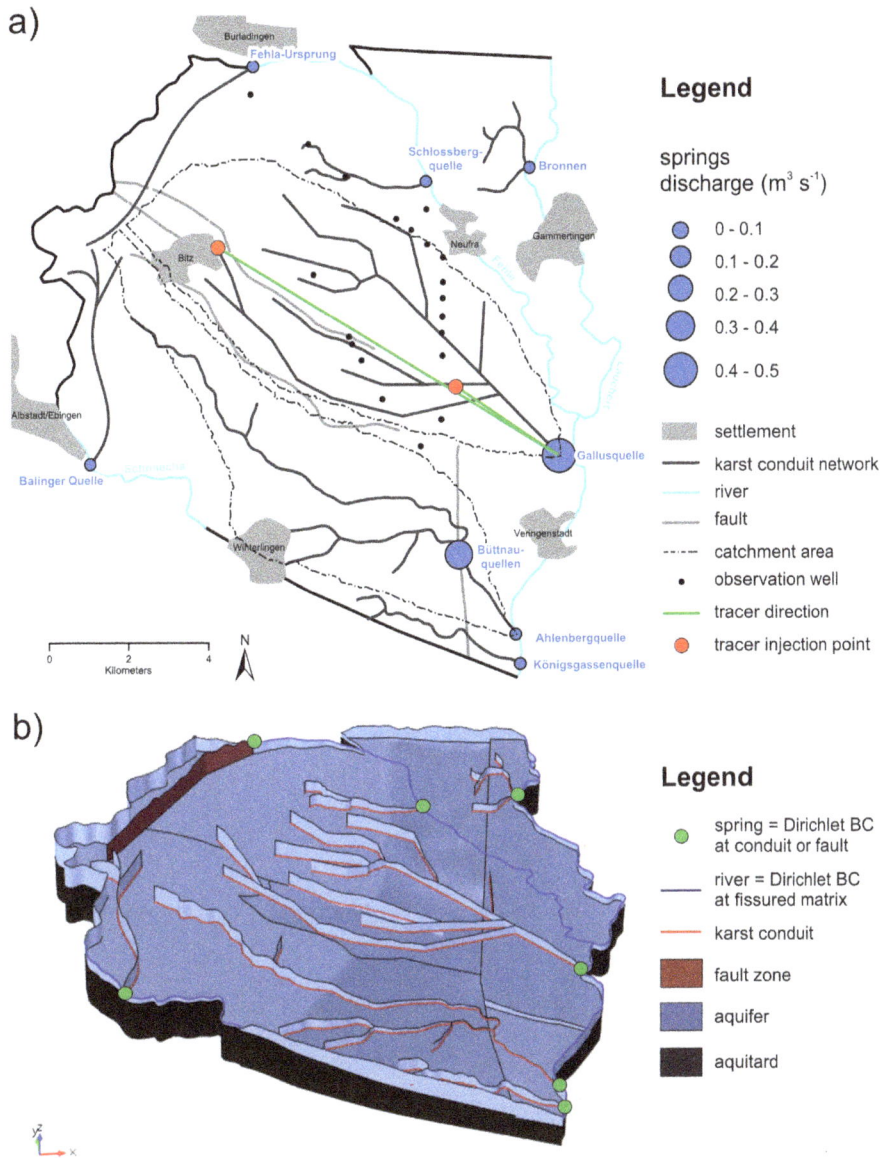

Figure 1. (a) Plan view of the model area. Settlements, fault zones and rivers in the area are plotted, as well as the 20 observation wells used for hydraulic head calibration, the six springs used for spring discharge calibration and the two tracer tests employed for flow velocity calibration. Catchment areas for the Gallusquelle spring and the Ahlenberg and Büttnauquellen springs were simulated according to Oehlmann et al. (2013). **(b)** Three-dimensional view of the model. The upper boundary is hidden to allow a view of the karst conduit system and the aquifer base. The abbreviation BC stands for boundary condition. At the hidden upper boundary, a constant recharge Neumann BC is applied.

al. (2013) it is more likely that the water crosses the graben structure by a transversal cross-fault (Strayle, 1970). Therefore, the one-dimensional conduit element was replaced by a two-dimensional fault element (Fig. 1b). This leads to a small adjustment in the catchment areas compared to the results of Oehlmann et al. (2013) (Fig. 1a). While the discharge data for the Fehla-Ursprung spring are not as extensive as for the other simulated springs, it is approximated to 0.1 m³ s⁻¹, the annual average ranging from 0.068 to 0.135 m³ s⁻¹. The fault zone aperture was calibrated accordingly (Sect. 5).

Due to a large number of studies conducted in the area during the last decades (e.g. Villinger, 1977; Sauter, 1992; Geyer et al., 2008; Kordilla et al., 2012; Mohrlok, 2014) many data for pattern matching are available even though the karst conduit network itself is not accessible. Since the groundwater flow simulation was performed for steady-state conditions, direct recharge, which is believed to play an important role during event discharge (Geyer et al., 2008), was neglected. It is not expected that recharge dynamics exhibit significant influence on the flow field during recession periods. From Sauter (1992) the long-term average annual recharge, ranges

Table 1. Calibrated and simulated parameters for the best-fit simulations. Literature values are given if available. TT 1 and TT 2 refer to the two tracer tests.

Parameter	Simulated values scenario 2	Simulated values scenario 5	Literature values
K_m (m s^{-1})	8×10^{-6}	1.5×10^{-5}	1×10^{-6}–2×10^{-5} (local scale)[e] 2×10^{-5}–1×10^{-4} (regional scale)[e]
m_h (m$^{-2/3}$ s^{-1})	0.3	0.3	–
b_h (m$^{1/3}$ s^{-1})	0.22	0.18	–
n (s m$^{-1/3}$)	1.04–4.55	1.05–5.56	0.03–1.07[a]
b (m)	0.01	0.01	–
m (–)	2.04×10^{-4}	1.42×10^{-4}	–
ε_1 (m) for TT 1	7.15	7.5	4.4–6.9[f], 10[e]
ε_2 (m) for TT 2	30	23	20[g]
A^h (m^2)	11.9	13.4	13.9[f]
V (m^3)	109351	89286	$\leq 200\,000$[b]
RMSE H (m)	5.61	5.91	–
Peak offset TT 1 (h)	-0.28[c]	-0.28[c]	–
Peak offset TT 2 (h)	2.5[d]	-1.39[d]	–

[a] Jeannin (2001); [b] Geyer et al. (2008); [c] measurement interval 1 min, simulation interval 2.7 h; [d] measurement interval 6 h, simulation interval 2.7 h; [e] Sauter (1992); [f] Birk et al. (2005); [g] Merkel (1991); [h] average for the interval between tracer test 1 and the spring.

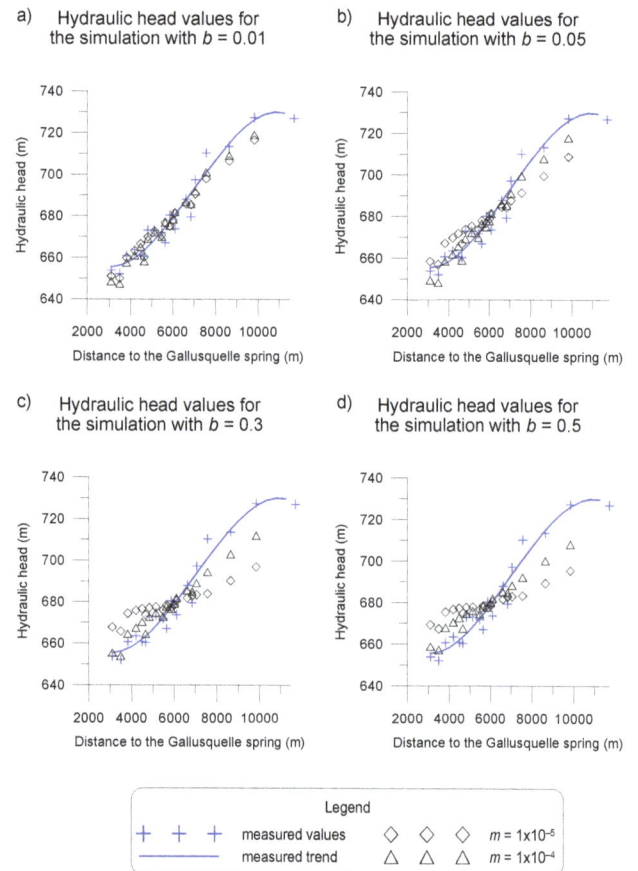

a) Hydraulic head values for the simulation with $b = 0.01$
b) Hydraulic head values for the simulation with $b = 0.05$
c) Hydraulic head values for the simulation with $b = 0.3$
d) Hydraulic head values for the simulation with $b = 0.5$

Legend: + measured values; — measured trend; ◇ $m = 1 \times 10^{-5}$; △ $m = 1 \times 10^{-4}$

Figure 2. Hydraulic head distributions for different combinations of geometric conduit parameters for scenario 1. b is the lowest conduit radius and m the radius increase along the conduit. For comparison, a trend line is fitted to the measured hydraulic head values showing the distribution of hydraulic gradients from the Gallusquelle spring to the western border of its catchment area.

of hydraulic parameters and the average annual hydraulic head distribution derived from 20 observation wells (Fig. 1a) are available. Villinger (1993) and Sauter (1992) provided data on the geometry of the aquifer base. Available literature values for the model parameters are given in Table 1.

The observed hydraulic gradients in the Gallusquelle area are not uniform along the catchment. Figure 2 shows a S-shaped distribution with distance to the Gallusquelle spring. The gradient at each point of the area depends on the combination of the respective transmissivity and total flow. The amount of water flowing through a cross sectional area increases towards the springs due to flow convergence. In the Gallusquelle area, the transmissivity rises in the vicinity of the springs leading to a low hydraulic gradient. In the central part of the area discharge is relatively high while the transmissivities are lower leading to the observed steepening of the gradient starting in a distance of 4000 to 5000 m from the Gallusquelle spring. Towards the boundary of the catchment area in the west the water divide reduces discharge in the direction of the Gallusquelle spring leading to a smoothing of hydraulic gradients.

Geyer et al. (2008) calculated the maximum conduit volume for the Gallusquelle spring V_c [L^3] with information from the tracer test that will be referred to as tracer test 2 in the following. Since the injection point of the tracer test is close to the catchment boundary, it is assumed that it covers the whole length of the conduit system. The authors calculated the maximum volume at 218 000 m^3. Their approach assumes the volume of the conduit corresponds to the total volume of water discharged during the time between tracer

input and tracer arrival neglecting the contribution of the fissured matrix.

The six springs that were monitored and therefore simulated are shown in Fig. 1. Except for the Balinger spring, their discharges were fitted to long-term average annual discharge data. For the Balinger spring discharge calibration was not possible due to lack of data. It was included as a boundary condition because several tracer tests provided a valuable basis for the conduit structure leading to the spring.

Tracer directions were available for 32 tracer tests conducted at 20 different tracer injection locations (Oehlmann et al., 2013). In all, 16 of the tracer tests were registered at the Gallusquelle spring. For this work two of them were chosen for pattern matching of transport parameters. Both of them were assumed to have a good and direct connection to the conduit network. Tracer test 1 (Geyer et al., 2007) has a tracer injection point at a distance of 3 km to the Gallusquelle spring. Tracer test 2 (MV746 in Merkel, 1991; Reiber et al., 2010) was conducted at 10 km distance to the Gallusquelle

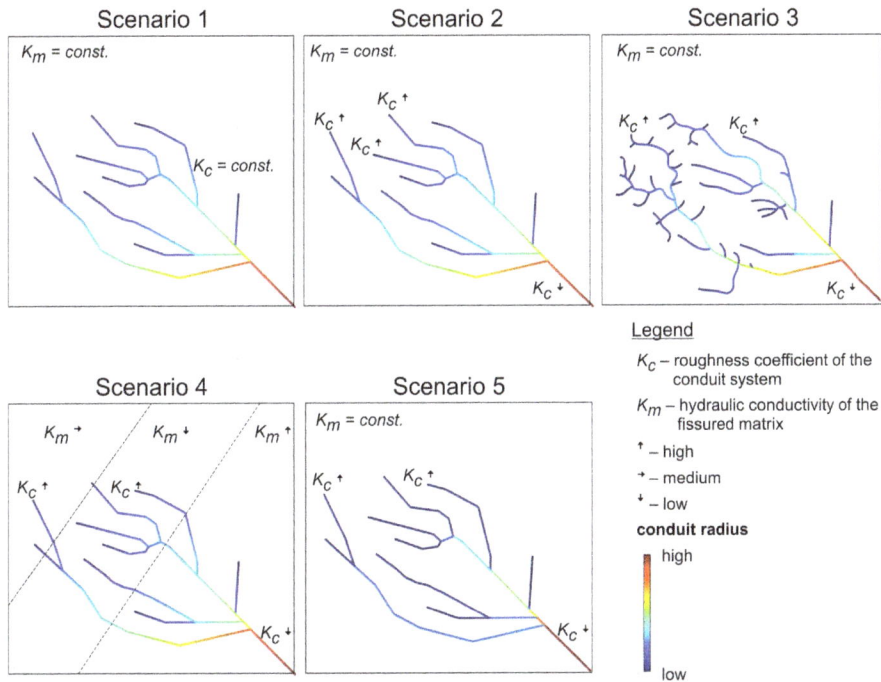

Figure 3. Conceptual overview of the simulated scenarios. The conduit geometry and the varying parameters are shown.

spring (Fig. 1a). Due to the flow conditions (Fig. 1a) it can be assumed that tracer test 2 covers the total length of the conduit network feeding the Gallusquelle spring. The recovered tracer mass was chosen as input for the tracer test simulation. The basic information about the tracer tests is given in Table 2.

Since the tracer tests were not performed at average flow conditions, the model parameters were calibrated first for the long-term average annual recharge of 1 mm d^{-1} and the long-term average annual discharge of 0.5 m^3 s^{-1}. For the transport simulations, the recharge was then adapted to produce the respective discharge observed during the tracer experiment (Table 2).

4 Parameter analysis

An extensive parameter analysis was performed in order to identify parameters determining the hydraulic parameter field in the model area, as well as their relative contributions to the discharge and conduit flow velocities. The fitting parameters include the parameters controlling the respective transmissivities of the fissured matrix and the karst conduit system, i.e. the geometry and roughness of the conduit system, the hydraulic conductivity of the fissured matrix and the fracture aperture for the Fehla-Ursprung spring. Furthermore, the apparent dispersivities for the two artificial tracer tests were calibrated (Table 1). Since all model runs were performed for steady-state conditions parameters controlling the temporal distribution of recharge were not con-

Table 2. Field data of the simulated tracer tests.

	Tracer test 1	Tracer test 2
Input mass (kg)	0.75	10
Recovery (%)	72	50
Distance to spring (km)	3	10
Spring discharge (m^3 s^{-1})	0.375	0.76
Sampling interval	1 min	6 h
Peak time (h)	47	79.5

sidered. The parameter analysis was performed with COMSOL Multiphysics® parametric sweep tool, which sweeps over a given parameter range. Parameter ranges were chosen according to literature values (Table 1). For the conduit geometry parameters, lowest conduit radius b and slope of radius increase m, no literature values are available. Therefore, the ranges were chosen so that conduit volumes ranged below the maximum volume given by Geyer et al. (2008). In addition to the variation of the fitting parameters, five basic scenarios were compared. They correspond to different conceptual representations of the area and are summarized in Fig. 3 and Table 3.

Three objective functions were employed for pattern matching: spring discharge, hydraulic head distribution and flow velocities of the two tracer tests (Sect. 3). The average spring discharge of the Gallusquelle spring was set by the difference between simulated and the measured discharge. A difference of 10 L s^{-1} was considered as acceptable. Param-

Table 3. Specifics of the different scenarios. The bold writing indicates the parameter that is analysed in the respective scenario. The results are indicated by comparative markers. "+" means good, "o" means average and "–" means bad compared to the other scenarios. Details to the scenarios and results evaluation can be found in Sect. 4.

Parameter	Scenario 1	Scenario 2	Scenario 3	Scenario 4	Scenario 5
K_c	constant	**linear increase**	linear increase	linear increase	linear increase
Lateral network	minimal	minimal	**extended**	minimal	minimal
K_m	constant	constant	constant	**variable**	constant
Intersection radius r_{c2}	r_{c0}	r_{c0}	r_{c0}	r_{c0}	$\sqrt{r_{c0}^2 + r_{c1}^2}$
Main results					
Hydraulic head fit	+	+	+	+	+
Fit of breakthrough	–	+	+	+	+
Model applicability	+	o	–	–	o

eter sets, which could not fulfil this criterion, were not considered for parameter analysis. The other low-discharge and less-investigated springs (Sect. 3) were used to inspect the flow field and water balance in the modelling area, i.e. they were only considered after parameter fitting to check the plausibility of the deduced parameter set.

The fit of the tracer tests was determined by comparing the arrival times of the highest peak concentration of the simulation with the measured value (peak offset). Since tracer experiments conducted in karst conduits usually display very narrow breakthrough curves, this procedure appears to be justified. The quality of the fit was judged as satisfactory if the peak offset was lower than either the simulation interval or the measurement interval.

The fit of the hydraulic head distribution was determined by calculating the root mean square error (RMSE) between the simulated and the observed values at the respective locations of the observation wells. Since the fit at local points with a large-scale modelling approach generally shows large uncertainties due to low-scale heterogeneities, an overall fit of < 10 m RMSE was accepted. Furthermore, a qualitative comparison with the hydraulic gradients in the area was performed (e.g. Fig. 2) to ensure that the general characteristics of the area were represented instead of only the statistical value.

4.1 Scenario 1 – standard scenario

In scenario 1 all features were implemented as described in Sects. 2 and 3. The parameter analysis shows that for each conduit geometry, defined by their smallest conduit radii b and their slopes of radius increase along the conduit length m (Eq. 1), only one value of the Manning coefficient n allows a simulated discharge for the Gallusquelle spring of $0.5\,\mathrm{m^3\,s^{-1}}$. The n value correlates well with that for the total conduit volume due to the fact that the spring discharge is predominantly determined by the transmissivity of the karst conduit system. The transmissivity of the conduit system at each point in space is the product of its hydraulic conduc-

tivity, which is proportional to $1/n$, and the cross sectional area of the conduit A. Thus, to keep the spring discharge at $0.5\,\mathrm{m^3\,s^{-1}}$ a higher conduit volume requires a higher calibrated n value (Eq. 4).

With scenario 1 it is possible to achieve a hydraulic head fit resulting in a RMSE of 6 m that can be judged as adequate on catchment scale. Regarding the conduit geometry, a good hydraulic head fit can be achieved with small b values independently of the chosen m value (Fig. 2a). The higher the b value, the higher the m value to reproduce the hydraulic gradients of the area (Fig. 2). This implies that the hydraulic head fit is independent of the conduit volume during steady-state conditions but depends on the b/m ratio. The influence of the b/m ratio on the hydraulic head fit depends on the hydraulic conductivity of the fissured matrix K_m. For low K_m values of ca. $1 \times 10^{-6}\,\mathrm{m\,s^{-1}}$ the hydraulic head fit is completely independent of the conduit geometry and the RMSE is very high (Fig. 4a). For high K_m values of ca. $5 \times 10^{-4}\,\mathrm{m\,s^{-2}}$ (Fig. 4a) the dependence is also of minor importance and the RMSE is relatively stable at ca. 11 m. Due to the high hydraulic conductivity of the fissured matrix the hydraulic gradients do not steepen in the vicinity of the spring even for high b/m ratios. For K_m values between the above values the RMSE significantly rises for b/m ratios above 1000 m. For the range of acceptable errors, i.e. lower than 10 m, it is apparent in Fig. 4a that the best-fit K_m value is approximately $1 \times 10^{-5}\,\mathrm{m\,s^{-1}}$ independent of the conduit geometry. However, no distinct best-fit conduit geometry can be derived. There are several parameter combinations providing a good fit for the Gallusquelle spring discharge and the hydraulic head distribution.

The goodness of the fit of the simulation of the tracer breakthrough is mainly determined by the conduit geometry. The influence of the hydraulic conductivity of the fissured matrix K_m on flow velocities inside the karst conduits is comparatively low and decreases even further in the vicinity of the springs (Fig. 4b and c) leading to minor influences on tracer travel times. Instead, the quality of the fit

Objective functions in relation to the hydraulic conductivity of the fissured matrix K_m

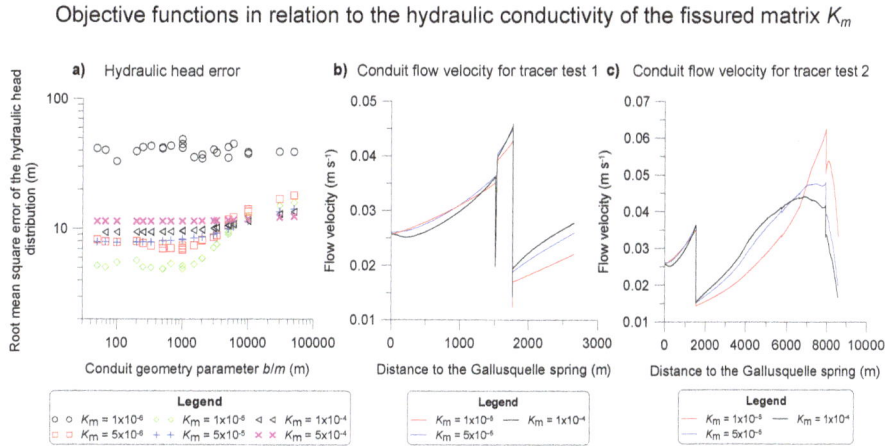

Figure 4. Influence of the hydraulic conductivity of the fissured matrix on the objective functions. (**a**) Influence on the root mean square error of the hydraulic head distribution in relation to the conduit geometry. The conduit geometry is represented by the parameter b/m (Eq. 1), which is the ratio of the smallest radius to the slope of radius increase along the conduit length. (**b**) Influence on the conduit flow velocity for tracer test 1. (**c**) Influence on the conduit flow velocity for tracer test 2.

mainly depends on the conduit volume and accordingly on the Manning coefficient n (Fig. 5). It is possible to simulate only one of the two tracer experiments with this scenario (Fig. 5). Given the broad range of geometries for which an adequate hydraulic head fit can be achieved (Figs. 2 and 4) it is possible to simulate one of the two tracer peak velocities and the hydraulic head distribution with the same set of parameters. While the simulation of the breakthrough of tracer test 1 requires relatively high n values, of ca. $2.5 \, \text{s m}^{-1/3}$, that of tracer test 2 can only be calibrated with lower values of ca. $1.7 \, \text{s m}^{-1/3}$ (cf. Fig. 5a and b). For every parameter set, where the travel time of the simulated tracer test 2 is not too long, that of tracer test 1 is too short. For the simulation of tracer test 2, the velocities at the beginning of the conduits must be relatively high. To avoid the flow velocities from getting too high in downgradient direction, the conduit size would have to increase drastically due to the constant additional influx of water from the fissured matrix. In the given geometric range, the conduit system has a dominant influence on spring discharge. Physically, this situation corresponds to the conduit-influenced flow conditions (Kovács et al., 2005). Thus, conduit transmissivity is a limiting factor for conduit–matrix exchange and a positive feedback mechanism is triggered, if the conduit size is increased. A higher conduit size leads to higher groundwater influx from the fissured matrix and spring discharge is overestimated. Therefore, parameter analysis shows that scenario 1 is too strongly simplified to correctly reproduce the complex nature of the aquifer.

4.2 Scenario 2 – conduit roughness coefficient K_c

In scenario 2 the Manning coefficient n was changed from constant to laterally variable. In the literature, n is generally

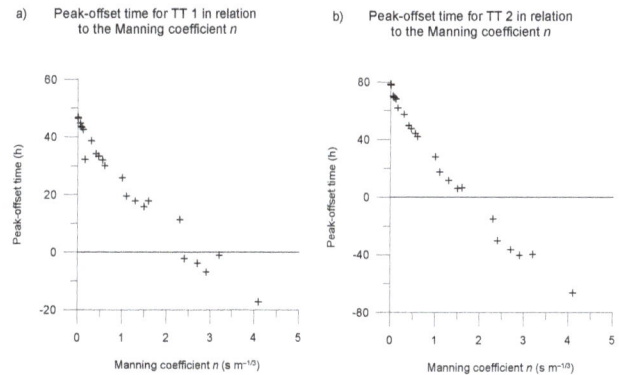

Figure 5. Difference between peak concentration times vs. the Manning n value for scenario 1. High n values correspond to high conduit volumes and high cross sectional areas at the spring (**a**) for tracer test 1 and (**b**) for tracer test 2.

kept constant throughout the conduit network (e.g. Jeannin, 2001; Reimann et al., 2011) for lack of information on conduit geometry. However, it is assumed that the Gallusquelle spring is not fed by a single large pipe. Rather there is some evidence in the spring area that a bundle of several small-interconnected pipes feed the spring. Since the number of individual conduits per bundle is unknown and the regional modelling approach limits the resolution of local details, the small diameter conduits, which the bundle consists of, cannot be simulated individually. Therefore, each single pipe in the model represents a bundle of conduits in the field.

It can be assumed that the increase in conduit cross section is at least partly provided by additional conduits added to the bundle rather than a single individual widening conduit. Therefore, while the cross section of the simulated conduit, i.e. the total effective cross section of the conduit bundle, increases towards the springs, it is not specified how much of this increase is due to the individual conduits widening and how much is due to additional conduits, not distinguishable in the simulation. If the simulated effective cross sectional area increase is mainly due to additional conduits being included in the bundle, the surface / volume ratio increases with the cross section, contrary to what would be observed, if a single conduit in the model would represent a single conduit in the field. The variation in surface area / volume ratio implicitly leads to a larger roughness in the simulation, even further enhanced by exchange processes between the individual conduits. This effect again leads to an increase in the Manning coefficient n in the downgradient direction towards the spring for a simulated single conduit. Since the number and size of the individual conduits is unknown, it is impossible to calculate the change of n directly from the geometry. Thus, a simple scenario was assumed where the roughness coefficient K_c, which is the reciprocal of n, was linearly and negatively coupled to the rising conduit radius (Eq. 13).

$$K_c = \frac{1}{n} = -m_h r_c + m_h r_{c,max} + b_h, \quad (13)$$

where r_c [L] is the conduit radius and $r_{c,max}$ [L] the maximum conduit radius simulated for the respective spring, which COMSOL® calculates from Eq. (1). m_h [$L^{-2/3} T^{-1}$] and b_h [$L^{1/3} T^{-1}$] are calibration parameters determining the slope and the lowest value of the roughness coefficient respectively.

For every conduit geometry several combinations of m_h and b_h lead to the same spring discharge. However, hydraulic head fit and tracer velocities are different for each m_h–b_h combination even if spring discharge is the same. With the new parameters a higher variation of velocity profiles is possible. This allows for the calibration of the tracer velocities of both tracer tests. The dependence of tracer test 2 on m_h is much higher than that of tracer test 1 since it is injected further upgradient towards the beginning of the conduit (Fig. 6). Therefore, tracer test 2 is influenced more strongly by the higher velocities far away from the spring introduced by high m_h values and always shows a significant positive correlation with m_h (Fig. 6).

Since the slope of K_c is negative with respect to the conduit length, the variable K_c leads to a slowing down of water towards the springs. As discussed in detail by Oehlmann et al. (2013) a rise of transmissivity towards the springs is observed in the Gallusquelle area. Therefore, adequate hydraulic head fits can only be obtained, if the decrease of K_c towards the spring is not too large and compensates the effect of the increase in conduit transmissivity due to the increasing conduit radius. This effect reduces the number of possible and plausible parameter combinations. From these considerations a best-fit model can be deduced capable of reproducing all objective functions within the given error ranges (Fig. 7a). According to the model simulations, karst groundwater discharge and flow velocities significantly depend on

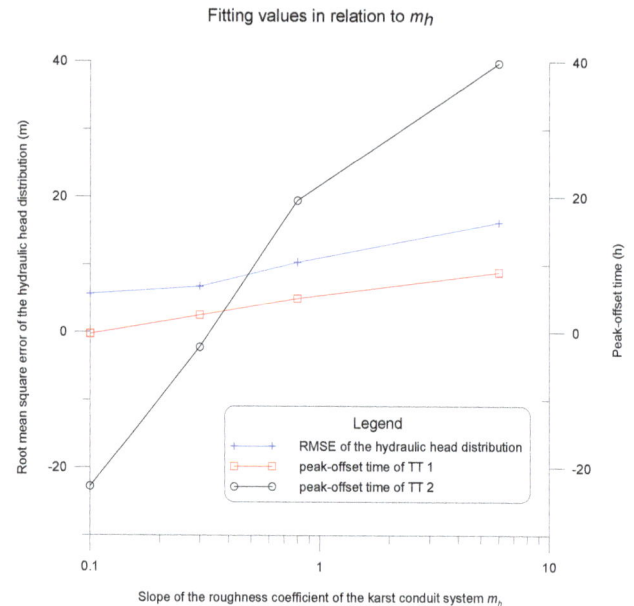

Figure 6. Hydraulic head errors and differences between peak concentration times for both tracer tests for scenario 1. The example is shown for a conduit geometry with a starting value $b = 0.01$ m and a radius increase of $m = 2 \times 10^{-4}$. Each m_h ($m^{-2/3}$ s^{-1}) value corresponds to a respective value of the highest conduit roughness b_h ($m^{1/3}$ s^{-1}) and each combination results in the same spring discharge.

the total conduit volume as is to be expected. It can be deduced from the parameter analysis that the conduit volume can be estimated at ca. 100 000 m^3 for the different parameters to match equally well (Fig. 7a).

4.3   Scenario 3 – extent of conduit network

In scenario 3, a laterally further extended conduit system was employed, assuming the same maximum conduit volume as in scenarios 1 and 2 but with different spatial distribution along the different total conduit lengths. The original conduit length for the Gallusquelle spring in scenarios 1 and 2 is 39 410 m, for scenario 3 it is 63 490 m; therefore, the total length was assumed to be larger by ca. 50 % (Fig. 8). The geometry of the original network was mainly constructed along dry valleys where point-to-point connections are observed based on qualitative evaluation from artificial tracer tests. Of the dry valleys without tracer tests, only the larger ones were included, where the assumption of a high karstification is backed up by the occurrence of sinkholes (Mohrlok and Sauter, 1997). Therefore, it represents the minimal extent of the conduit network. For scenario 3 the network was extended along all dry valleys within the catchment, where no tracer tests were conducted.

The results of the parameter variations are comparable to those of scenario 2 (cf. Fig. 7a and b). While the hydraulic head contour lines are smoother than for the original conduit

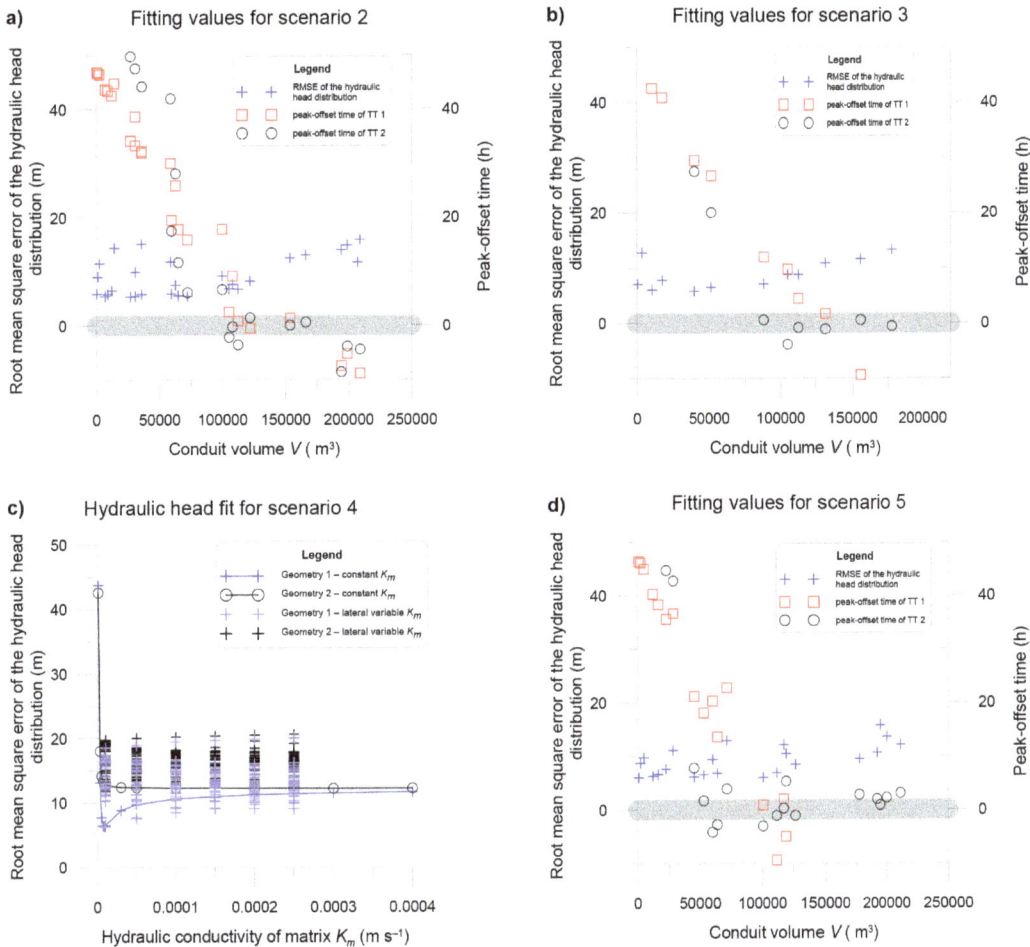

Figure 7. Calibrated values for the simulated scenarios. For scenarios 2, 3 and 5 (**a, b, d**) hydraulic head fit and the peak-offset times of both tracer tests (referred to as TT 1 and TT 2) are shown in relation to conduit volume. The thick grey bar marks the target value of zero. For scenario 4 (**c**) the root mean square error of the hydraulic heads is given for two different conduit geometries in relation to the hydraulic conductivity of the fissured matrix K_m. For the version with laterally variable matrix conductivity the axis shows as an example the hydraulic conductivity of the north-western part. The parameters for the two geometries are given in Table 3.

length the general hydraulic head fit is the same (Fig. 7b). It seems possible to obtain a good fit for all model parameters but the scenario is more difficult to handle numerically. Calculation times are up to 10 times larger compared to the other scenarios and goodness of convergence is generally lower. Since the calibrated parameters are not significantly different from those deduced in scenario 2 it is concluded that the ambiguity introduced by the uncertainty in total conduit length is small if hydraulic conduit parameters and total conduit volumes are the aim of investigation.

4.4 Scenario 4 – matrix hydraulic conductivity K_m

In scenario 4, the homogeneously chosen hydraulic conductivity of the fissured matrix K_m was changed into a laterally variable conductivity based on different types of lithology and the spatial distribution of the groundwater potential. Sauter (1992) found from field measurements that the area

can be divided into three parts with different hydraulic conductivities. Oehlmann et al. (2013) discussed that the major influence is the conduit geometry leading to higher hydraulic transmissivities close to the springs in the east of the area. It is also possible that not only the conduit diameters change towards the spring but the hydraulic conductivity of the fissured matrix as well, since the aquifer cuts through three stratigraphic units (Sect. 3). These geologic changes are likely to affect the lateral distribution of hydraulic conductivities (Sauter, 1992). Figure 9 shows the division into three different areas. K_m values were varied in the range of the values measured by Sauter (1992).

It was expected that a laterally variable K_m value has a major influence on the hydraulic head distribution. All variations of scenario 2 that produce good results for both tracer tests and have a high total conduit volume above $100\,000\,\text{m}^3$ yield poor results for hydraulic head errors and spatial dis-

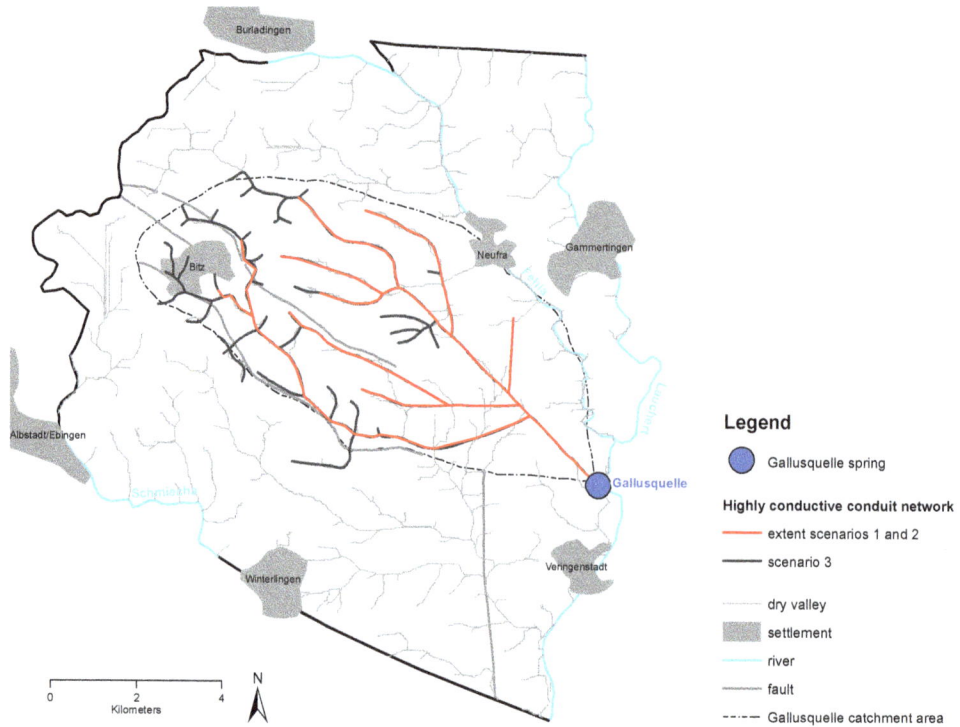

Figure 8. Extended conduit system for scenario 3. The conduit configuration (extent) that is used for the other scenarios is marked in red.

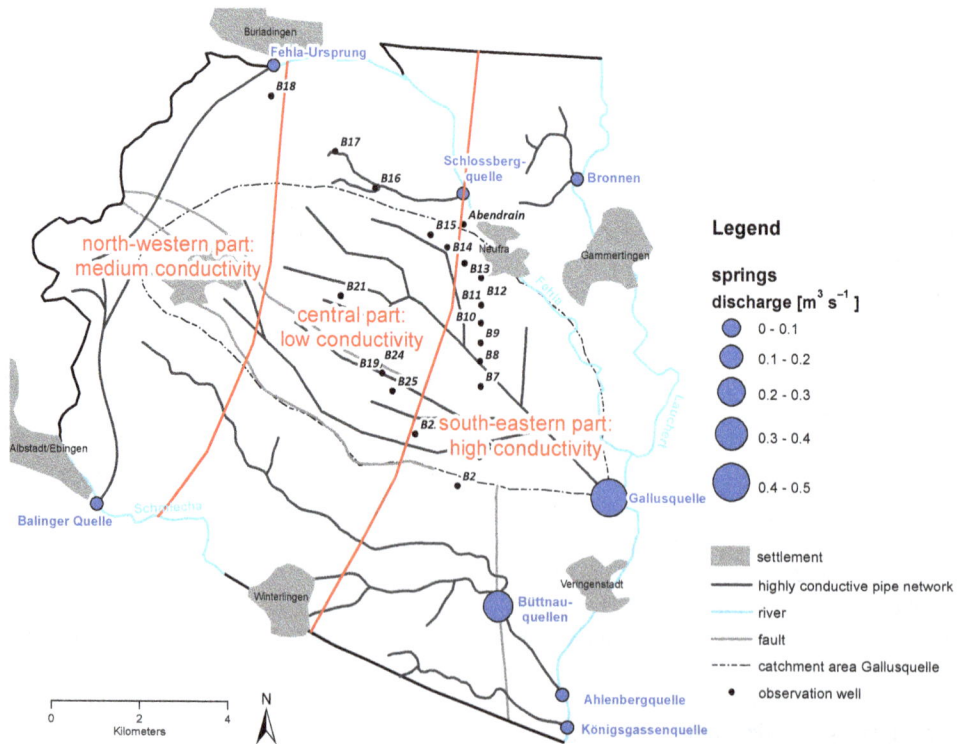

Figure 9. Model catchment with spatially distributed hydraulic conductivities. The model area is divided into three parts after geologic aspects. For each segment different values of the hydraulic conductivity were examined during parameter analysis in scenario 4.

tributions of the hydraulic heads (Fig. 7a). For scenario 4, two different conduit configurations (geometries) were chosen that achieve good results with respect to conduit flow velocities. Geometry G1 has a conduit volume of $112\,000\,\mathrm{m}^3$. G2 has a higher b value which leads to the maximum conduit volume of ca. $150\,000\,\mathrm{m}^3$. All parameters for the two simulations are given in Table 4.

It was found that while the maximum root mean square error of the hydraulic head fit is similar for both geometries, the minimum RMSE for the hydraulic head is determined by the conduit system. It is not possible to compensate an unsuitable conduit geometry with suitable K_m values (Fig. 7c), which assists in the independent conduit network and fissured matrix calibration. This observation increases the confidence in the representation of the conduits and improves the possibility to deduce the conduit geometry from field measurements. For an adequate conduit geometry, laterally variable matrix conductivities do not yield any improvement. The approach introduces additional parameters and uncertainties because the division of the area into three parts is not necessarily obvious without detailed investigation. From the distribution of the exploration and observation wells (Fig. 1a) it is apparent that especially in the south and west the boundaries are not well defined.

4.5 Scenario 5 – conduit intersections

In scenario 5, the effect of the conduit diameter change at intersections was investigated. In the first four scenarios the possible increase in cross sectional area at intersecting conduits was neglected. In nature, however, the influx of water from another conduit is likely to influence conduit evolution and therefore its diameter. In general, higher flow rates lead to increased dissolution rates because dissolution products are quickly removed from the reactive interface. If conditions are turbulent the solution is limited by a diffusion dominated layer that gets thinner with increasing flow velocities (Clemens, 1998). Clemens (1998) simulated karst evolution in simple Y-shaped conduit networks and found higher diameters for the downstream conduit even after short simulation times. Preferential conduit widening at intersections could further be enhanced by the process of mixing corrosion (Dreybrodt, 1981). However, Hückinghaus (1998) found during his karst network evolution simulations that the water from other karst conduits has a very high saturation with respect to Ca^{2+} compared to water entering the system through direct recharge. Thus, if direct recharge is present, the mixing with nearly saturated water from an intersecting conduit could hamper the preferential evolution of the conduit downstream slowing down the aforementioned processes. In scenario 5 the influence of an increase in diameter at conduit intersections was investigated. Since the amount of preferential widening at intersections is unknown, the cross sections of two intersecting conduits were added and used as starting cross section for the downstream conduit. The new conduit

Table 4. Parameters for the two different conduit configurations compared in scenario 4. b is the minimum conduit radius, m the slope of radius increase towards the springs, b_h the highest conduit roughness, m_h the slope of roughness decrease away from the spring and V the conduit volume.

	Geometry 1	Geometry 2
b (m)	0.01	0.5
m (–)	2.07×10^{-4}	1.5×10^{-4}
b_h ($\mathrm{m}^{1/3}\,\mathrm{s}^{-1}$)	0.17	0.15
m_h ($\mathrm{m}^{-2/3}\,\mathrm{s}^{-1}$)	0.4	0.6
V (m^3)	112 564	153 435

radius was then calculated according to Eq. (14) at each intersection.

$$r_{\mathrm{c}2} = \sqrt{r_{\mathrm{c}0}^2 + r_{\mathrm{c}1}^2}, \tag{14}$$

where $r_{\mathrm{c}2}$ is the conduit radius downstream of the intersection and $r_{\mathrm{c}0}$ and $r_{\mathrm{c}1}$ the conduit radii of the two respective conduits before their intersection.

Results are very similar to those of scenario 2 (cf. Fig. 7a and d). Both simulations result in nearly the same set of parameters (Table 1). The estimated conduit volume is even a little smaller for scenario 5 since larger cross sections in the last conduit segment near the spring are reached for a lower total conduit volume. The drastic increase of conduit cross sections at the network intersections leads to higher variability in the cross sections along the conduit segments. The differences between the peak offsets of both tracer tests are higher compared to those of scenario 2. While the peak time of tracer test 2 can be calibrated for large conduit volumes, i.e. conduit volumes above $120\,000\,\mathrm{m}^3$ (Fig. 7d), the peak time of tracer test 1 is too late for large conduit volumes. This is due to the fact that the injection point for tracer test 1 is much closer to the spring than that for tracer test 2. In scenario 5 the conduit volume is spatially differently distributed from that of scenario 2 for the identical total conduit volume. The drastic increase in conduit diameters downgradient of conduit intersections leads to rather high conduit diameters in the vicinity of the spring. Therefore, while tracer transport in tracer test 2 occurs in relatively small conduits with high flow velocities and larger conduits with lower velocities, the tracer in tracer test 1 is only transported through the larger conduits whose flow velocities are restricted by the spring discharge. In Fig. 7d the parameter values for the best fit would lie well below the lower boundary of the diagram at negative values below $-10\,\mathrm{h}$. However, since the fit for conduit volumes around $100\,000\,\mathrm{m}^3$ is similar to that of scenario 2, the two scenarios can in this case not be distinguished based on field observations.

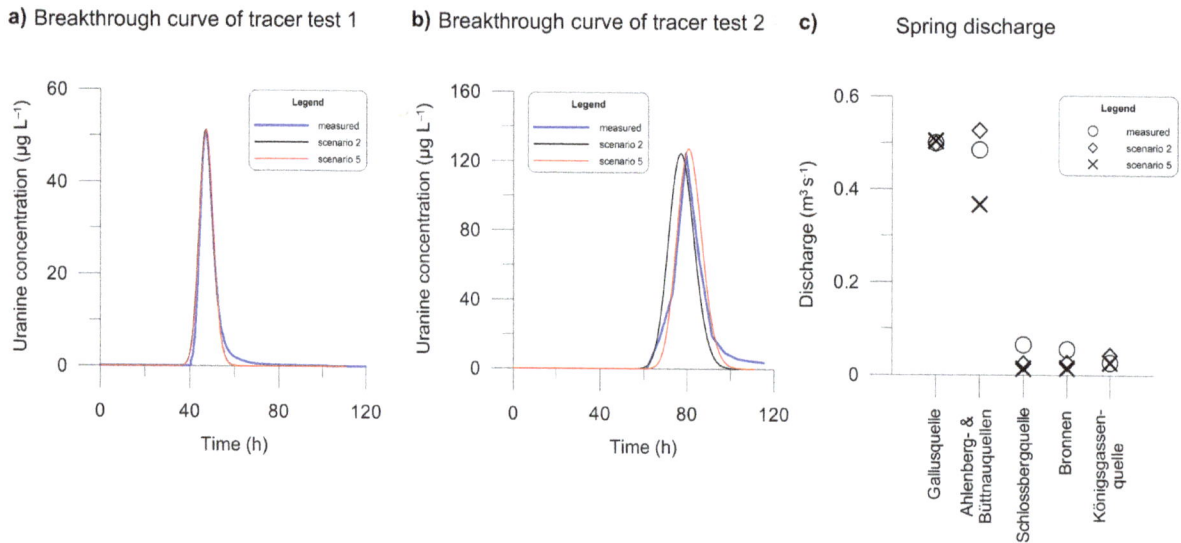

Figure 10. Comparison of the best-fit simulations with field data for scenarios 2 and 5. (**a**) Breakthrough curve of tracer test 1, (**b**) breakthrough curve of tracer test 2 and (**c**) spring discharge.

5   Conclusions of the parameter analysis

Table 3 provides a comparison, i.e. the characteristics for all scenarios. The parameter analysis shows that there is only a limited choice of parameters with which the spring discharges (water balance), the hydraulic head distribution and the tracer velocities can be simulated. Scenario 1 is the only scenario that cannot reproduce the peak travel times observed in both tracer tests simultaneously (Sect. 4.1). It underestimates the complexity of the geometry and internal surface characteristics (e.g. roughness) of the conduit system.

Scenario 4 introduces two additional model parameters. The best fit for this scenario is, however, still achieved with all three K_m values being equal, which basically results in the parameter set of scenario 2. This implies that the major influence leading to the differences in hydraulic gradients observed throughout the area is the conduit system and not the variability of the fissured matrix hydraulic conductivity. It was also shown that for the Madison aquifer (USA), by Saller et al. (2013), a better representation of the hydraulic head distribution can be achieved by including a discrete conduit system even for reduced variability in the hydraulic conductivity of the fissured matrix. Their conclusion complies very well with the findings for scenario 4.

Scenario 3 simulates the presence of a couple of additional smaller dendritic branches. The deduced parameter values and the fit of the objective functions are similar to those of scenarios 2 and 5. Because of long calculation times without additional advantage for the presented study, scenario 3 is not considered for further analysis.

Scenarios 2 and 5 are both judged as suitable. Their parameters and the quality of the fit are similar. Therefore, it is not possible to decide which one is the better representation

of reality. Regarding the different processes interacting during karst evolution (Sect. 4.5) it is most likely that the actual geometry ranges somewhat in between these two scenarios. Table 1 summarizes all parameters of both simulations and Fig. 10 shows the simulated tracer breakthrough curves and spring discharges.

6   Discussion

6.1   Plausibility of the best-fit simulations

The main objective of the model simulation is not only to reproduce the target values but also to provide insight into dominating flow and transport processes, sensitive parameters and to check the plausibility of the model set-up. Possible ambiguities in parameterizations can also be checked, i.e. different combinations of parameters producing identical model output.

For these aims model parameters and aquifer properties simulated with scenarios 2 and 5 are compared to those observed in the field. As seen in Table 1 most of the calibrated parameters range well within values provided in the literature. The calibrated Manning coefficients are relatively high compared to other karst systems. Jeannin (2001) lists effective conductivities for several different karst networks that translate into n values of between 0.03 and $1.07\,\mathrm{s\,m^{-1/3}}$, showing that the natural range of n values easily extends across 2 orders of magnitude and the minimum n values of the simulation lie within the natural range. The maximum n values are significantly higher than those given by Jeannin (2001). This is not surprising since the calibrated n value reflects the total roughness of the conduit bundles and therefore includes geometric conduit properties in addition to the wall roughness that it was originally defined for. This effect is specific for the Gallusquelle area but it might be important to consider for other moderately karstified areas as well where identification of conduit geometries is especially difficult.

The total conduit volume of the Gallusquelle spring derived from scenarios 2 and 5 is only 50 % of that estimated with traditional methods (Geyer et al., 2008). Since the conduit transmissivity increases towards the spring water enters the conduits preferably in the vicinity of the spring in the Gallusquelle area. Therefore, the matrix contribution is high. In addition, the travel time at peak concentration of tracer test 2, which was used for the volume estimation by Geyer et al. (2008), is longer than 3 days, during which time matrix–conduit water exchange can readily take place. Based on the results of a tracer test conducted in a distance of 3 km to the Gallusquelle spring Birk et al. (2005) estimated the error incurred by deducing the conduit volume without taking conduit–matrix exchange fluxes into account with a very simple numerical model. The authors found a difference in conduit volumes of approximately 50 %. This fits well with the results of the present simulation. Birk et al. (2005) also the simulated equivalent conduit cross sectional area between their tracer injection point and the spring to be $13.9 \, m^2$. For scenario 2 the simulated average cross sectional area is $11.9 \, m^2$ and for scenario 5 $13.4 \, m^2$, which compares very well with the results of Birk et al. (2005).

It was not possible to match the shape of both breakthrough curves with the same dispersivity. The apparent dispersion in the tracer test 2 breakthrough is much higher compared to that of tracer test 1, while the breakthrough of tracer test 1 shows a more expressed tailing (Fig. 10a and b). This corresponds to the effect observed by Hauns et al. (2001). The authors found scaling effects in karst conduits: the larger the distance between input and observation point, the more mixing occurred. The tailing is generally induced by matrix diffusion or discrete geometric changes such as pools, where the tracer can be held back and released more slowly. Theoretically, every water drop employs medium and slow flow paths if the distance is large enough, leading to a more or less symmetrical, but broader, distribution and therefore a higher apparent dispersion (Hauns et al., 2001). To quantify this effect, exact knowledge of the geometric conduit shape such as the positions and shapes of pools would be necessary. Furthermore, an additional unknown possibly influencing the observed retardation and dispersion effects is the input mechanism. The simulation assumes that all introduced tracers immediately and completely enter the conduit system, which neglects effects of the unsaturated zone on tracer breakthrough curves. In addition, the shape of the breakthrough curve of tracer test 2 is difficult to deduce since the 6 h sampling interval can be considered as rather low leading to a breakthrough peak which is described by only seven measurement points. Therefore, the apparent dispersivity was calibrated for both breakthrough curves separately. Calibrated dispersivity ranges well within those quoted in literature (Table 1). The mass recovery during the simulation was determined to range between 98.4 and 99.9 % in all simulations. The slight mass difference results from a combination of diffusion of the tracer into the fissured matrix and numerical inaccuracies.

The spring discharge of the minor springs in the area (Sect. 3) was slightly underestimated in most cases (Fig. 10c). For most springs the models of scenarios 2 and 5 provide similar results. The underestimation of discharge is in the order of $< 0.05 \, m^3 \, s^{-1}$ and is not expected to significantly influence the general flow conditions. It probably results from the unknown conduit geometry in the catchments of the different minor springs. The only case in which the two scenarios give significantly different results is the spring discharge of the spring group consisting of the Ahlenberg and Büttnauquellen springs (Fig. 10c). Scenario 2 overestimates and scenario 5 underestimates the discharge. This is due to the fact that the longest conduit of the Ahlenberg and Büttnauquellen springs is longer than the longest one of the Gallusquelle spring but the conduit network has less intersections (Fig. 1). Therefore, the conduit volume of the Ahlenberg and Büttnauquellen springs is $134\,568 \, m^3$ in scenario 2 and only $75\,085 \, m^3$ in scenario 5 leading to the different discharge values. It is reasonable to assume that a better fit for the spring group can be achieved, if more variations of conduit intersections are tested. An adequate fit for the Fehla-Ursprung spring of $0.1 \, m^3 \, s^{-1}$ was achieved for both scenarios with a fault aperture of 0.005 m.

6.2 Uncertainties and limitations

The most important uncertainties regarding the reliability of the simulation include the assumptions that were made prior to modelling. First, flow dynamics were neglected. This approach was chosen because tracer tests are supposed to be conducted during quasi-steady-state flow conditions. However, this is only the ideal case. During both tracer tests spring discharge declined slightly. The influence of transient flow on transport velocities inside the conduits was estimated by a very simple transient flow simulation for the best-fit models in which recharge and storage coefficients were calibrated to reproduce the observed decline in spring discharges. The transient flow only slightly affected peak velocities but lead to a larger spreading of the breakthrough curves and therefore lower calibrated dispersion coefficients. This effect occurred because the decline in flow velocities is not completely uniform inside the conduits and depending on where the tracer is at which time it experiences different flow velocities in the different parts of the conduits, which leads to a broader distribution at the spring. The same breakthrough curves can be simulated under steady-state flow conditions with slightly higher dispersivity coefficients. So, the calibrated dispersivities do not only represent geometrical heterogeneities but also temporal effects as is the case for all standard evaluations of dispersion from tracer breakthrough curves.

Flow velocities inside the karst conduits
with and without a direct recharge component

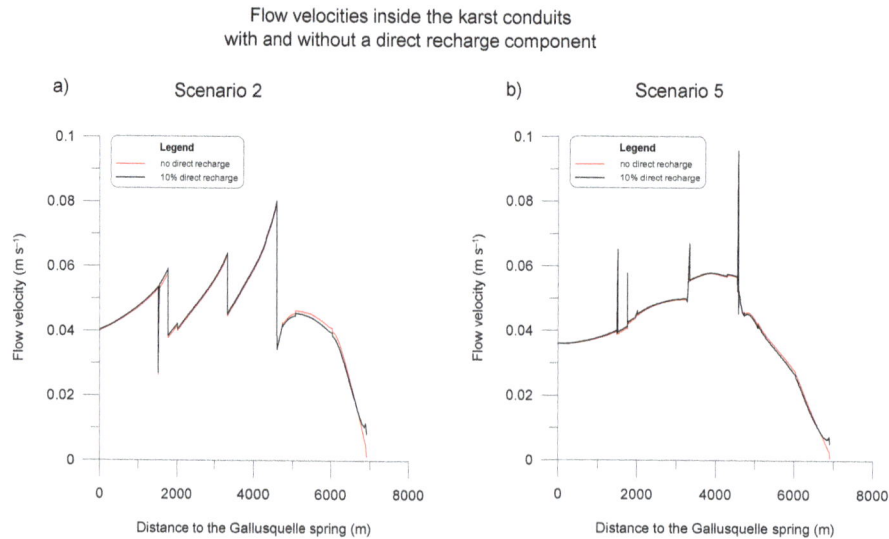

Figure 11. Flow velocities inside the main conduit branch of the Gallusquelle spring during the simulation of tracer test 2. The best-fit simulations for scenarios 2 and 5 are compared to simulations where a direct recharge of 10 % is introduced.

The influence of rapid recharge is not to considered in the simulation of baseflow conditions. However, there might be an influence on flow velocities during the actual recharge events, i.e. if rapid recharge is intensive and strong enough to lead to a reversal of the flow gradients between conduit and fissured matrix. Therefore, an alternative simulation was performed for tracer test 2, which was conducted during high flow conditions (Table 2) after a recharge event. The maximum percentage of direct recharge of 10 % estimated by Sauter (1992) and Geyer et al. (2008) was used for this simulation. Neither for scenario 2 nor for scenario 5 a gradient reversal between conduit and matrix occurred and the influence on flow velocities was negligible (Fig. 11).

Furthermore, flow in all karst conduits was simulated for turbulent conditions. Turbulent conditions can be generally assumed in karst conduits (Reimann et al., 2011) and also apply to all calibrated model conduit cross sections. Since the conduit cross section presents the total cross section of the conduit bundle, the cross sections of the individual tubes are uncertain, though. The high n values suggest that the surface / volume ratio is relatively high, which implies that the individual conduit cross sections are rather small. Therefore, laminar flow in some conduits is likely. While laminar flow conditions in the conduits influence hydraulic gradients considerably, this fact is believed not to influence the overall results and conclusions of this study, i.e. the relative significance of the parameters deduced from parameter analysis and the deduced conduit volume, especially since flow is simulated for steady-state conditions.

For all distributed numerical karst simulations, uncertainties regarding the exact positions and interconnectivities of the conduit branches still remain. Due to the extensive investigations already performed in previous work (Sect. 3) these

uncertainties are reduced in the Gallusquelle area and the above scenarios include the most probable ones. However, the flexibility of the modelling approach allows for the integration of any future information that might enhance the numerical model further.

6.3 Calibration strategy

For a successful calibration of a distributed groundwater flow and transport model for a karst area on catchment scale certain constraints have to be set a priori. The geometry of the model area, i.e. locations/types of boundary conditions and aquifer base, fixed during calibration, has to be known with sufficient certainty. Furthermore, the objective functions for calibration have to be defined, i.e. the hydraulic response of the system and transport velocities. In a karst groundwater model, these consist of measurable variables, i.e. spring discharges, hydraulic heads in the fissured matrix and two tracer breakthrough curves. The hydraulic head measurements should be distributed across the entire catchment and preferably close to the conduit system, should geometric conduit parameters be calibrated for as well. It is expected that the influence of the conduits on the hydraulic head decreases and the influence of matrix hydraulic conductivities increases with distance to the conduit system. In the design of the tracer experiment, the following criteria should be observed: for a representative calibration, the dye should be injected at as large a distance to each other as possible with one of them including the length of the whole conduit system. Each tracer test gives integrated information about its complete flow path. If the injection points lie close together, no information about the development of conduit geometries from water divide to spring can be obtained. Further, the dye should be injected as directly as possible into the conduit system, e.g. via a flushed sinkhole, to obtain information on the conduit flow regime and to minimize matrix interference. To ease interpretation a constant spring discharge during the tests is desirable.

In this study, the flow field was simulated not only for the catchment area of the Gallusquelle spring, but also for a larger area including the catchment areas of several smaller springs (Fig. 1). This is in general not essential for deducing conduit volumes and setting up a flow and transport model. Simulating several catchments, however, helps to increase the reliability of the simulation. The positions of water divides are majorly determined by the hydraulic conductivity of the fissured matrix K_m, so that the simulated catchment areas of the different springs can be used to estimate how realistic the simulated flow field is and decrease the range of likely K_m values. In this study, high K_m values above ca. $3 \times 10^{-5} \, \text{m s}^{-1}$ made the simulation of the spring discharge of the Fehla-Ursprung spring (Fig. 1) impossible because the water divide in the west could not be simulated and most of the water in the area discharged to the east towards the river Lauchert resulting in a very narrow and long catchment area for the Gallusquelle spring.

There are eight parameters available for model calibration in this study. Two of these parameters define the conduit geometry: b is the lowest conduit radius and m the slope with which the conduit radius increases. One parameter, d_f, defines the aperture of the fault zone. The hydraulic conductivity of the fissured matrix is represented by the parameter K_m and the roughness of the conduit system by two parameters: b_h represents the highest roughness and m_h the slope of roughness decrease in upgradient direction from the spring. The last two parameters ε_1 and ε_2 are the respective conduit dispersivities obtained from the two artificial tracer experiments (Table 1).

For efficiency reasons it is important to know which of these parameters can be calibrated independently. The apparent transport dispersivities ε_1 and ε_2 are pure transport parameters, which influence only the shape of the breakthrough curves and not the flow field. The hydraulic model parameters influence the shape of the tracer breakthrough curves as well. Therefore, dispersivities ε_1 and ε_2 should be calibrated separately after calibrating the hydraulic model parameters.

Only for hydraulically dominant fault zones knowledge of the fault zone aperture d_f is required. For the model area this parameter was required for one fault zone lying in the west of the area feeding the Fehla-Ursprung spring (Fig. 1). Since the Fehla-Ursprung spring has its own catchment area the fault zone has only minor influence on the flow regime in the Gallusquelle catchment. Its hydraulic parameters were calibrated at the beginning of the simulation procedure to reproduce the catchment and the discharge of the Fehla-Ursprung spring adequately and kept constant throughout all the simulations. In the final calibrated models it was rechecked, but the calibrated value was still acceptable.

The hydraulic conductivity of the fissured matrix K_m can be calibrated independently in principle as well. The influence on spring discharge is relatively small. The best-fit K_m value depends on the conduit parameters, i.e. geometry and roughness, since the hydraulic conductivities of the conduit system and of the fissured matrix define the total transmissivity of the catchment area together. Nonetheless, the best-fit value lies in the same range for different conduit geometries (Figs. 4a and 7c). The greater the difference between the simulated conduit geometries, the more likely is a slight shift of the best-fit K_m value. Therefore, it is advisable to calibrate it anew for significant model changes, e.g. different scenarios, but to keep it constant during the rest of the calibrations. For the best-fit configuration, potentially used as a prognostic tool, the K_m value needs to be checked and adapted if necessary. This observation is, however, only valid for steady-state flow conditions. The dynamics of the hydraulic head and spring discharge might be highly sensitive to the matrix hydraulic conductivity, the conduit–matrix exchange coefficient and the lateral conduit extent. This work focuses on the conduits as highly conductive pathways for e.g. contaminant transport, but the calibration of matrix velocities, e.g. by use of environmental tracers, would likely be sensitive to the K_m values as well. Therefore, the choice of the flow regime and the objective functions determines the strength of the interdependencies between fissured matrix and conduit system parameters and therefore whether K_m can be calibrated independently.

The conduit parameters for geometry and roughness, here four parameters (lowest conduit radius b, slope of radius increase m, highest roughness b_h and slope of roughness decrease m_h), have to be varied simultaneously. All of them have a major influence on spring discharge and cannot be varied separately without introducing discharge errors. For each conduit geometry, there are a number of possible b_h–m_h combinations that result in the observed spring discharge. In general, the slowest transport velocities are achieved with a m_h value of zero. So, to deduce the range of geometric parameters that reproduce the objective functions, it is advisable to check the minimum conduit volume for which the tracer tests are not too fast for a value of m_h equal to zero. For the Gallusquelle area, transmissivities significantly increase towards the springs, which is characteristic for most karst catchments. Therefore, low b_h values oppose the general hydraulic head trend: they increase the conduit roughness at the spring leading to slower flow and higher gradients. The higher the conduit volume, the higher b_h is required to reproduce the observed transport velocities. Therefore, the best-fit model likely has the smallest conduit volume for which both tracer tests can be reproduced. In Fig. 7 this condition can be seen to clearly range in the order of $100\,000 \, \text{m}^3$ for the Gallusquelle area. While the four conduit parameters allow for a good model fit, they are pure calibration parameters. They show that the karst conduit system has a high complexity, which cannot be neglected for distributed velocity and hydraulic head representation. A systematic simulation of the heterogeneities, e.g. with a karst genesis approach, would be a process-based improvement to the current method and give more physical meaning to the parameters.

7 Conclusions

The study presents a large-scale catchment-based distributed hybrid karst groundwater flow model capable of simulating groundwater flow and solute transport. For flow recession conditions this model can be used as a predictive tool for the Gallusquelle area with relative confidence. The approach of simultaneous pattern matching of flow and transport parameters provides new insight into the hydraulics of the Gallusquelle conduit system. The model ambiguity was significantly reduced to the point where an estimation of the actual karst conduit volume for the Gallusquelle spring could be made. This would not have been possible simulating only one or two of the three objective functions, i.e. the spring discharge, the hydraulic head distribution and two tracer tests.

The model allows for the identification of the relevant parameters affecting karst groundwater discharge and transport in karst conduits and the examination of the respective overall importance in a well-investigated karst groundwater basin for steady-state flow conditions. While a differentiated representation of the roughness values in the karst conduits is substantial for buffering the lack of knowledge of the exact conduit geometry, e.g. local variations in cross section and the number of interacting conduits, variable matrix hydraulic conductivities cannot improve the simulation. It was shown that the effect of the unknown exact lateral extent of the conduit system and the change in conduit cross section at conduit intersections is of minor importance for the overall karst groundwater discharge. This is important since these parameters are usually unknown and difficult to measure in the field.

For calibration purposes, this study demonstrates that for a steady-state flow field and the observed objective functions the hydraulic conductivities of the fissured matrix can practically be calibrated independently of the conduit parameters. Furthermore, a strategy for the simultaneous calibration of conduit volumes and conduit roughness in a complex karst catchment was developed.

As discussed in Sect. 5 the major limitation of the simulation is the neglect of flow dynamics, which limits the applicability to certain flow conditions. Therefore, transient flow simulation is the focus of on-going work. This will enhance the applicability of the model as a prognostic tool to all essential field conditions and lead to further conclusions regarding the important karst system parameters, their influences on karst hydraulics and their interdependencies. It can be expected that some parameters, which are of minor importance in a steady-state flow field, e.g. the lateral conduit extent and the percentage of recharge entering the conduits directly, will exhibit significant influence for transient flow conditions.

Acknowledgements. The presented study was funded by the German Federal Ministry of Education and Research (promotional reference no. 02WRS1277A, AGRO, Risikomanagement von Spurenstoffen und Krankheitserregern in ländlichen Karsteinzugsgebieten).

This open-access publication is funded by the University of Göttingen.

Edited by: M. Giudici

References

Barenblatt, G. I., Zheltov, I. P., and Kochina, I. N.: Basic concepts in the theory of seepage in fissured rocks (strata), J. Appl. Math. Mech.-USS, 24, 1286–1303, 1960.

Bauer, S., Liedl, R., and Sauter, M.: Modeling of karst aquifer genesis: Influence of exchange flow, Water Resour. Res., 39, 1285, doi:10.1029/2003WR002218, 2003.

Birk, S., Geyer, T., Liedl, R., and Sauter, M.: Process-Based Interpretation of Tracer Tests in Carbonate Aquifers, Ground Water, 43, 381–388, 2005.

Birk, S., Liedl, R., and Sauter, M.: Karst Spring responses examined by process-based modeling, Ground Water, 44, 832–836, 2006.

Clemens, T.: Simulation der Entwicklung von Karstaquiferen, PhD thesis, Eberhard-Karls-Universität zu Tübingen, Tübingen, 1998.

Clemens, T., Hückinghaus, D., Sauter, M., Liedl, R., and Teutsch, G.: A combined continuum and discrete network reactive transport model for the simulation of karst development, in: Proceedings of the ModelCARE 96 Conference, 24–26 September 1996, Golden, Colorado, USA, 309–318, 1996.

COMSOL AB: COMSOL Multiphysics® User's Guide v4.3, 1292 pp., 2012.

Coronado, M., Ramírez-Sabag, J., and Valdiviezo-Mijangos, O.: On the boundary conditions in tracer transport models for fractured porous underground formations, Rev. Mex. Fís., 53, 260–269, 2007.

Covington, M. D., Luhmann, A. J., Wicks, C. M., and Saar, M. O.: Process length scales and logitudinal damping in karst conduits, J. Geophys. Res., 117, F01025, doi:10.1029/2011JF002212, 2012.

Doummar, J., Sauter, M., and Geyer, T.: Simulation of flow processes in a large scale karst system with an integrated catchment model (Mike She) – Identification of relevant parameters influencing spring discharge, J. Hydrol., 426, 112–123, doi:10.1016/j.jhydrol.2012.01.021, 2012.

Dreybrodt, W.: Mixing in $CaCO_3$–CO_2–H_2O systems and its role in the karstification of limestone areas, Chem. Geol., 32, 221–236, 1981.

Ford, D. C. and Williams, P. W.: Karst geomorphology and hydrology, Wiley, West Sussex, 562 pp., 2007.

Geyer, T., Birk, S., Licha, T., Liedl, R., and Sauter, M.: Multi-tracer test approach to characterize reactive transport in karst aquifers, Ground Water, 45, 36–45, 2007.

Geyer, T., Birk, S., Liedl, R., and Sauter, M.: Quantification of temporal distribution of recharge in karst systems from spring hydrographs, J. Hydrol., 348, 452–463, 2008.

Golwer, A.: Erläuterungen zu Blatt 7821 Veringenstadt, Geologische Karte 1 : 25 000 von Baden-Württemberg, Geologisches Landesamt Baden-Württemberg, Stuttgart, 151 pp., 1978.

Gwinner, M. P.: Erläuterungen zu Blatt 7721 Gammertingen, Geologische Karte 1 : 25 000 von Baden-Württemberg, Geologisches Landesamt Baden-Württemberg, Freiburg, Stuttgart, 78 pp., 1993.

Hartmann, A., Weiler, M., Wagener, T., Lange, J., Kralik, M., Humer, F., Mizyed, N., Rimmer, A., Barberá, J. A., Andreo, B., Butscher, C., and Huggenberger, P.: Process-based karst modelling to relate hydrodynamic and hydrochemical characteristics to system properties, Hydrol. Earth Syst. Sci., 17, 3305–3321, doi:10.5194/hess-17-3305-2013, 2013.

Hartmann, A., Goldscheider, N., Wagener, T., Lange, J., and Weiler, M.: Karst water resources in a changing world: Review of hydrological modeling approaches, Rev. Geophys., 52, 1–25, doi:10.1002/2013RG000443, 2014.

Hauns, M., Jeannin, P.-Y., and Atteia, O.: Dispersion, retardation and scale effect in tracer breakthrough curves in karst conduits, J. Hydrol., 241, 177–193, 2001.

Hu, R.: Hydraulic tomography: A new approach coupling hydraulic travel time, attenuation and steady shape inversions for high-spatial resolution aquifer characterization, PhD thesis, University of Göttingen, Göttingen, 116 pp., 2011.

Hubinger, B. and Birk, S.: Influence of initial heterogeneities and recharge limitations on the evolution of aperture distributions in carbonate aquifers, Hydrol. Earth Syst. Sci., 15, 3715–3729, doi:10.5194/hess-15-3715-2011, 2011.

Hückinghaus, D.: Simulation der Aquifergenese und des Wärmetransports in Karstaquiferen, C42, Tübinger Geowissenschaftliche Arbeiten, Tübingen, 1998.

Hunter, N. M., Bates, P. D., Horritt, M. S., De Roo, A. P. J., and Werner, M. G. F.: Utility of different data types for calibrating flood inundation models within a GLUE framework, Hydrol. Earth Syst. Sci., 9, 412–430, doi:10.5194/hess-9-412-2005, 2005.

Jeannin, P.-Y.: Modeling flow in phreatic and epiphreatic karst conduits in the Hölloch cave (Muotatal, Switzerland), Water Resour. Res., 37, 191–200, 2001.

Jeannin, P.-Y. and Sauter, M.: Modelling in karst systems, Bulletin d'Hydrogéologie, 16, Université de Neuchâtel, Neuchâtel, 1998.

Khu, S.-T., Madsen, H., and di Pierro, F.: Incorporating multiple observations for distributed hydrologic model calibration: An approach using a multi-objective evolutionary algorithm and clustering, Adv. Water Resour., 31, 1387–1398, 2008.

Kordilla, J., Sauter, M., Reimann, T., and Geyer, T.: Simulation of saturated and unsaturated flow in karst systems at catchment scale using a double continuum approach, Hydrol. Earth Syst. Sci., 16, 3909–3923, doi:10.5194/hess-16-3909-2012, 2012.

Kovács, A. and Sauter, M.: Modelling karst hydrodynamics, in: Methods in karst hydrogeology, edited by: Goldscheider, N. and Drew, D., Taylor and Fracis, London, 201–222, 2007.

Kovács, A., Perrochet, P., Király, L., and Jeannin, P.-Y.: A quantitative method for the characterisation of karst aquifers based on spring hydrograph analysis, J. Hydrol., 303, 152–164, 2005.

Li, H. T., Brunner, P., Kinzelbach, W., Li, W. P., and Dong, X. G.: Calibration of a groundwater model using pattern information from remote sensing data, J. Hydrol., 377, 120–130, doi:10.1016/j.jhydrol.2009.08.012, 2009.

Liedl, R., Sauter, M., Hückinghaus, D., Clemens, T., and Teutsch, G.: Simulation of the development of karst aquifers using a coupled continuum pipe flow model, Water Resour. Res., 39, 1057, doi:10.1029/2001WR001206, 2003.

Luhmann, A. J., Covington, M. D., Alexander, S. C., Chai, S. Y., Schwartz, B. F., Groten, J. T., and Alexander, E. C.: Comparing conservative and nonconservative tracers in karst and using them to estimate flow path geometry, J. Hydrol., 448–449, 201–211, doi:10.1016/j.jhydrol.2012.04.044, 2012.

Madsen, H.: Parameter estimation in distributed hydrological catchment modelling using automatic calibration with multiple objectives, Adv. Water Resour., 26, 205–216, 2003.

Merkel, P.: Karsthydrologische Untersuchungen im Lauchertgebiet (westl. Schwäbische Alb), Diplom thesis, University of Tübingen, Tübingen, 108 pp., 1991.

Mohrlok, U.: Numerische Modellierung der Grundwasserströmung im Einzugsgebiet der Gallusquelle unter Festlegung eines Drainagesystems, Grundwasser, 19, 73–85, doi:10.1007/s00767-013-0249-x, 2014.

Mohrlok, U. and Sauter, M.: Modelling groundwater flow in a karst terraine using discrete and double-continuum approaches: importance of spatial and temporal distribution of recharge, in: Proceedings of the 12th International Congress of Speology, 2/6th Conference on Limestone Hydrology and Fissured Media, 10–17 August 1997, La Chaux-de-Fonds, Switzerland, 167–170, 1997.

Oehlmann, S., Geyer, T., Licha, T., and Birk, S.: Influence of aquifer heterogeneity on karst hydraulics and catchment delineation employing distributive modeling approaches, Hydrol. Earth Syst. Sci., 17, 4729–4742, doi:10.5194/hess-17-4729-2013, 2013.

Ophori, D. U.: Constraining permeabilities in a large-scale groundwater system through model calibration, J. Hydrol., 224, 1–20, 1999.

Perrin, C., Andréassian, V., Serna, C. R., Mathevet, T., and Le Moine, N.: Discrete parameterization of hydrological models: Evaluating the use of parameter sets libraries over 900 catchments, Water Resour. Res., 44, W08447, doi:10.1029/2007WR006579, 2008.

Rehrl, C. and Birk, S.: Hydrogeological characterisation and modelling of spring catchments in a changing environment, Aust. J. Earth Sci., 103/2, 106–117, 2010.

Reiber, H., Klein, F., Selg, M., and Heidland, S.: Hydrogeologische Erkundung Baden-Württemberg – Mittlere Alb 4 – Markierungsversuche, Abwassereinleitungen, Landesamt für Umwelt, Messungen und Naturschutz Baden-Württemberg, Tübingen, 71 pp., 2010.

Reimann, T., Rehrl, C., Shoemaker, W. B., Geyer, T., and Birk, S.: The significance of turbulent flow representation in single-continuum models, Water Resour. Res., 47, W09503, doi:10.1029/2010WR010133, 2011.

Saller, S. P., Ronayne, M. J., and Long, A. J.: Comparison of a karst groundwater model with and without discrete conduit flow, Hydrogeol. J., 21, 1555–1566, doi:10.1007/s10040-013-1036-6, 2013.

Sauter, M.: Quantification and Forecasting of Regional Groundwa-
ter Flow and Transport in a Karst Aquifer (Gallusquelle, Malm,
SW Germany), C13, Tübinger Geowissenschaftliche Arbeiten,
Tübingen, 1992.

Schmidt, S., Geyer, T., Marei, A., Guttman, J., and Sauter, M.:
Quantification of long-term wastewater impacts on karst groundwater resources in a semi-arid environment by chloride mass balance methods, J. Hydrol., 502, 177–190, 2013.

Strayle, G.: Karsthydrologische Untersuchungen auf der Ebinger
Alb (Schwäbischer Jura), in: Jahreshefte des Geologischen Lan-
desamtes Baden-Württemberg, Freiburg im Breisgau, 109–206,
1970.

Teutsch, G. and Sauter, M.: Groundwater Modeling in karst terranes: scale effects, data aquisition and field validation, in: Proceedings of the 3rd Conference on Hydrogeology, Ecology, Monitoring and Management of Ground Water in Karst Terranes, 4–
6 December 1991, Nashville, USA, 17–34, 1991.

Villinger, E.: Über Potentialverteilung und Strömungssysteme
im Karstwasser der Schwäbischen Alb (Oberer Jura, SW-
Deutschland), Geologisches Jahrbuch, C18, Bundesanstalt für
Geowissenschaften und Rohstoffe und Geologische Landesämter
der Bundesrepublik Deutschland, Hannover, 1977.

Villinger, E.: Hydrogeologie, in: Erläuterungen zu Blatt 7721
Gammertingen, Geologische Karte 1 : 25 000 von Baden-
Württemberg, edited by: Gwinner, M. P., Geologisches Lan-
desamt Baden-Württemberg, Freiburg, Stuttgart, 30–57, 1993.

Worthington, S. R. H.: Diagnostic hydrogeologic characteristics of
a karst aquifer (Kentucky, USA), Hydrogeol. J., 17, 1665–1678,
doi:10.1007/s10040-009-0489-0, 2009.

10

Using groundwater age and hydrochemistry to understand sources and dynamics of nutrient contamination through the catchment into Lake Rotorua, New Zealand

U. Morgenstern[1], C. J. Daughney[1], G. Leonard[1], D. Gordon[2], F. M. Donath[3,4], and R. Reeves[4]

[1]GNS Science, P.O. Box 30368, Lower Hutt, New Zealand
[2]Hawke's Bay Regional Council, Private Bag 6006, Napier, New Zealand
[3]Department of Applied Geology, Georg-August-Universität Göttingen, Goldschmidtstr. 3, 37077 Göttingen, Germany
[4]GNS Science, Private Bag 2000, Taupo, New Zealand

Correspondence to: U. Morgenstern (u.morgenstern@gns.cri.nz)

Abstract. The water quality of Lake Rotorua has steadily declined over the past 50 years despite mitigation efforts over recent decades. Delayed response of the groundwater discharges to historic land-use intensification 50 years ago was the reason suggested by early tritium measurements, which indicated large transit times through the groundwater system. We use the isotopic and chemistry signature of the groundwater for detailed understanding of the origin, fate, flow pathways, lag times and future loads of contaminants. A unique set of high-quality tritium data over more than four decades, encompassing the time when the tritium spike from nuclear weapons testing moved through the groundwater system, allows us to determine detailed age distribution parameters of the water discharging into Lake Rotorua.

The Rotorua volcanic groundwater system is complicated due to the highly complex geology that has evolved through volcanic activity. Vertical and steeply inclined geological contacts preclude a simple flow model. The extent of the Lake Rotorua groundwater catchment is difficult to establish due to the deep water table in large areas, combined with inhomogeneous groundwater flow patterns.

Hierarchical cluster analysis of the water chemistry parameters provided evidence of the recharge source of the large springs near the lake shore, with discharge from the Mamaku ignimbrite through lake sediment layers. Groundwater chemistry and age data show clearly the source of nutrients that cause lake eutrophication, nitrate from agricultural activities and phosphate from geologic sources. With a naturally high phosphate load reaching the lake continuously via all streams, the only effective way to limit algae blooms and improve lake water quality in such environments is by limiting the nitrate load.

The groundwater in the Rotorua catchment, once it has passed through the soil zone, shows no further decrease in dissolved oxygen, indicating an absence of bioavailable electron donors along flow paths that could facilitate microbial denitrification reactions. Nitrate from land-use activities that leaches out of the root zone of agricultural land into the deeper part of the groundwater system must be expected to travel with the groundwater to the lake.

The old age and the highly mixed nature of the water discharges imply a very slow and lagged response of the streams and the lake to anthropogenic contaminants in the catchment, such as nitrate. Using the age distribution as deduced from tritium time series data measured in the stream discharges into the lake allows prediction of future nutrient loads from historic land-use activities 50 years ago. For Hamurana Stream, the largest stream to Lake Rotorua, it takes more than a hundred years for the groundwater-dominated stream discharge to adjust to changes in land-use activities. About half of the currently discharging water is still pristine old water, and after this old water is completely displaced by water affected by land use, the nitrogen load of Hamurana Stream will approximately double. These timescales apply to activities that cause contamination, but also to remediation action.

1 Introduction

Detailed information on groundwater age distribution is required for the Lake Rotorua catchment to understand the agricultural contaminant loads that travel from land to the lake with the groundwater and discharge via springs and streams into the lake, with a large lag time. The water quality of Lake Rotorua has declined continuously over the past 50 years, despite cessation of direct-to-lake sewage discharge in 1991 (Burger et al., 2011) and the fencing-off of streams in grazing land in parts of the lake catchment.

Land use in the catchment has intensified significantly over the past 60 years and is now predominantly forest (39 %), pasture (27 %) and dairy (9 %) (Burger et al., 2011; Rutherford et al., 2009). Increasing nitrate concentrations had been observed in virtually all of the major streams flowing into the lake during the period 1968–2003 (Hoare, 1987; Rutherford, 2003). We measured nitrate concentrations of 6–10 mg L^{-1} NO$_3$–N in three young groundwater samples under dairy farms in the SE catchment. In the absence of significant overland runoff, nutrients from land use are transported with the water through the groundwater system to the lake. Early tritium measurements indicated large transit times through the groundwater system (the subject of this study). With a time lag > 50 years in the groundwater system, nitrate loads to the lake may be expected to increase further in the future due to delayed arrival of nutrients from historic land use as they ultimately discharge from the groundwater system via the springs and streams into the lake. This trend will be exacerbated by any further intensification of land use within the catchment over recent decades, as this recently recharged water has largely not yet reached the streams (Morgenstern and Gordon, 2006).

Groundwater age is a crucial parameter for understanding the dynamics of the groundwater and the contaminants that travel with the water. Determining water age, and hence transit times, allows identification of delayed impacts of past and present land-use practices on water quality (Böhlke and Denver, 1995; Katz et al., 2001, 2004; McGuire et al., 2002; MacDonald et al., 2003; Broers, 2004; Moore et al., 2006), and for identification of anthropogenic versus geologic impacts on groundwater quality (Morgenstern and Daughney, 2012). Understanding the dynamics of groundwater is fundamental for most groundwater issues. Water age is defined by the transit time of water through catchments and hence is vital for conceptual understanding of catchment processes such as response to rainfall, stream flow generation, recharge source and rate (McGuire and McDonnell, 2006; Morgenstern et al., 2010, 2012; Stewart et al., 2010; Cartwright and Morgenstern, 2012). Water age, being directly related to fluid flux, is also very useful for calibrating numerical surface water and groundwater transport models (Goode, 1996; Burton et al., 2002; Molson and Frind, 2005; Bethke and Johnson, 2008). Water age provides important information on vulnerability to contamination and can therefore be used to assess the se-

curity of drinking water supplies, particularly from groundwater bores (Darling et al., 2005; Morris et al., 2005; New Zealand Ministry of Health, 2008). Water age measurements can also be used to quantify rates of hydrochemical evolution resulting from water–rock interaction (Katz et al., 1995; Burns et al., 2003; Glynn and Plummer, 2005; Daughney et al., 2010; Beyer et al., 2014). These applications of water dating cover the spectrum from applied water resource management to fundamental scientific research.

In all of the above-mentioned applications it is important to constrain not only the mean age of water, but also the distribution of ages within a sample from the groundwater discharge. Transit time determinations in catchment hydrology typically identify a range of water ages contributing to stream flow, and the time- and location-dependent distribution of transit times provides insight into the processes that generate runoff (Maloszewski and Zuber, 1982; McGuire and McDonnell, 2006; Stewart et al., 2007; McDonnell et al., 2010). Use of water age determinations for calibration of numerical transport models must also account for the full distribution of age and its variation in space and time (Goode, 1996; Cornaton et al., 2011; Cornaton, 2012). Assessment of the security of drinking water from groundwater bores also requires an understanding of the water's age distribution (Eberts et al., 2012; Morgenstern, 2004). For example, New Zealand legislation states that a water supply bore is considered secure (unlikely to have a risk of contamination by pathogenic organisms) when less than 0.005 % of the water has been present in the aquifer for less than 1 year (New Zealand Ministry of Health, 2008).

For the Lake Rotorua catchment study, tritium is the tracer of choice. Tritium dating can be applied to both river/stream water and groundwater, whereas gas tracers are less suitable for surface waters that are in contact with air. Tritium ages, in contrast to gas tracer ages, include travel through the unsaturated zone (Zoellmann et al., 2001; Cook and Solomon, 1995); travel times can be > 40 years through the thick unsaturated zones of the Rotorua catchment ignimbrite aquifers (Morgenstern et al., 2004). Tritium is not subject to transformation, degradation or retardation during water transport through the catchment. Tritium dating is applicable to water with mean residence times of up to about 200 years (Cook and Solomon, 1997; Morgenstern and Daughney, 2012), as is typical of New Zealand's dynamic surface waters and shallow groundwaters. In addition, monitoring the movement of the pulse-shaped bomb-tritium through groundwater systems is an excellent opportunity to obtain information about the age distribution parameters of the groundwater. This is particularly useful in groundwater systems, such as the Rotorua system, that have high uncertainties within flow models due to a deep water table and preferential flow paths. Finally, tritium is a particularly sensitive marker for study of the timing of nitrate contamination in groundwater, because the main anthropogenic nitrate contamination of groundwater systems started coincidentally with the bomb-tritium peak from the atmospheric nuclear weapons testing after WWII; water recharged before this post-war upsurge in intensive agriculture has low tritium and low nitrate concentrations.

For the Rotorua catchment we have an extensive data set available over time and space. Tritium time series data for the main lake inflows cover more than 4 decades, and data covering the last decade are available with an extremely high spatial resolution of about 100 sites in the Lake Rotorua catchment. Tritium concentration can be measured at GNS Science with the required extremely high accuracy using 95-fold electrolytic enrichment prior to ultralow-level liquid scintillation spectrometry (Morgenstern and Taylor, 2009). Tritium is highly applicable for groundwater dating in the post-bomb low-tritium environment of the Southern Hemisphere, as bomb tritium from atmospheric thermonuclear weapons testing has now been washed out from the atmosphere for 20 years, as is described in detail in Morgenstern and Daughney (2012).

The objective of this study is to understand the origin, fate, flow pathways, lag times, and future loads of contaminants that cause lake eutrophication in the Lake Rotorua catchment, central North Island, New Zealand. This will assist in mitigating the deterioration in lake water quality since the 1960s (Rutherford et al., 1989) that threatens the lake's significant cultural and tourist value. Environmental hydrochemistry tracers and age tracers are used to identify the recharge source of the main water discharges to Lake Rotorua, to identify the source of the contaminants (anthropogenic versus geologic), and to evaluate the water age distributions in order to understand groundwater processes, lag times, and the groundwater flow dynamics. The Rotorua groundwater system is complicated due to the catchment's highly complex geology, which has evolved through volcanic activity, and due to the deep water table of > 50 m in large areas, which prevents detailed groundwater studies and introduces uncertainty in catchment boundaries and flow patterns. The complex geology leads to inhomogeneous groundwater flow patterns, as indicated by large parts of the catchment having particularly large positive or negative specific water yields (White et al., 2004). The groundwater discharge in the northern catchment is unusually large for the size of the surface water catchment, probably due to preferential flow paths that route groundwater towards the north across the surface slope in this part of the catchment.

The tritium and age distribution data are currently being used to calibrate a numeric groundwater transport model. The use of this rich data set for groundwater transport model calibration is part of a larger investigation that evaluates the calibration of hydrological and hydrogeological models using hydrochemical data, including tracers of water age, with the aim of using tracer-calibrated groundwater models for nutrient transport and economic modelling (e.g. Lock and Kerr, 2008; Rutherford et al., 2009), ultimately supporting optimal and sustainable land and water management in catchments. The broader findings from the Rotorua investigation will be applied to the many other New Zealand catchments for which time series age tracer data are available (Stewart and Morgenstern, 2001; Morgenstern, 2004; Stewart and Thomas, 2008; Gusyev et al., 2014).

2 Hydrogeological setting

2.1 Geology

Lake Rotorua is located in a roughly circular caldera basin in the central North Island, New Zealand (Fig. 1), situated in the Taupo Volcanic Zone (TVZ), an area of silicic volcanism with NW–SE extension and geothermal activity roughly 60 km wide by 300 km long that is related to subduction of the Pacific plate beneath the Australian plate off New Zealand's east coast (Wilson et al., 1995; Spinks et al., 2005). Figure 1 shows the surficial geology. Mesozoic greywacke, which outcrops to the east and west of the TVZ, forms the basement rocks in the area. Younger formations are predominantly rhyolite ignimbrites, rhyolite and dacite lava domes, and lacustrine and alluvial sediments derived from these volcanic lithologies. Note that "rhyolite" is sometimes used colloquially to refer to rhyolite lava, but the use of "rhyolite" is here applied only as a formal compositional definition (volcanic rocks $> 69\%$ SiO_2). Deposits of rhyolite composition may be pyroclastic (explosively formed, including airfall deposits and the pyroclastic flow deposit termed "ignimbrite") or they may be lavas (effusively erupted without explosion). Rhyolite lavas are viscous and often push up into high lava domes over the eruption vent. The geological formations and processes of greatest relevance to the hydrology and hydrogeology of the Lake Rotorua catchment area are shown in the three-dimensional geological model of White et al. (2004) (Fig. 2) and summarized in the following paragraphs.

From 2 million to 240 thousand years ago (ka), a number of rhyolite lava domes were emplaced and volcanic activity from TVZ calderas resulted in pyroclastic deposits across the area, including the highly welded Waiotapu ignimbrite (ca. 710 ka; "older ignimbrite" in Fig. 1) and a range of variably welded, variably altered, sometimes-jointed ignimbrites of which the Matahina, Chimp and Pokai formations (ca. 320–270 ka) are most significant and mapped within "undifferentiated rhyolite pyroclastics" in Fig. 1. These ignimbrites are expected to be the main basal units for groundwater aquifers in the study area (White et al., 2004).

The period from 240 to 200 ka is defined by the eruption that deposited the Mamaku Plateau Formation (240 ka) and formed the Rotorua Caldera. The caldera collapse downfaulted parts of older lava domes positioned across the western and northern edge of the caldera. The Mamaku Plateau Formation is predominantly composed of ignimbrite (hereafter "Mamaku ignimbrite"), which is variably welded, variably jointed and very permeable. Several rhyolite lava domes (mainly Ngongotaha and its neighbours) began to develop

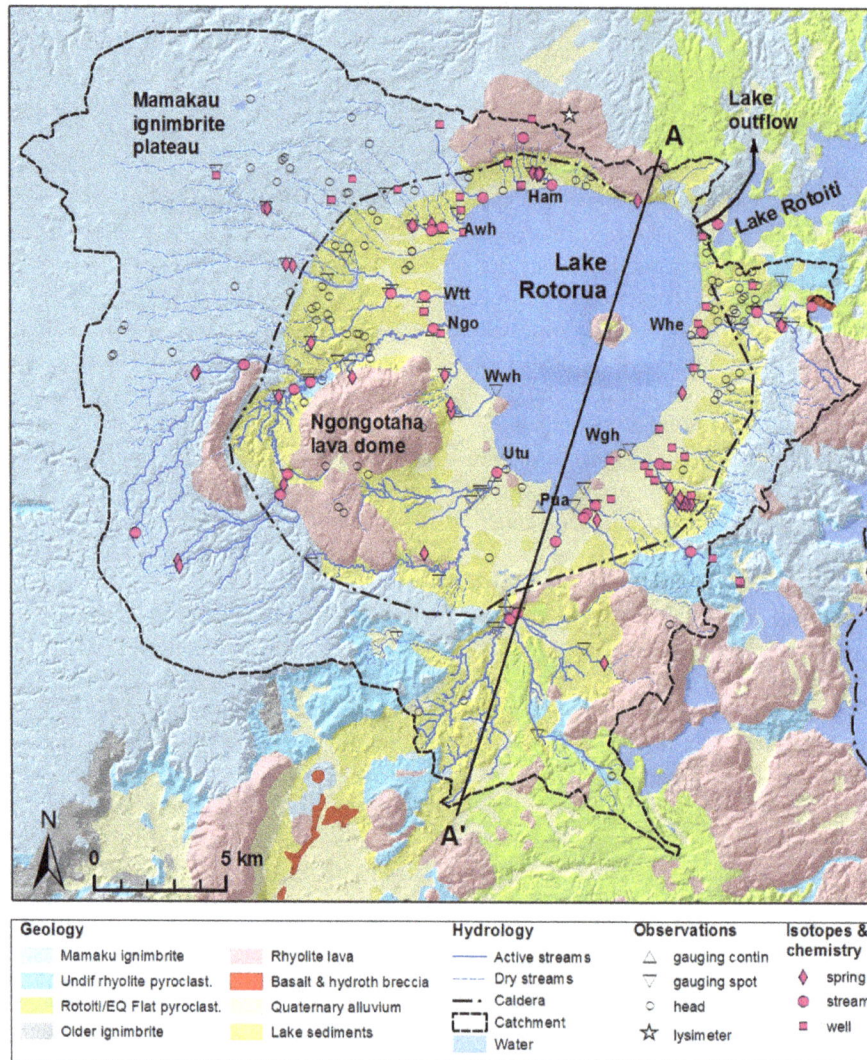

Figure 1. Location and geology of the Lake Rotorua catchment, with sampling sites. The assumed groundwater catchment is from White and Rutherford (2009). Surficial geology is based on the 1 : 250 000 map of Leonard et al. (2010). For abbreviations of the names of the major streams refer to Table 1. The approximate trace of the caldera is shown. Cross-section A′–A is shown on the cut-away face of Fig. 2.

soon after the caldera collapse. Also, soon after the eruption, a lake began to form in the collapsed depression, leading to the deposition of lacustrine fine ash and pumice, commonly referred to as lacustrine sediments (Leonard et al., 2010); note that these sediments have sometimes been referred to as Huka or Huka Group sediments throughout the TVZ, but this definition formally refers only to specific units near Taupo City and the term Huka is avoided here.

From 200 to 61 ka, volcanic activity in the vicinity of Lake Rotorua was relatively subdued. A number of eruptions from the Okataina Volcanic Centre (OVC), located to the east of the Rotorua caldera, produced widely dispersed but relatively thin airfall deposits. These pyroclastic materials caused periodic damming of drainage pathways and led to fluctuations in lake level that in turn resulted in widespread and variably

thick sediments being deposited in the Rotorua Caldera. This period of relatively quiet volcanic activity ended with the Rotoiti and Earthquake Flat eruptions from the OVC, which produced widespread pyroclastic deposits, including the non-welded ignimbrites of the Rotoiti Formation and the Earthquake Flat Formation.

From 61 ka to present, numerous eruptions from the OVC (the most recent of which was in 1886) deposited airfall layers in the Lake Rotorua catchment area. Numerous rhyolite lava units were also emplaced during this period. The periodic deposition of pyroclastic materials, along with activity on faults of the Taupo Rift (Leonard et al., 2010), presumably caused fluctuations in the lake level, with current lake level being reached sometime within the last few thousand years (White et al., 2004). Due to the decline of lake level, Holocene alluvial sand and gravel deposits are found in stream channels and around the current lake shoreline.

The southern Rotorua basin hosts a vigorous geothermal system producing many hot water, hot mud, steam and geyser features, along with gas emission, between the southern edge of the Lake and about the southern edge of the caldera (Fig. 1). There is hot local groundwater flow in this area, generally flowing down-hill northwards into the lake. Beyond this relatively confined area the groundwater system does not appear to interact with fluids from this geothermal system.

2.2 Hydrology

Lake Rotorua has a surface area of $79\,km^2$ and a mean depth of 10.8 m (Burger et al., 2011), with a total water volume of $0.85\,km^3$. The assumed total catchment area is ca. $475\,km^2$ (White and Rutherford, 2009) (Fig. 1).

Annual rainfall in the catchment is strongly affected by topography and varies from more than 2200 mm northwest of the lake to less than 1400 mm southeast of the lake (Hoare, 1980; White et al., 2007; Rutherford et al., 2008). Approximately 50 % of rainfall infiltrates into the groundwater system. This is based on two sources of information: (1) comparisons of rainfall and actual evapotranspiration that have been made for various parts of the catchment (Hoare, 1980; Dell, 1982; White et al., 2004, 2007; Rutherford et al., 2008); and (2) data from paired lysimeters, a standard rain gauge and a ground-level rain gauge installed at Kaharoa (White et al., 2007) (Fig. 1). With 50 % of rainfall recharge, total infiltration into the groundwater system is estimated to be $14\,500\,L\,s^{-1}$, based on the catchment shown in Fig. 1, excluding rainfall inputs direct to the lake, and assuming recharge is 50 % of rainfall. This rainfall recharge supports stream flow and potentially direct inputs of groundwater to the lake.

There are nine major streams (Fig. 1, for abbreviations refer to Table 1) and several minor streams that flow into the lake; the remainder of the inflows are provided from direct inputs of rainfall and lake-front features, and potentially from groundwater seepage through the lake bed. The major streams are baseflow-controlled and characterized by very constant water flow (Hoare, 1980) and temperature, and groundwater-derived baseflow accounts for approximately 90 % of the average flow in the typical Rotorua stream (Hoare, 1987). Baseflows in the nine major streams entering Lake Rotorua cumulatively amount to $11\,800\,L\,s^{-1}$, and total inflows to the lake from minor streams and lake-front features amounts to $350\,L\,s^{-1}$ (Hoare, 1980; White et al., 2007) (Table 1).

With the lake water volume of $0.85\,km^3$, the lake water turnover time via the groundwater-fed streams is 2.2 years. The only surface outflow occurs through Ohau Channel via Lake Rotoiti (Fig. 1). Water balance calculations suggest that the total catchment area exceeds the surface water catchment area (White et al., 2007); in other words, groundwater

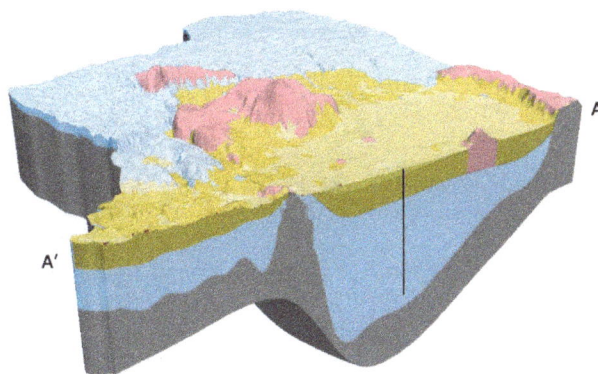

Figure 2. Three-dimensional geological model of the Lake Rotorua catchment area (from White et al., 2007). The location of the vertical cut-away face A′–A is shown in Fig. 1. The vertical exaggeration is 5 ×, with a 1 km vertical scale bar shown in the centre. The colour scheme is similar to that in Fig. 1 (bottom to top): pre-Mamaku formations (grey), Mamaku Plateau Formation (light blue), sediment formation (dark yellow), lava domes (red) and Holocene alluvial deposits (light yellow).

from outside the surface water catchment is flowing through the aquifer system to Lake Rotorua (White and Rutherford, 2009).

2.3 Hydrogeology

Groundwater flow in the Lake Rotorua catchment is influenced by several fundamental geological characteristics. First, the Mamaku ignimbrite, the dominant hydrogeological feature in the catchment, is assumed to be up to 1 km thick in the centre of the caldera depression, and from about 200 to tens of metres thick outside of the caldera, decreasing generally with distance from the caldera (Fig. 2). Ignimbrite tends to fill in pre-existing valleys and landforms, so its thickness can be quite variable over horizontal distances of as little as hundreds of metres. Transit times of groundwater through such a thick aquifer may be lengthy compared to times in the shallow alluvial aquifers used for water supply in many other parts of the world and New Zealand. Second, the ignimbrites in the Lake Rotorua catchment are known to be variably welded, altered and jointed, with the potential for preferential groundwater flow paths. Groundwater may be routed from its recharge area along lengthy preferential flow paths and discharge in neighbouring surface catchments, leading to water ages that vary substantially even within a localized area (the presence of such preferential flow paths can therefore be demonstrated with age tracers). Third, the broadly circular collapse faults of the caldera constitute a major structural feature that may influence the flow of groundwater within the catchment.

The major water contribution to Lake Rotorua is from the western catchment that drains the eastern flanks of the Mamaku Plateau (Fig. 1). The Mamaku ignimbrite formation

Table 1. Estimates of baseflow from White et al. (2007). Abbreviations refer to Fig. 1.

Stream	Abbreviation	Baseflow $(\mathrm{L\,s^{-1}})$
Hamurana	Ham	2750
Awahou	Awh	1700
Waiteti	Wtt	1300
Ngongotaha	Ngo	1700
Waiowhiro	Wwh	370
Utuhina	Utu	1600
Puarenga	Pua	1700
Waingaehe	Wgh	250
Waiohewa	Whe	390
Minor streams	n/a	350

serves as a major source of groundwater in the area (Gordon, 2001). A large area ($\sim 250\,\mathrm{km^2}$) is drained by several major springs ($> 1000\,\mathrm{L\,s^{-1}}$) emerging from the ignimbrite on the western side of Lake Rotorua. Given the large extent and thickness of the ignimbrite aquifer, a large groundwater reservoir exists, with long water residence times expected in the aquifer. Taylor and Stewart (1987) estimated the mean residence time of the water of some of the springs as 50–100 years.

The post-240 ka ignimbrites in this area (and some lava domes) are extremely porous; they sustain hardly any overland water flow (Dell, 1982), with most of the stream beds dry throughout most of the year except during heavy rain, and they allow the infiltrated water to percolate down to an extremely deep groundwater table $> 50\,\mathrm{m}$ (the ignimbrite formations around Ngongotaha Dome are an exception).

Little is known about the hydrogeology of the groundwater system; borehole data collected by drillers is often not of sufficient quality to identify and correlate aquifer units. Rosen et al. (1998) developed a schematic model for the Mamaku ignimbrite, with a lower and upper ignimbrite aquifer sheet considered permeable, and a middle sheet considered impermeable but fractured and not acting as an aquitard for the other two sheets. This is probably an over-simplification in many areas (see Milner et al., 2003) but does point at horizontally planar discontinuities within the formation that appear to influence groundwater flow. The water, after easy passage through the large aquifer, is forced to the surface only 1–2 km before the lake shore via large groundwater springs, feeding large streams that drain into Lake Rotorua.

The water-bearing lava dome formations that predate the Mamaku ignimbrite are likely to have fracture flow, based on spring discharge permeability analysis, varying depending on fracture sizes and linkages. Faulting associated with the Rotorua caldera has offset several of the rhyolite domes and groundwater may flow through these faults.

The palaeo-lake sediments that post-date the Mamaku ignimbrite and rhyolite lava formations comprise silt, sand and gravel (ignimbrite, obsidian and rhyolite pumice) and are considered permeable, with lenses of low permeability that can also act as confining layers.

Overall, understanding of the Rotorua groundwater system is complicated due to the highly complex geology that has evolved through volcanic activity. Vertical and steeply inclined geological contacts are common, precluding a simple horizontal-layer-based succession model throughout the catchment usually applicable in sedimentary basins. Aquifers have not been well determined due to insufficient bore log data, and also the extent of the Lake Rotorua groundwater catchment is difficult to establish, due to the deep water table of $> 50\,\mathrm{m}$ in large areas at the catchment boundaries, combined with an inhomogeneous groundwater flow pattern, as indicated by the groundwater discharge in the northern catchment being too large compared to the size of the surface water catchment.

3 Methods

3.1 Determination of water age

The age of the groundwater at the discharge point characterizes the transit time of the water through a groundwater system. For groundwater dating, we use tritium time series (repeated sampling after several years), and the complementary tracers tritium, CFCs and SF$_6$ together where possible. The method of dating young groundwater with mean ages of less than 200 years for current New Zealand Southern Hemispheric conditions is described in detail in Morgenstern and Daughney (2012, Sects. 2.3 and 2.4). In short: tritium dating, in previous decades problematic due to interference from the artificial tritium produced by atmospheric nuclear weapons testing in the early 1960s, has now become very efficient and accurate due to the fading of the bomb-tritium.

For groundwater dating, one or more tracer substances are measured that have a time-dependent input into the groundwater system or a well-defined decay-term (e.g. radioactive decay). The tracer concentration data are then fitted using a lumped-parameter model (Maloszewski and Zuber, 1982; Zuber et al., 2005). For dating young groundwater – i.e. water less than about 100 years – the most commonly used tracers are tritium, chlorofluorocarbons (CFCs) and sulfur hexafluoride (SF$_6$) (Cook and Solomon, 1997; Edmunds and Smedley, 2000; Stewart and Morgenstern, 2001; Morgenstern et al., 2010). The measured output tracer concentration in the groundwater (C_{out}) is then compared to its tracer concentration at the time of rainfall input (C_{in}) using the convolution integral

$$C_{\mathrm{out}}(t) = \int_0^\infty C_{\mathrm{in}}(t - \tau) e^{-\lambda \tau} g(\tau)\,\mathrm{d}\tau, \qquad (1)$$

where t is the time of observation, τ is the transit time (age), $e^{-\lambda\tau}$ is the decay term with $\lambda = \ln(2)/T_{1/2}$ (e.g. radioactive decay of tritium with a half-life $T_{1/2}$ of 12.32 years) and $g(\tau)$ is the system response function (Cook and Herczeg, 1999; Zuber et al., 2005).

The system response function accounts for the distribution of ages within the water sample, for example from mixing of groundwater of different ages within the aquifer, or at the well (Maloszewski and Zuber, 1982, 1991; Goode, 1996; Weissman et al., 2002; Zuber et al., 2005). The two response functions most commonly used are the exponential piston flow model and the dispersion model (Zuber et al., 2005). The exponential piston flow model combines the piston flow model, assuming piston flow within a single flow tube in which there is minimal mixing of water from different flow lines at the discharge point (e.g. confined aquifer), and the exponential model, assuming full mixing of water from different flow paths with transit times at the groundwater discharge point that are exponentially distributed (e.g. mixing of stratified groundwater at an open well in an unconfined aquifer). The response functions of the various models are described in Maloszewski and Zuber (1982) and Cook and Herczeg (1999). To interpret the ages of the Lake Rotorua catchment data set, the exponential piston flow model was used, given by

$$g = 0 \text{ for } \tau < T(1 - f) \tag{2}$$

$$g = \frac{1}{Tf}e^{(-\frac{\tau}{Tf} + \frac{1}{f} - 1)} \text{ for } \tau \geq T(1 - f), \tag{3}$$

where T is the mean residence time (MRT), f is the ratio of the volume of exponential flow to the total flow volume at the groundwater discharge point, and $T(1 - f)$ is the time it takes the water to flow through the piston flow section of the aquifer (Maloszewski and Zuber (1982) use the variable η; $\eta = 1/f$). When $f = 0$ the model becomes equivalent to the piston flow model, and when $f = 1$ it becomes equivalent to the exponential model.

The two parameters of the response functions, the MRT and the distribution of transit times (f), are determined by convoluting the input (tritium concentration in rainfall measured over time) to model water passage through the hydrological system in a way that matches the output (e.g. tritium concentrations measured in wells or springs). Because of its pulse-shaped input, tritium is a particularly sensitive tracer for identifying both of these two parameters, which can be deduced uniquely by comparing the delay and the dispersion of the bomb-pulse tritium in the groundwater to that from tritium in the original rain input. This method is particularly useful for interpretation of ages of groundwater in the Lake Rotorua catchment, where most of the groundwater discharges lack any other information on mixing of groundwater with varying flow path lengths and of different age, such as ratio of confined to unconfined flow volume, or screen depth for wells.

For tracer age interpretation, the integral (Eq. 1) was used to convolute the historical rainfall tracer input to an output that reflects mixing in a groundwater system, with the best match of the simulated output to the measured output time series data (Fig. 3). The TracerLPM workbook (Jurgens et al., 2012) was used. The tritium input function is based on concentrations of tritium in rainfall measured monthly since the 1960s at Kaitoke, near Wellington, New Zealand (Morgenstern and Taylor, 2009). The Kaitoke rainfall input function is multiplied by a scaling factor of 0.87 to account for variation in atmospheric tritium concentrations due to latitude and orographic factors, as deduced from measurements from rain at various locations in New Zealand (e.g. Morgenstern et al., 2010). For the prevailing New Zealand climatic conditions there is no need for correction of the tritium input for seasonal infiltration (Morgenstern et al., 2010).

The problem of ambiguity in tritium dating over the last decades is demonstrated in Fig. 3. Hangarua Spring discharges old water with a mean residence time of about 90 years (see below), but during the late 1980s its tritium concentration was similar to that of very young water (rain curve in Fig. 3). At that time, the tritium concentration in Hangarua Spring would have been in agreement with both very young water and old water with a mean residence time of 90 years. Tritium data covering several decades, however, clearly distinguish this old water (low tritium concentration) from young rain water. Figure 3 also shows that due to the fading of the bomb-tritium in recent decades (tritium decay over four tritium half-lives since the bomb spike), in recent years the tritium concentration of old water is clearly distinguishable (lower) from that of young water, without ambiguity. The tritium time series data allow also for constraining groundwater mixing models. Figure 3 shows the model output curves that match the measured tritium data. Given sufficient analytical accuracy, this is also possible for extremely low tritium concentrations; the data for Hamurana water intake spring (blue in Fig. 3) are all below 0.4 TU, which is below the detection limit of many tritium laboratories (http://www-naweb.iaea.org/napc/ih/IHS_programmeihltric.html).

The application of mixing models is described in Morgenstern and Daughney (2012, Sect. 2.6). Throughout New Zealand, for springs and wells in almost all hydrogeological situations, the exponential piston flow model, with its age distribution, has produced good matches to most (about a hundred) tritium time series data. It was not, however, possible to obtain adequate matches in the ignimbrite area of the Rotorua catchment using such a simple exponential piston flow model. Alternatively, using the dispersion model did not improve the matches. The complex volcanic aquifers of the Lake Rotorua catchment, which have evolved through volcanic activity, require a more complex system response function. A combination of two exponential piston flow models was used.

3.2   Sample collection and analysis

Samples were collected from 41 springs, from 31 groundwater-dominated stream flow sites, and from 26 groundwater wells. To obtain the residence times of the water discharging into the lake after passage through the entire groundwater system, sampling focused on the naturally flowing groundwater discharges, the springs and streams. Samples were collected at times of base flow conditions.

All nine major streams were sampled multiple times near the inflow into the lake, typically 3–4 times (Figs. 3 and 4). Most of these tritium time series go back to the early 1970s and encompass the passage of the "bomb" tritium peak through the groundwater system, allowing determination of detailed age distribution parameters for these major inflows to the lake. These "historic" samples had been collected sporadically for various projects over the decades to study the transfer of the bomb-tritium through the hydrologic cycle. Over the recent decade, the streams have also been sampled for tritium at various points upstream, at various main confluences, or at main springs to obtain a detailed spatial distribution of water ages. Springs and wells were also once sampled for CFCs, SF_6, argon and nitrogen, to obtain complementary age information.

Sampling locations are shown in Fig. 1. Many of the sites have no road access, with some of them in remote steep gullies. A portable sampling system was required for the gas samples to allow fresh water from the well or spring to be pumped into the sample bottles from below the water surface without air contact. We used a pneumatic Bennet pump, powered from a cylinder of compressed air at the remote locations, and from a compressor powered by the car battery at sites with car access. Sampling from streams (tritium only) involved simply dipping the bottle under the water surface and filling the bottle.

Sampling methods for hydrochemistry and nutrients were according to Daughney et al. (2007). Age tracer samples were collected without filtration or preservation. For tritium, a 1 L plastic bottle was filled to the top. For CFC samples, two 125 mL glass bottles with aluminium liner cap were filled, rigorously excluding air contact by filling from the bottom via a nylon tube and three times volume replacement below the surface of the overflowing sample water. 1 L bottles were filled for SF_6.

Analytical details for hydrochemistry are described in Daugney et al. (2007). Details of the tritium analysis procedure are described in Morgenstern and Taylor (2009). While the early tritium measurements in the 1970s were performed with a detection limit of approximately 0.1 Tritium Units (TU), we now achieve significantly lower detection limit of 0.02 (TU) via tritium enrichment by a factor of 95 and reproducibility of tritium enrichment of 1 % via deuterium calibration. Analysis procedures for CFC-11, CFC-12, and SF_6 are described in van der Raaij and

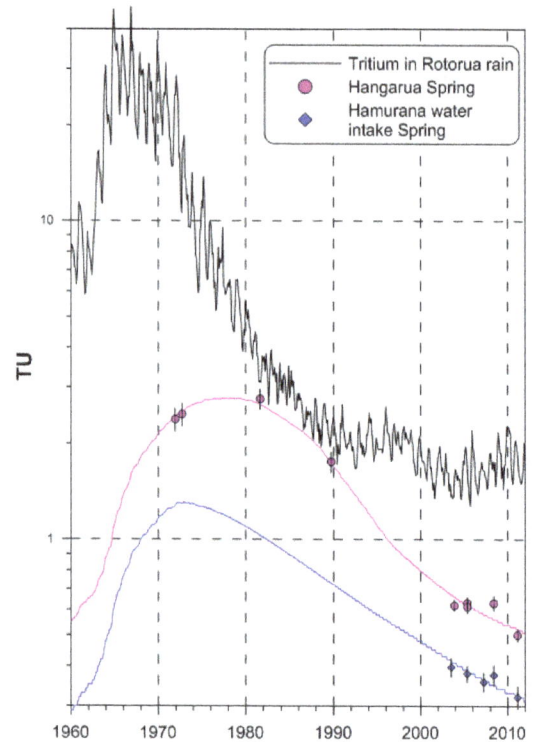

Figure 3. Tritium rain input for the Rotorua catchment, and measured tritium output at Hangarua Spring and at Hamurana water intake spring. The input curve is based on monthly measurements in Kaitoke near Wellington, New Zealand, scaled to the latitude of Rotorua with a factor 0.87, and smoothed by an exponential piston flow model with 0.3 years mean residence time and 50 % exponential flow within the total flow volume. One TU = one tritium atom per 10^{18} hydrogen atoms. For the spring samples one-sigma measurement errors are shown. Note the logarithmic scale of the TU axis.

Beyer (2014). Detection limits are 3×10^{-15} mol kg^{-1} for CFCs and 2×10^{-17} mol kg^{-1} for SF_6. Dissolved argon and nitrogen concentrations were measured for estimating the temperature at the time of recharge, and the excess air concentration, as described by Heaton and Vogel (1981), for calculation of the atmospheric partial pressure (ppt) of CFCs and SF_6 at the time of recharge.

4   Results and discussion

In the following section hydrochemistry cluster analysis and hydrochemistry evolution are discussed to assess the geographic sources of groundwater and groundwater processes in the aquifer. The nutrients nitrate, sulfate, potassium and phosphate are discussed to evaluate their source (anthropogenic versus geologic), lag time, fate and impact on lake eutrophication. The age distributions of the groundwater discharges to Lake Rotorua are discussed to understand the conceptual groundwater flow pattern and the lag time in the

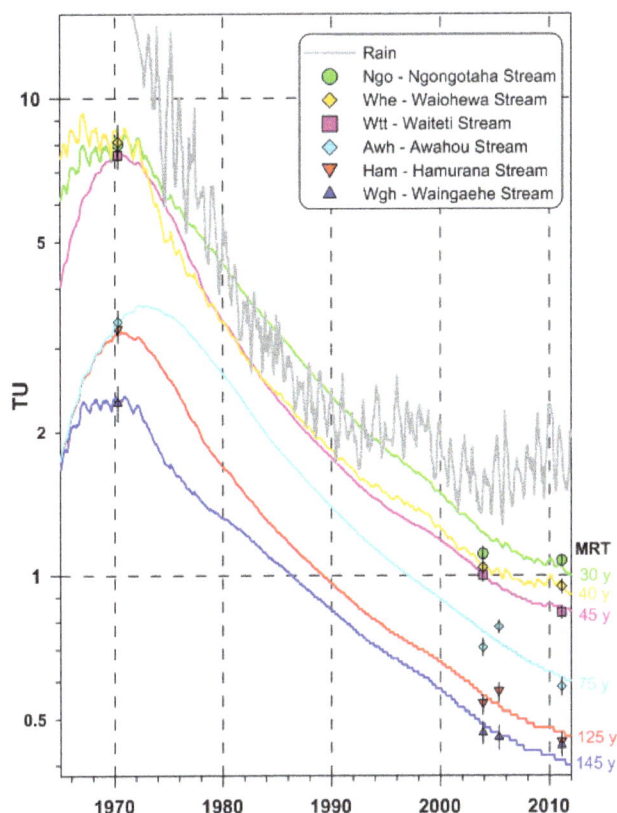

Figure 4. Tritium time series data, together with their matching lumped-parameter model outputs, for six major streams. Grey line is tritium input via rain from Fig. 3. The locations of the streams are shown in Fig. 1.

Table 2. Age distribution parameters for the two binary exponential piston flow models (EPMs) for the major stream discharges to Lake Rotorua. Average MRT is the mean residence time between the two EPMs.

Stream	EPM1		Fraction	EPM2		Average
	MRT1	f1	of EPM1	MRT2	f2	MRT [year]
Hamurana	185	0.82	0.65	12	0.77	125
Awahou	80	1.00	0.92	6	0.91	75
Waiteti	60	1.00	0.78	3	0.90	45
Ngongotaha	35	1.00	0.82	1	0.91	30
Waiowhiro	40	0.63	1.00	n/a	n/a	40
Utuhina	85	0.60	0.70	1	1.00	60
Puarenga	44	1.00	0.95	2	1.00	40
Waingaehe	160	0.94	0.90	3	1.00	145
Waiohewa	55	1.00	0.75	1	1.00	40

groundwater system. The ultimate goal of this project is the use of the hydrochemistry and groundwater age parameters for calibration of a groundwater transport model for improved management of the nutrient loads to the lake – the subject of follow-up papers.

4.1 Groundwater age interpretation

To obtain the unique solution for both parameters of the age distribution for a specific model, time series data are required (Sect. 3.1). Most of the large water inflows into Lake Rotorua have long time series data available (up to over four decades), allowing for well constrained age distribution parameters for both MRT and the fraction f between different flow models (or Péclet number for the dispersion model). The tritium time series data, together with the matching lumped-parameter model simulations, are shown in Fig. 3 for two of the large springs, and in Fig. 4 for six of the large streams (covering 2/3 of stream baseflow to Lake Rotorua). For the sites with shorter time series data (subcatchment stream discharges, groundwater wells), most of the sites have at least sufficient time series or multi-tracer data for unambiguous robust age interpretations. If fraction f cannot be established

uniquely from the tritium time series data, we applied mixing models that matched long tritium time series data from other sites with similar hydrogeologic settings to these sites. All 96 sites with tritium time series or tritium and complementary CFC and SF_6 data have unambiguous age interpretation. For the tritium time series data shown in Figs. 3 and 4, the lumped-parameter models, with their respective age distribution parameters that match the measured data, are listed in Table 2.

Throughout New Zealand, and including all hydrogeologic situations (but mainly groundwater wells), we have measured approximately a hundred long tritium time series covering several decades. A simple lumped-parameter model, the exponential piston flow model, usually can match these time series data well (e.g. Morgenstern and Daughney, 2012). The long-term tritium data from most of the large stream discharges shown in Fig. 4, however, cannot be matched by a simple model such as the exponential piston flow or dispersion model and require a more complex groundwater flow model combination. Using a binary mixing model, with parallel contributions from two exponential piston flow models, resulted in excellent matches. We justify this binary mixing model by inferring two different flow contributions in the catchment to stream and spring flow – from deep old groundwater, as indicated by very deep groundwater tables in the area (generally > 50 m), and from younger groundwater from shallow aquifers, as indicated by minor stream flows maintained by shallow aquifers. In Table 2 are also listed the average mean residence times between the two parallel models, weighted by their fraction within the total flow. For the MRTs, errors caused by our tritium measurement error and uncertainty in tritium input are typically ± 1 year for MRTs < 5 years, ± 2 years for MRTs between 5 and 10 years, ± 3 years for MRTs between 10 and 50 years, ± 5 years for MRTs between 50 and 100 years, and larger errors for older water towards the detection limit.

For convenience, the average MRTs are also listed in Fig. 4 next to their model output curve. It is obvious from Fig. 4 that all the main streams discharge very old water into Lake Rotorua. The tritium response of the streams is clearly distinguishable from that of young rain water (grey line). The youngest water discharge, Ngongotaha Stream (green), has an average MRT of 30 years. All other main streams discharge significantly older water, up to MRT of 145 years for Waingaehe Stream (dark blue). Note that even though bomb-tritium from atmospheric nuclear weapons testing in the 1960s has decayed enough to no longer cause ambiguous age interpretations, it is still possible to detect the tail of the bomb-tritium for matching model parameters if tritium analyses have sufficient accuracy; all data for the Hamurana water intake spring (blue in Fig. 3) are below 0.4 TU.

Most of the streams and springs discharge very old groundwater into Lake Rotorua, with MRTs typically between 50–150 years, indicating discharge from a large groundwater system with large water residence (turn-over) times. Only a few small subcatchments with minor flow rates discharge young water (MRT < 20 years), indicating local geologic units below the surface that do not allow water to infiltrate into and flow through larger, deeper groundwater systems.

Substantial fractions of that long residence time in the groundwater system may occur during passage through the thick unsaturated zones (50–100 m) as indicated by CFC and SF_6 results measured at groundwater wells and springs (Morgenstern et al., 2004). CFCs and SF_6 in groundwater are still exchanged with the atmosphere during passage through the unsaturated zone, therefore CFC and SF_6 ages represent travel time through the saturated zone only. Large observed differences between CFC and SF_6 ages, compared to tritium ages of up to 40 years and greater for the older waters, therefore indicate travel time of the groundwater through the unsaturated zone of > 40 years for the older groundwater discharges.

The old age of the majority of the Lake Rotorua water inflows and the highly mixed nature of the water discharges (note the high fractions of exponential flow, up to 100 %, in Table 2) implies a very slow and lagged response of the streams and the lake to anthropogenic contaminants in the catchment, such as nitrate. The majority of the nitrate load currently discharging into the lake is thus from land-use activities 30 and more years ago.

About a hundred stream and groundwater well samples have been dated in the Lake Rotorua catchment. The groundwater age distributions are used in the following sections to identify hydrochemistry evolution, sources of contaminants, and to predict future nitrate loads that will enter Lake Rotorua from the large contaminated groundwater system. In a future paper, the conceptual groundwater flow model in the Lake Rotorua catchment will be inferred from the groundwater age distribution data. The data will subsequently be used for calibration of a groundwater transport model.

4.2 Hydrochemistry and recharge source

The hydrochemical composition of the groundwater and surface waters in the Rotorua catchment have been investigated by Morgenstern et al. (2004) and Donath et al. (2014); the following section summarizes the results relevant to the present study.

Hydrochemistry is driven by interaction between water and the different major lithologies and can be used to track the origin of the groundwater. Hydrochemistry reflects the rhyolite ignimbrite and lava aquifer lithologies that dominate the Lake Rotorua catchment, with much lower concentrations of Ca, Mg and SO_4 and much higher concentrations of F, PO_4 and SiO_2, compared to groundwater in other parts of New Zealand.

Several statistical and graphical techniques were applied to characterize the variations in hydrochemistry across parts of the catchment. Hierarchical cluster analysis (HCA) was shown to be a useful technique to identify samples with similar hydrochemical composition, and to relate the groundwater samples to their origin from one of the main aquifer lithologic units. HCA conducted with Ward's linkage rule allowed the samples to be partitioned into four hydrochemical clusters. Three of the clusters, accounting for the majority of the samples, were inferred to reflect water–rock interaction with the dominant lithologies in the catchment, namely Mamaku ignimbrite, lava or palaeo-lake sediments. Hydrochemistry inferred to indicate interaction between water and the Mamaku ignimbrite had Na and HCO_3 as the dominant cation and anion, respectively, and among the highest concentrations of Mg, PO_4 and SiO_2 and among the lowest concentrations of F, K and SO_4 observed. Hydrochemistry inferred to indicate interaction between water and rhyolite lava also had Na and HCO_3 as the dominant cation and anion, respectively, but had relatively low concentrations of PO_4 and among the highest concentrations of K. Hydrochemistry inferred to indicate interaction between water and sediments had Na–Ca–HCO_3–Cl water type and relatively low concentrations of SiO_2. The remaining cluster was inferred to represent geothermal influences on the hydrochemistry (e.g. elevated concentrations of Na, Cl, SO_4, SiO_2 and NH_4).

Figure 5 shows the hydrochemical clusters of the water samples inferred to indicate interaction between water and Mamaku ignimbrite (light blue), lava (red) or lacustrine sediments (dark yellow). Note that samples assigned to the cluster inferred to indicate geothermal origin are not displayed in Fig. 5 or discussed further in the present study because geothermal influence is not the subject of this study.

Samples with hydrochemistry indicative of interaction with the Mamaku ignimbrite occur predominantly in the north and northwest portion of the catchment (Fig. 5, blue). All of the large springs discharging into Hamurana (Ham), Awahou (Awh), and Utuhina (Utu) streams have a Mamaku ignimbrite hydrochemistry signature (blue circles). The stream reaches in the Mamaku ignimbrite area upstream

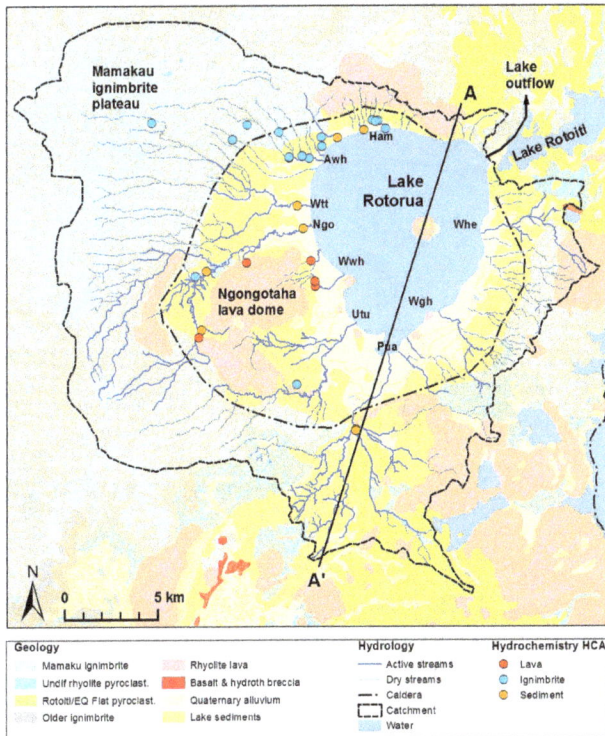

Figure 5. Distribution of hierarchical cluster analysis (HCA) clusters in Lake Rotorua catchment, together with the geological units (Leonard et al., 2010) and stream reaches. Stream reaches shown as dotted lines are usually dry. HCA clusters relate to origin of the groundwater from one of the three main geologic formations: Mamaku ignimbrite (light blue), lava (red) and lacustrine sediment (yellow).

from these large springs are usually dry. This, together with the Mamaku ignimbrite hydrochemistry signature, implies that these large springs drain the large Mamaku ignimbrite areas upstream that have negligible surface runoff. In the northwest in the Hamurana and Awahou catchments, these springs emerge close to the lake shore within the sediment area (Fig. 5, yellow), indicating that close to the lake shore, due to the more impermeable nature of the intra-caldera sediment, the deeper groundwater flow from the Mamaku ignimbrite is forced to the surface. All of these large springs emerge within the slopes of the sediment formation where sediment layers are thinner and weaker compared to the level area closer to the present lake shore; we infer that the thin nature of the sediments on these slopes allows the water from the underlying ignimbrite to flow to the surface. No large spring occurs in the level area closer to the lake, where sediments are thicker. Also the large Utuhina Spring in the southeast emerges within the slopes of the sediment (Fig. 5), indicating the more impermeable nature of the sediment forcing the groundwater from the Mamaku ignimbrite to the surface in the area of thin sediment layers. The Utuhina Spring emerges below a small local lava dome feature, but the ig-

nimbrite signature of the water indicates that this spring is the discharge from the large ignimbrite area southwest of the lava dome feature. The small lava feature may be fractured, discontinuous, or act as a water conduit, allowing water discharge from the ignimbrite behind.

Shallow wells and streams that gain most of their recharge and flow within the lacustrine sediments display a characteristic hydrochemical signature (Fig. 5, yellow circles). Such samples originate from the downstream parts of Waiteti (Wtt) and Ngongotaha (Ngo) streams. The study by Donath et al. (2014) also detected this characteristic hydrochemistry in samples collected with higher spatial resolution in parts of the Ngongotaha subcatchment that are not discussed here.

Hydrochemistry of the water draining the Ngongotaha lava dome west of the lake (Fig. 5, red circles) is inferred to indicate interaction with lava formations. The Ngongotaha dome, similar to the Mamaku ignimbrite, has no drainage via surface flow (stream beds are dry), indicating a highly porous nature, likely due to fractures and pumiceous zones within the dome. Only where the rhyolite dome intercepts the palaeo-lake sediments is the groundwater flow from the lava forced to the surface due to the low permeability of the sediments.

The investigated water discharges from the eastern catchment of Lake Rotorua entirely show geothermal influence in their hydrochemistry composition (not the subject of this study).

The above HCA results give, for the first time, consistent evidence of the link between the main recharge areas in the Mamaku ignimbrite, and the main groundwater discharges into the lake.

4.3 Hydrochemistry evolution

In the following two sections hydrochemistry data versus groundwater age is discussed for a better understanding of groundwater processes and geologic versus anthropogenic origin of contaminants.

The groundwater of the Rotorua rhyolite ignimbrite and lava dome aquifers (Fig. 6a) displays high dissolved oxygen (DO), between 5 and $11\,\mathrm{mg}\,\mathrm{L}^{-1}$ (50–100 % of equilibrium with air). There is no trend of decreasing DO with increasing age, indicating that microbial reduction reactions are insignificant in this volcanic aquifer within timescales of the water residence time in the aquifer. Microbial reduction reactions, facilitated by the presence of organic matter or other electron donors (e.g. pyrite), would usually consume the dissolved oxygen in the groundwater. Reduction of oxygen is energetically the most favourable reaction that micro-organisms use in a series of reactions, with the result that other reduction reactions (e.g. denitrification) typically do not occur until most of the dissolved oxygen has been consumed. These reduction reactions take time, and if these reactions are supported by the presence of electron donors in the geologic formation, it is expected that old waters be-

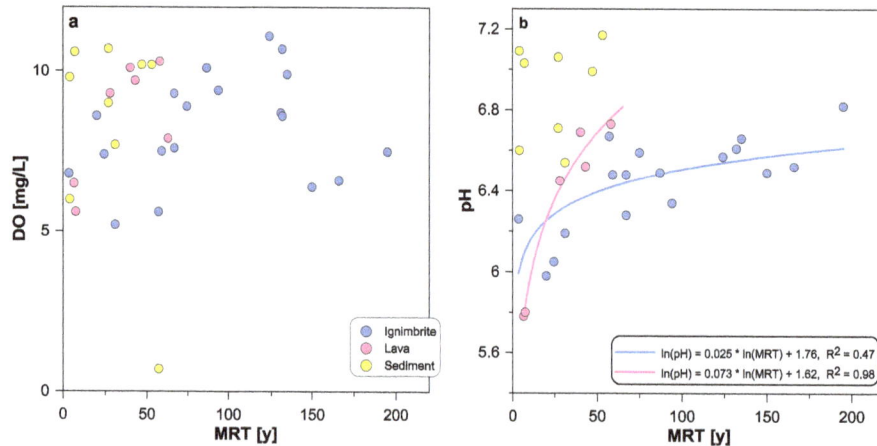

Figure 6. (a) Dissolved oxygen (DO) and **(b)** pH versus mean residence time (MRT). The colour codes of the samples indicate water from the relevant geologic formation, as indicated by hierarchical cluster analysis.

come increasingly anoxic (e.g. Böhlke et al., 2002; Tesoriero and Puckett, 2011). No trend of decreasing DO with increasing groundwater age was observed, suggesting an absence of significant amounts of electron donors such as organic matter or pyrite in this ignimbrite formation. This is supported by its depositional history as a single large ignimbrite formation without any organic matter involved. Absence of oxygen reduction indicates that there is no potential for significant denitrification reactions in this aquifer system.

Absence of a trend of decreasing DO with increasing groundwater age, but rather constant DO in very young and old groundwater of between 50 and 100 %, suggests that the partial oxygen reduction that is observed has occurred in the soil zone, which does contain organic matter, and that once the water has passed the soil zone, no further oxygen reduction processes occur. Only one groundwater sample in the Rotorua catchment was depleted in oxygen (below $1 \, \text{mg} \, \text{L}^{-1}$). This is related to the palaeo-lake sediments, suggesting localized deposits of reactive organic matter in these sediments, as would be expected in lake sediments.

The pH of groundwater usually increases over time due to ongoing hydrogeochemical reactions, resulting in an increasing pH of groundwater with age. In New Zealand we observed in groundwater an increase from about pH 6 for very young groundwater (< 1 year) to about pH 8 in very old groundwater (> 10 000 years) (Morgenstern and Daughney, 2012). For the Rotorua catchment, the groundwater pH data from lava formations (Fig. 6b) show a sharp increase from pH 5.8 to 6.7 over the age range from 5 to 50 years, with a power law fit of $\ln(\text{pH}) = 0.073 \times \ln(\text{MRT}) + 1.62$, $R^2 = 0.98$. For the groundwater from the Mamaku ignimbrite, the pH increases from just 6.3 to 6.6 over the age range from 0 to 100 years, with a power law fit of $\ln(\text{pH}) = 0.025 \times \ln(\text{MRT}) + 1.76$, $R^2 = 0.47$. The groundwater from the sediment formation shows no clear trend in pH with groundwater age, but displays higher pH for rela-

tively young water, pH 6.5–7.2 for water with MRT between 1 and 60 years.

As groundwater becomes more evolved over time due to water–rock interaction, concentrations of phosphorus, silica, bicarbonate and fluoride typically increase due to dissolution of volcanic glass, silicate minerals, carbonates and fluoride likely deposited from the volatile phases in magma exsolved during eruption (Morgenstern and Daughney, 2012). With increasing groundwater age, ion concentrations are expected to increase up to a maximum equilibrium concentration. Groundwater from the sediment formation often follows different or unclear trends compared to the rhyolite ignimbrite and lava formations.

Dissolved reactive phosphate (PO_4–P) in groundwater from all three formations, the rhyolite lava and ignimbrite aquifers, and the sediments originating from the same formations, shows excellent correlation with groundwater age (Fig. 7a, black curve), with $\ln(PO_4-P) = 0.458 \times \ln(\text{MRT}) - 4.72$, $R^2 = 0.94$.

Silica (SiO_2) also shows good correlation with groundwater age for the rhyolite ignimbrite and lava formations (Fig. 7b). The silica concentration of groundwater in lava formations (red circles) increases faster compared to ignimbrite (blue circles). The power fit to the lava data is $\ln(SiO_2) = 0.310 \times \ln(\text{MRT}) + 2.96$, $R^2 = 0.88$ (red curve), and to the ignimbrite data is $\ln(SiO_2) = 0.238 \times \ln(\text{MRT}) + 3.05$, $R^2 = 0.83$ (blue curve). The correlation between silica and groundwater age for the lacustrine sediment aquifers (yellow circles) is rather erratic; high silica concentration can also occur in very young groundwater.

For bicarbonate (HCO_3), only groundwater samples from the Mamaku ignimbrite show a reasonable correlation with groundwater age, with a power fit of $\ln(HCO_3) = 0.206 \times \ln(\text{MRT}) + 2.58$, $R^2 = 0.71$ (Fig. 7c).

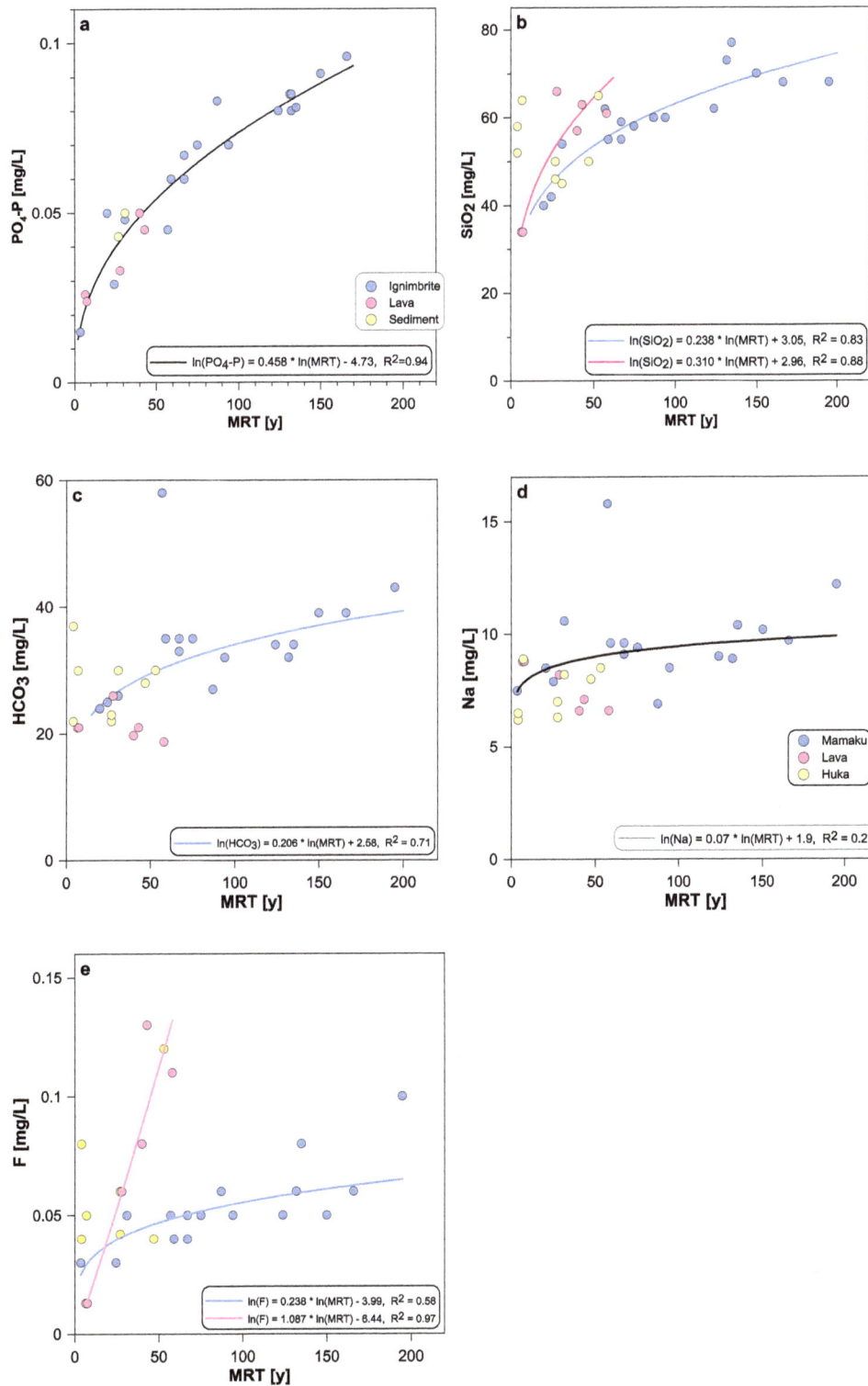

Figure 7. (a) Dissolved reactive phosphate (PO_4–P), (b) silica (SiO_2), (c) bicarbonate (HCO_3), (d) sodium (Na) and (e) fluoride (F) versus mean residence time (MRT). The sample colour code for all panels is shown in (a), and indicates water origin from the relevant geologic formation, as indicated by hierarchical cluster analysis.

The high data point at $58 \, mg \, L^{-1}$ was considered an outlier and not included in the fit.

Sodium (Na) in general also shows increasing concentration with groundwater age because it is part of common minerals and leached from these. But the correlation is poor (Fig. 7d). Note that elevated Na in young groundwater can also be caused by land-use impacts, as observed in other parts of New Zealand (Morgenstern and Daughney, 2012). Considering the high data point at $15.8 \, mg \, L^{-1}$ an outlier (it is from the same site considered as an outlier for HCO_3), the correlation for the data from all three formations is $\ln(Na) = 0.07 \times \ln(MRT) + 1.9$, $R^2 = 0.27$.

Fluoride concentrations (F) show good correlations with age for the rhyolite ignimbrite and lava formations (Fig. 7e), even though the trends are masked by the fact that the concentrations are close to the detection limit. Concentrations increase in lava formations significantly faster, with a power fit of $\ln(F) = 1.087 \times \ln(MRT) - 6.44$, $R^2 = 0.97$, compared to ignimbrite with $\ln(F) = 0.238 \times \ln(MRT) - 3.99$, $R^2 = 0.58$.

In groundwater of the Rotorua catchment (excluding groundwater from the eastern catchment indicating geothermal influence, which is not the subject of this study), the hydrochemistry parameters phosphate, silica, bicarbonate, sodium and fluoride are purely of geologic origin, because they do not display elevated concentrations in young water that was recharged during the time of anthropogenic high intensity land-use activities. The groundwater samples show, for the rhyolite Mamaku ignimbrite and lava formations, excellent correlations across the western and northern Lake Rotorua catchment. The samples in each of these geologic units follow similar trends of hydrochemistry concentration versus mean residence time, indicating the relatively homogeneous nature of these aquifers. Rather erratic trends in water originating from the sediments suggest that these are not a homogeneous formation but rather finely layered lensoidal geologic deposits that vary spatially and support complex or fragmented groundwater systems. Good trends of hydrochemistry versus groundwater age may be an indirect indication of robust age interpretations.

In rhyolite lava formations, geochemical reactions lead to increased pH, Si, and F in groundwater significantly faster than in ignimbrite, indicating higher reaction rates for dissolution of these elements from lava formations. While this is important for understanding water–rock interaction, we do not yet have sufficient information on the lithogeochemistry to develop a mechanistic understanding of the reaction processes.

4.4 Nutrients

Elevated nutrient levels in surface water cause poisonous algal blooms and lake eutrophication. Presence of both phosphate and nitrate, above a threshold concentration, triggers algae blooms in lakes. Limitation of one of these, P or N, can limit algae blooms. In New Zealand, increasing nutrient loads from high-intensity animal farming and fertilizers have triggered lake eutrophication. In the absence of significant overland runoff, nitrate travels from the land to the lake via the groundwater, which eventually discharges into streams and lakes.

Nutrient concentrations in New Zealand groundwaters from agricultural sources have increased steadily after European settlement in the early 19th century and with development of the meat industry after 1880 (Morgenstern and Daughney, 2012). In a national context, for groundwater recharged before 1880 at pre-anthropogenic pristine conditions, low nutrient concentrations prevailed (e.g. nitrate $< 0.2 \, mg \, L^{-1} \, NO_3$–N). In groundwater recharged between 1880 and 1955, nutrient concentrations are slightly elevated due to low-intensity land use. In groundwater recharged after 1955 a sharp increase of nutrient concentrations is observed due to the impact of high-intensity land use after World War II (Morgenstern and Daughney, 2012).

The main nutrients derived from land use in the Rotorua catchment, as indicated by elevated concentrations in young groundwater, are nitrate (NO_3), sulfate (SO_4) and potassium (K). These nutrient concentrations are shown in Figs. 8 and 9a and b versus mean residence time, also correlated to recharge year (upper x-axes). The majority of the chemistry data of the Rotorua data set are from calendar year 2003, therefore mean residence times of about 50 and 125 years correspond to mean groundwater recharge years 1955 and 1880, respectively. Homogeneous nitrate concentrations in discharges from within subcatchments of typically $0.7 \pm 0.2 \, mg \, L^{-1} \, NO_3$–N indicate that nutrient inputs are derived from diffuse rather than a small number of point sources, pointing to agricultural sources.

Figure 8 also includes data (labelled "other") from the sites in the Lake Rotorua catchment that could not be assigned to one of the HCA clusters because these sites had not been analysed for the full suite of hydrochemical parameters required for input into HCA. In several surveys only nitrate was measured to obtain a higher spatial resolution of the nitrate distribution. The analysis of all hydrochemical parameters, as required for HCA, was mainly undertaken at the large discharges into the lake that contain old water, and only few of these sites contain water young enough to show the impact of recent land-use intensification. Therefore the "other" samples were added to Fig. 8 to better represent younger waters. In addition, samples from the eastern catchment having a geothermal signature are also included in the cluster "other". The geothermal influence is minimal and does not affect the nitrate signature, and hence does not bias the display of results in Fig. 8.

Nitrate concentrations (Fig. 8) in oxic groundwaters with MRT > 125 years (recharged prior to 1880) in the Rotorua catchment are higher, with up to about $0.7 \, mg \, L^{-1} \, NO_3$–N (dotted line in Fig. 8) compared to other regions in New Zealand with $0.2 \, mg \, L^{-1} \, NO_3$–N. The reason for elevated nitrate in water despite a high mean residence time is the

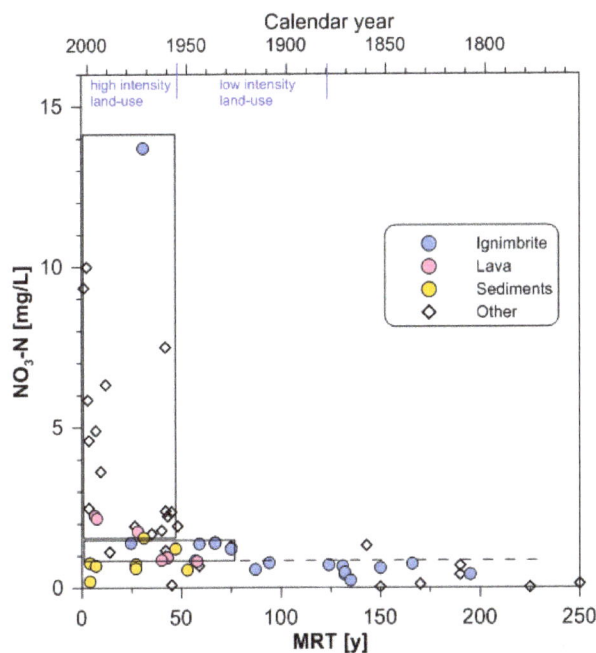

Figure 8. Nitrate (NO_3) versus mean residence time (MRT). The sample colour code indicates water origin from the relevant geologic formation, as indicated by hierarchical cluster analysis.

high degree of mixing in the groundwater discharges from the highly porous unconfined Rotorua ignimbrite aquifers (see next section). In such aquifer conditions, groundwater from short and long flow paths converge at the groundwater discharges, causing a high degree of mixing of young and old water. For example, groundwater with a MRT as high as 170 years, using an exponential piston flow model with 95% exponential flow volume within the total flow volume (see next section), contains over 20 % of water recharged after 1955. This post-1955 water can contribute significant amounts of nitrate from high-intensity land use, raising the nitrate concentration of the water mix considerably, despite such a long MRT.

A significant increase in nitrate occurred only recently (Fig. 8). Apart from one data point, an increase up to 1.5 mg L^{-1} NO_3–N was observed only in water with MRTs of less than 75 years, and a dramatic increase up to 14 mg L^{-1} NO_3–N was observed in water with MRTs of less than 50 years. Note the dramatic increase of nitrate in water with MRT < 20 years, reflecting the increased conversions to dairy farming during the 1980s and 1990s (Rutherford et al., 2011). As the majority of the water discharges into Lake Rotorua are significantly older than a few decades, with MRTs of up to 145 years, the impact of the dairy conversions and their nitrate loads over the recent decades has to a large extent not yet reached the lake. Increased nitrate loads to the lake over the next decades must be expected as these nitrate loads work their way through the large groundwater system and eventually discharge into the streams and lake.

Sulfate and potassium are part of fertilizers and also show elevated concentrations in young groundwater (Fig. 9). Note that sulfate in groundwater in the eastern lake catchment has much elevated concentrations, up to 40 mg L^{-1} SO_4, due to geothermal influence. Groundwater with indications of geothermal influence is not discussed in this study. Also note that sulfate can be biased due to anoxic SO_4 reduction. The data shown are, however, not from anoxic groundwater environments. Sulfate and potassium show slightly elevated concentrations, up to a factor of 3, only in water with MRT < 50 years, corresponding to water recharged after approximately 1950. Sulfate concentrations (in mg L^{-1} SO_4) in the Rotorua volcanic aquifers are significantly lower compared to a national survey: pre-anthropogenic concentration of 2 versus 12, and high-intensity land-use concentrations of up to 6 versus 94 for the Rotorua volcanic aquifers and the national survey (Morgenstern and Daughney, 2012), respectively.

Phosphate, in conjunction with nitrate the cause for lake eutrophication, is not elevated in young groundwater (Fig. 7a) despite its frequent application as super-phosphate fertilizers. Absence of elevated PO_4 in young groundwater implies that fertilizer phosphate from non-point sources has not yet reached the saturated groundwater systems and is still retained in the soil. This finding is consistent with the usually high P-retention scores for ashfall soils and thick unsaturated zones across this region, which are very efficient at buffering P loss. P-retention in soils was also observed in the New Zealand National Groundwater Monitoring Programme across other soil types (Morgenstern and Daughney, 2012).

The presence of elevated PO_4 only in old groundwater indicates that its source is purely due to geological factors, because these waters were recharged before land-use intensification. PO_4 concentrations up to 0.1 mg L^{-1} PO_4–P are observed, due to phosphate leaching from the rhyolite ignimbrite and lava formations. With most groundwater discharging into Lake Rotorua being very old (MRT > 50 years), the water has naturally high PO_4 concentrations, well above the threshold for primary algae production of ca. 0.03 mg L^{-1} total phosphate (Dodds, 2007).

The high phosphate load to the lake via groundwater is natural. As the turn-over time of the lake water is only 2.2 years via the high PO_4-bearing streams, there is a constantly high PO_4 load reaching the lake via all streams. Therefore, the only effective way to limit algae blooms and improve lake water quality in such environments is by limiting the nitrate load.

4.5　Prediction of future nitrate load

The water quality of Lake Rotorua has declined continuously over the past 60 years, responding very slowly to historical agricultural and urban development in the catchment, and large amounts of groundwater have insidiously become contaminated over the last 60 years because of the long travel

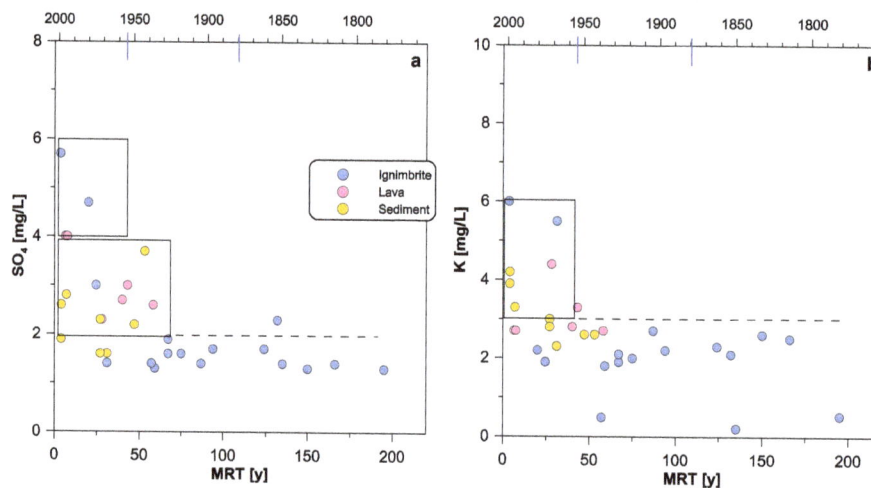

Figure 9. (a) Sulfate (SO4) and (b) potassium (K) versus mean residence time (MRT). The sample colour code indicates water origin from the relevant geologic formation, as indicated by hierarchical cluster analysis. The upper axis indicates calendar year.

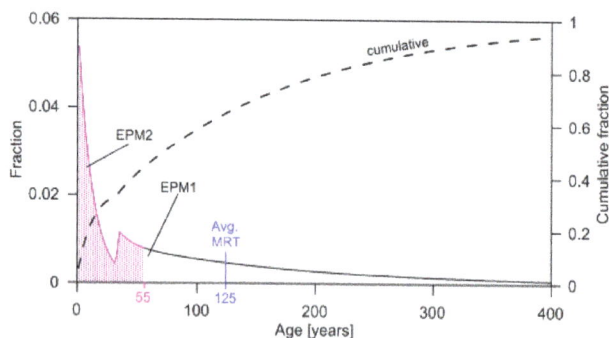

Figure 10. Age distribution for Hamurana Stream (at inflow into Lake Rotorua). The red shaded area indicates the fraction of water that was recharged after land-use intensification. EPM1 and EPM2 are exponential piston flow models.

times through the groundwater system of the Lake Rotorua catchment. The response time of the groundwater system to mitigation action will also be lengthy; it will take similar time frames until the contaminated water is flushed out of the aquifers. To improve lake water quality and define reduction targets for nutrients that affect lake water quality, prediction of future contaminant loads from current and historic activities in the Lake Rotorua catchment are required.

In the previous section we have shown that of the two main contaminants that together cause lake eutrophication, phosphorus is naturally present in the volcanic lake environment, but nitrate from anthropogenic sources has been leaching into the groundwater since the onset of industrial agriculture, delivering increasing nitrate loads to the lake. Figure 8 shows significantly elevated nitrate concentrations in groundwater recharged after 1955.

Due to the large lag time in the groundwater system, these younger groundwaters, with their higher nitrate load, have

not yet worked their way fully through the groundwater system. Significant fractions of the groundwaters discharging to the lake are older (Figs. 3 and 4), with MRT > 50 years, and were recharged before land-use intensification. Therefore the water discharges into the lake are currently still diluted by old pristine water. With the delayed arrival of nitrate from historic land use, which ultimately will discharge from the groundwater system via the springs and streams into the lake, nitrate loads to the lake from historic land-use activities must be expected to increase further in the future. No significant denitrification can be expected in the Rotorua groundwater system (Fig. 6a).

The age distributions functions derived from the tritium time series data in the stream discharges to Lake Rotorua (Table 2) can be used to project the future arrival to the lake of water that was recharged since land-use development in the catchment (Morgenstern and Gordon, 2006). The age distribution function for Hamurana Stream, the largest stream (Table 1), which discharges some of the oldest water to the lake (Fig. 2), is shown in Fig. 10.

Figure 10 shows the two superimposed age distributions of exponential piston flow models: EPM2 with younger water of MRT = 12 years and EPM1 with significantly older water of MRT = 185, together with the average MRT = 125 years between the two models (blue). Only the water younger than 55 years has been recharged after land-use intensification (red shaded) and contains elevated nitrate. The cumulative fraction of land-use impacted water is about 45%, implying that more than half of the water is still pristine old water. After this old water is completely displaced by land-use impacted water, the nitrogen load of Hamurana Stream will approximately double. The projected increase in nitrogen load over time, as derived from the age distribution, is shown in Fig. 11.

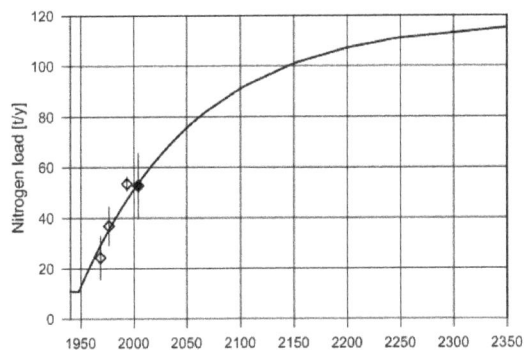

Figure 11. Projected increase over time of nitrogen load to Lake Rotorua from Hamurana Stream.

Nitrate, as opposed to other nitrogen fractions, is clearly the major component of nitrogen in the Rotorua groundwater system (Morgenstern et al., 2004). Concentrations of nitrate in the catchment were low (0.14 mg L^{-1} NO$_3$–N) prior to catchment development, as determined from old groundwater (Morgenstern and Gordon, 2006). The prediction of nitrogen load increase is calculated by scaling the nitrogen load currently measured in the stream (full symbol, Fig. 11) according to its fraction of land-use impacted water to the various years over time, using the age distribution (Fig. 10).

Good agreement with average historic monitoring results of nitrogen loads (Rutherford, 2003, hollow symbols in Fig. 11) confirms that the assumptions regarding the baseline concentration and timing of nitrogen input in the catchment are reasonable. Using the age distributions derived for all stream discharges to the lake, we also projected the total nitrogen load increase to Lake Rotorua (Morgenstern and Gordon, 2006). In regard to phosphorus, there are no elevated phosphorus concentrations in young groundwater (Fig. 7a) and the phosphorus load to Lake Rotorua is projected to stay constant, as long as fertilizer phosphate does not break through the soil into the groundwater.

The timescales necessary for the Hamurana Stream to adjust to changes in land-use activities in the catchment are long. Due to the long residence time of the water in the large aquifer system, it takes more than a hundred years for the groundwater discharge to the lake to adjust to changes in land-use activities. These timescales apply to activities that cause contamination, but also to remediation action.

This projection of nitrogen load via the stream is based on actual nitrogen concentrations in the stream (combined with the age of the water) and accounts only for the nitrogen from land-use activities that leaches out of the root zone of agricultural land into the deeper part of the groundwater system. Any nitrogen uptake in the soil is already taken into account.

The above nitrogen prediction is based on constant nitrogen input since catchment development. This trend will, however, be exacerbated by any further intensification of land

use within the catchment over recent decades, as this recently recharged water has largely not yet reached the streams.

5 Conclusions

This study shows how the isotopic and chemistry signature of groundwater can be used to help determine the sources and the dynamics of groundwater and contaminants that travel with it, in particular in complex groundwater systems that are difficult to characterize using conventional hydrogeologic methods, such as that of the Lake Rotorua catchment. The isotopic and chemistry signatures of the major groundwater-dominated stream discharges to the lake, after passing through the large aquifer system of the catchment, allow us to understand groundwater processes and lag time on a catchment scale.

Tritium time series data and complementary age tracers SF$_6$ and CFCs can be used to establish age distribution parameters, allowing for understanding of groundwater processes and dynamics, and the timing of groundwater contamination. This is particularly useful in catchments where little information is available on historic land-use activities.

After long-standing controversies (e.g. White et al., 2004; Rutherford et al., 2011), hierarchical cluster analysis of the water chemistry parameters has provided evidence about the recharge areas and hydraulic connections of the large springs near the northern shore of Lake Rotorua. Streams and shallow wells that gain most of their flow and recharge within the lacustrine sediments display a characteristic hydrochemical signature. Hydrochemistry of the water draining the Ngongotaha lava dome also has a characteristic signature due to interaction with lava formations. Only where the lava dome intercepts the palaeo-lake sediments is the groundwater flow from the lava formation forced to the surface due to the low permeability of the sediments. The water from the ignimbrite also displays a characteristic hydrochemical signature. Similarly to the discharges from the lava formation, the water from the ignimbrite discharges near the intercept of the ignimbrite formation with the palaeo-lake sediments, indicating that the groundwater flow from the ignimbrite is forced to the surface due to the low permeability of the sediments. The largest springs, discharging in the northwest of the lake, emerge close to the lake shore within the sediment area, but the ignimbrite signature of these water discharges implies that these springs drain the Mamaku ignimbrite plateau, which has negligible surface runoff, through the lake sediment layers in slope areas where the sediments are thinner and weaker.

Groundwater chemistry and age data show clearly the source of nutrients that discharge with the groundwater into the lake and cause lake eutrophication. Low nitrate concentration in old oxic groundwater and high nitrate concentration in young groundwater recharged after catchment development in the 1950s implies an anthropogenic source of nitrate from agricultural activities, while low phosphate (PO_4) concentrations in young groundwater but high PO_4 concentrations in old groundwater imply a geologic source. High PO_4 is a natural constituent of the groundwater that discharges via the streams into the lake, and with a turn-over time of the lake water of only 2.2 years, there is a constantly high PO_4 load reaching the lake via all streams. Therefore, the only effective way to limit algae blooms and improve lake water quality in such environments is by limiting the nitrate load.

The groundwater in the Rotorua catchment, once it has passed through the soil zone, shows no further decrease in dissolved oxygen over the full range of residence time of the water in the aquifer, indicating an absence of significant microbial reactions due to limitation of electron donors in the aquifer (e.g. organic matter) that could facilitate microbial denitrification reactions (Kendall and McDonnell, 1998; Tesoriero et al., 2007). Nitrate from land-use activities that leaches out of the root zone of agricultural land into the deeper part of the groundwater system is unlikely to undergo any significant degree of reduction through denitrification and must be expected to travel with the groundwater to the lake.

The old age of the water, with mean residence time of > 50 years for most water discharges to the lake, implies that there is a large lag time for transmission of the nitrate through the groundwater system. Younger groundwaters, with their higher nitrate load, have not yet worked their way fully through the groundwater system. With increasing arrival of this nitrate from historic land uses, a further increase of the nitrate load to the lake must be expected in the future.

The old age and the highly mixed nature of the water discharges imply a very slow and lagged response of the streams and the lake to anthropogenic contaminants in the catchment, such as nitrate. Using the age distribution as deduced from tritium time series data measured in the stream discharges to the lake allows extrapolation of the nutrient load from historic land-use activities into the future. For Hamurana Stream, the largest stream to Lake Rotorua, it takes more than a hundred years for the groundwater-dominated stream discharge to adjust to changes in land-use activities. These time scales apply to activities that cause contamination, but also to remediation action.

Without age information on the groundwater-dominated streams, it would be difficult to obtain such an understanding of groundwater process, groundwater dynamics, and contaminant loads that travel with the groundwater.

Acknowledgements. We thank personnel from Bay of Plenty Regional Council for assistance with sample collection and for provision of some information used in this study, and Eileen McSaveney for editing the paper. This research was supported by funding from the Bay of Plenty Regional Council and the Ministry of Science and Innovation (Contract C05X1002).

Edited by: C. Harman

References

Bethke, C. M. and Johnson, T. M.: Groundwater age and age dating, Annu. Rev. Earth Pl. Sc., 36, 121–152, 2008.

Beyer, M., Morgenstern, U., and Jackson, B.: Review of dating techniques for young groundwater (< 100 years) in New Zealand, J. Hydrol., 53, 93–111, 2014.

Böhlke, J. K. and Denver, J. M.: Combined use of groundwater dating, chemical, and isotopic analyses to resolve the history and fate of nitrate contamination in two agricultural watersheds, Atlantic coastal plain, Maryland, Water Resour. Res., 31, 2319–2339, 1995.

Böhlke, J. K., Wanty, R., Tuttle, M., Delin, G., and Landon, M.: Denitrification in the recharge area and discharge area of a transient agricultural nitrate plume in a glacial outwash sand aquifer, Minnesota, Water Resour. Res., 38, 1105, doi:10.1029/2001WR000663, 2002.

Broers, H. P.: The spatial distribution of groundwater age for different geohydrological situations in the Netherlands: implications for groundwater quality monitoring at the regional scale, J. Hydrol., 299, 84–106, 2004.

Burger, D. F., Hamilton, D. P., and Pilditch, C. A.: Modelling the relative importance of internal and external nutrient loads on water column nutrient concentrations and phytoplankton biomass in a shallow polymictic lake, Ecol. Model., 211, 411–423, 2011.

Burns, D. A., Plummer, L. N., McDonnell, J. J., Busenberg, E., Casile, G. C., Kendall, C., Hooper. R. P., Freer. J. E., Peters, N. E., Beven, K., and Schlosser, P.: The geochemical evolution of riparian ground water in a forested piedmont catchment, Ground Water, 41, 913–925, 2003.

Burton, W. C., Plummer, L. N., Busenberg, E., Lindsey, B. D., and Gburek, W. J.: Influence of fracture anisotropy on ground water ages and chemistry, Valley and Ridge Province, Pennsylvania, Ground Water, 40, 242–257, 2002.

Cartwright, I. and Morgenstern, U.: Constraining groundwater recharge and the rate of geochemical processes using environmental isotopes and major ion geochemistry: Ovens Catchment, southeast Australia, J. Hydrol., 475, 137–149, 2012.

Cook, P. G. and Herczeg, A. L.: Environmental Tracers in Subsurface Hydrology, Kluwer Academic Publishers, Boston, Dordrech, London, 1999.

Cook, P. G. and Solomon, D. K.: Transport of atmospheric trace gases to the water table: Implication for groundwater dating with CFCs and Krypton 85, Water Resour. Res., 31, 263–270, 1995.

Cook, P. G. and Solomon, D. K.: Recent advances in dating young groundwater: Chlorofluorocarbons, $^3H/^3He$, and ^{85}Kr, J. Hydrol., 191, 245–265, 1997.

Cornaton, F. J.: Transient water age distributions in environmental flow systems: The time-marching Laplace transform solution technique, Water Resour. Res., 48, W03524, doi:10.1029/2011WR010606, 2012.

Cornaton, F. J., Park, Y.-J., and Deleersnijder, E.: On the biases affecting water ages inferred from isotopic data, J. Hydrol., 410, 217–225, 2011.

Darling, W. G., Morris, B., Stuart, M. E., and Gooddy, D. C.: Groundwater age indicators from public supplies tapping the chalk aquifer of Southern England, J. Chart. Inst. Water Environ. Man., 19, 30–40, 2005.

Daughney, C. J., Jones, A., Baker, T., Hanson, C., Davidson, P., Reeves, R. R., Zemansky, G. M., and Thompson, M.: A national protocol for state of the environment groundwater sampling in New Zealand, Miscellaneous Series 5, Institute of Geological & Nuclear Sciences, Wellington, New Zealand, 2007.

Daughney, C. J., Morgenstern, U., van der Raaij, R., and Reeves, R. R.: Discriminant analysis for estimation of groundwater age from hydrochemistry and well construction: Application to New Zealand aquifers, Hydrogeol. J., 18, 417–428, 2010.

Dell, P. M.: The effect of afforestation on the water resources of the Mamaku Plateau region, MSc Thesis, University of Waikato, New Zealand, 1982.

Dodds, W. K.: Trophic state, eutrophication and nutrient criteria in streams, Trends Ecol. Evol., 22, 669–676, 2007.

Donath, F. M., Daughney, C. J., Morgenstern, U., Cameron, S. G. and Toews, M. W.: Hydrochemical interpretation of groundwater-surface water interactions at regional and local scales, Lake Rotorua catchment, New Zealand, J. Hydrol. (NZ), in review, 2014.

Eberts, S. M., Böhlke, J. K., Kauffman, L. J., and Jurgens, B. C.: Comparison of particle-tracking and lumped-parameter age-distribution models for evaluating vulnerability of production wells to contamination, Hydrogeol. J., 20, 263–282, 2012.

Edmunds, W. M. and Smedley, P. L.: Residence time indicators in groundwater: the East Midlands Triassic sandstone aquifer, Appl. Geochem., 15, 737–752, 2000.

Glynn, P. D. and Plummer, L. N.: Geochemistry and the understanding of ground-water systems, Hydrogeol. J., 13, 263–287, 2005.

Goode, D. J.: Direct simulation of groundwater age. Wat. Resour. Res., 32, 289–296, 1996.

Gordon, D.: Bay of Plenty, in: Groundwaters of New Zealand, edited by: Rosen, M. R. and White, P. A., New Zealand Hydrological Society, Wellington, New Zealand, 327–354, 2001.

Gusyev, M. A., Abrams, D., Toews, M. W., Morgenstern, U., and Stewart, M. K.: A comparison of particle-tracking and solute transport methods for simulation of tritium concentrations and groundwater transit times in river water, Hydrol. Earth Syst. Sci., 18, 3109–3119, doi:10.5194/hess-18-3109-2014, 2014.

Heaton, T. H. E. and Vogel, J. C.: Excess air in groundwater, J. Hydrol., 50, 201–216, 1981.

Hoare, R. A.: Inflows to Lake Rotorua, J. Hydrol., 19, 49–59, 1980.

Hoare, R. A.: Nitrogen and phosphorus in the catchment of Lake Rotorua. Publication No. 11, Water Quality Centre, Ministry of Works and Development, Hamilton, New Zealand, 1987.

Jurgens, B. C., Böhlke, J. K., and Eberts, S. M.: TracerLPM (Version 1): An Excel® workbook for interpreting groundwater age distributions from environmental tracer data, US Geological Survey Techniques and Methods Report 4-F3, 60 pp., http://ca.water.usgs.gov/user_projects/TracerLPM/ (last access: 4 February 2015), 2012.

Katz, B. G., Böhlke, J. K., and Hornsby, H. D.: Timescales for nitrate contamination of spring waters, northern Florida, USA, Chem. Geol., 179, 167–186, 2001.

Katz, B. G., Chelette, A. R., and Pratt, T. R.: Use of chemical and isotopic tracers to assess nitrate contamination and ground-water age, Woodville Karst Plain, USA, J. Hydrol., 289, 36–61, 2004.

Katz, B. G., Plummer, L. N., Busenberg, E., Revesz, K. M., Jones, B. F., and Lee, T. M.: Chemical evolution of groundwater near a sinkhole lake, northern Florida 2. Chemical patterns, mass transfer modelling, and rates of mass transfer reactions, Water Resour. Res., 31, 1565–1584, 1995.

Kendall, C. and McDonnell, J. J. (Eds.): Isotope Tracers in Catchment Hydrology, Elsevier Science B. V., Amsterdam, 519–576, 1998.

Leonard, G. S., Begg, J. G., Wilson, C. J. N. (Compilers): Geology of the Rotorua area, Institute of Geological and Nuclear Sciences 1 : 250,000 geological map 5, GNS Science, Lower Hutt, New Zealand, 2010.

Lock, K. and Kerr, S.: Nutrient trading in Lake Rotorua: Social, cultural, economic and environmental issues around a nutrient trading system, Motu Manuscript MEL0319, Motu Economic and Public Policy Research, Wellington, New Zealand, 2008.

MacDonald, A. M., Darling, W. G., Ball, D. F., and Oster, H.: Identifying trends in groundwater quality using residence time indicators: An example from the Permian aquifer of Dumfries, Scotland, Hydrogeol. J., 11, 504–517, 2003.

Maloszewski, P. and Zuber, A.: Determining the turnover time of groundwater systems with the aid of environmental tracers, 1. Models and their applicability, J. Hydrol., 57, 207–231, 1982.

Maloszewski, P. and Zuber, A.: Influence of matrix diffusion and exchange reactions on radiocarbon ages in fissured carbonate aquifers, Water Resour. Res., 27, 1937–1945, 1991.

McDonnell, J. J., McGuire, K., Aggarwal, P., Beven, K., Biondi, D., Destouni, G., Dunn, S., James, A., Kirchner, J., Kraft, P., Lyon, S., Maloszewski, P., Newman, B., Pfister, L., Rinaldo, A., Rodhe, A., Sayama, T., Seibert, J., Solomon, K., Soulsby, C., Stewart, M., Tetzlaff, D., Tobin, C., Troch, P., Weiler, M., Western, A., Wörman, A., and Wrede, S.: How old is streamwater? Open questions in catchment transit time conceptualization, modelling and analysis, Hydrol. Process., 24, 1745–1754, 2010.

McGuire, K. J., DeWalle, D. R., and Gburek, W. J.: Evaluation of mean residence time in subsurface waters using oxygen-18 fluctuations during drought conditions in the mid-Appalachians, J. Hydrol., 261, 132–149, 2002.

Milner, D. M., Cole, J. W., and Wood, C. P.: Mamaku Ignimbrite: a caldera-forming ignimbrite erupted from a compositionally zoned magma chamber in Taupo Volcanic Zone, New Zealand, J. Volcanol. Geoth. Res., 122, 243–264, 2003.

Molson, J. W. and Frind, E. O.: How old is the water? Simulating groundwater age at the watershed scale, IAHS Publ., 297, 482–488, 2005.

Moore, K. B., Ekwurkel, B., Esser, B. K., Hudson, G. B., and Moran, J. E.: Sources of groundwater nitrate revealed using residence time and isotope methods, Appl. Geochem., 21, 1016–1029, 2006.

Morgenstern, U.: Assessment of age distribution in groundwater, in: Proceedings of the 2nd Asia Pacific Association of Hydrology and Water Resources Conference, vol. 1, 5–8 July 2004, Singapore, 580–587, 2004.

Morgenstern, U. and Daughney, C. J.: Groundwater age for identification of baseline groundwater quality and impacts of land-use intensification – The National Groundwater Monitoring Programme of New Zealand, J. Hydrol., 456–457, 79–93, 2012.

Morgenstern, U. and Gordon, D.: Prediction of Future Nitrogen Loading to Lake Rotorua, GNS Science Report 2006/10, Lower Hutt, New Zealand, GNS Science, 28 pp., available at: http://www.boprc.govt.nz/media/33280/Report-060600-PredictionofFutureNloadLRotorua.pdf (last access: 4 February 2015), 2006.

Morgenstern, U. and Taylor, C. B.: Ultra Low-level tritium measurement using electrolytic enrichment and LSC, Isotop. Environ. Health Stud., 45, 96–117, 2009.

Morgenstern, U., Reeves, R., Daughney, C., Cameron, S., and Gordon, D.: Groundwater age and Chemistry, and Future Nutrient Load for Selected Rotorua Lakes Catchments, Institute of Geological & Nuclear Sciences Science Report 2004/31, GNS Science, Lower Hutt, New Zealand, 74 pp., available at: http://www.boprc.govt.nz/media/32425/GNS-091118 (last access: 4 February 2015), 2004.

Morgenstern, U., Stewart, M. K., and Stenger, R.: Dating of streamwater using tritium in a post nuclear bomb pulse world: continuous variation of mean transit time with streamflow, Hydrol. Earth Syst. Sci., 14, 2289–2301, doi:10.5194/hess-14-2289-2010, 2010.

Morgenstern, U., van der Raaij, R., and Baalousha, H.: Groundwater flow pattern in the Ruataniwha Plains as derived from the isotope and chemistry signature of the water, GNS Science Report 2012/23, GNS Science, Lower Hutt, New Zealand, 50 pp., available at: http://www.gns.cri.nz/static/pubs/2012/SR_2012-023.pdf (last access: 5 February 2015), 2012.

Morris, B., Stuart, M. E., Darling, W. G., and Gooddy, D. C.: Use of groundwater age indicators in risk assessment to aid water supply operational planning, J. Chart. Inst. Water Environ. Man., 19, 41–48, 2005.

New Zealand Ministry of Health: Drinking-water standards for New Zealand 2005 (revised 2008), Ministry of Health, Wellington, New Zealand, 2008.

Rosen, M. R., Milner, D., Wood, C. P., Graham, D., and Reeves, R.: Hydrogeologic investigation of groundwater flow in the Taniwha Springs area. Institute of Geological and Nuclear Sciences Client Report 72779C.10, GNS Science, Lower Hutt, New Zealand, 1998.

Rutherford, J. C., Pridmore, R. D., and White, E.: Management of phosphorus and nitrogen inputs to Lake Rotorua, New Zealand, J. Water Resour. Pl. Manage., 115, 431–439, 1989.

Rutherford, K.: Lake Rotorua Nutrient Load Targets, NIWA Client Report HAM2003-155, National Institute of Water and Atmospheric Research, Hamilton, New Zealand, 2003.

Rutherford, K., Tait, A., Palliser, C., Wadhwa, S., and Rucinski, D.: Water balance modelling in the Lake Rotorua catchment, NIWA Client Report HAM2008-048, National Institute of Water and Atmospheric Research, Hamilton, New Zealand, 2008.

Rutherford, K., Palliser, C., and Wadhwa, S.: Nitrogen exports from the Lake Rotorua catchment – calibration of the ROTAN model, NIWA Client Report HAM2009-019, National Institute of Water and Atmospheric Research, Hamilton, New Zealand, 2009.

Rutherford, K., Palliser, C., and Wadhwa, S.: Prediction of nitrogen loads to Lake Rotorua using the ROTAN model, NIWA Client Report HAM2010-134, National Institute of Water and Atmospheric Research, Hamilton, New Zealand, 2011.

Spinks, K. D., Acocella, V., Cole, J. W., and Bassett, K. N.: Structural control of volcanism and caldera development in the transtensional Taupo Volcanic Zone, New Zealand, J. Volcanol. Geoth. Res., 144, 7–22, 2005.

Stewart, M. K. and Morgenstern, U.: Age and source of groundwater from isotope tracers, in: Groundwaters of New Zealand, edited by: Rosen, M. R. and White, P. A., New Zealand Hydrological Society, Wellington, New Zealand, 161–183, 2001.

Stewart, M. K. and Thomas, J. T.: A conceptual model of flow to the Waikoropupu Springs, NW Nelson, New Zealand, based on hydrometric and tracer (^{18}O, Cl, ^{3}H and CFC) evidence, Hydrol. Earth Syst. Sci., 12, 1–19, doi:10.5194/hess-12-1-2008, 2008.

Stewart, M. K., Mehlhorn, J., and Elliott, S.: Hydrometric and natural tracer (^{18}O, silica, ^{3}H and SF$_6$) evidence for a dominant groundwater contribution to Pukemanga Stream, New Zealand, Hydrol. Process., 21, 3340–3356, 2007.

Stewart, M. K., Morgenstern, U., and Mc Donnell, J. J.: Truncation of stream residence time: how the use of stable isotopes has skewed our concept of streamwater age and origin, Hydrol. Process., 24, 1646–1659, 2010.

Taylor, C. B. and Stewart, M. K.: Hydrology of the Rotorua Geothermal Aquifer, NZ, in: Isotope Techniques in Water Resources Development, IAEA STI/PUB/757, IAEA, Vienna, 25–45, 1987.

Tesoriero, A. J. and Puckett, L. J.: O$_2$ reduction and denitrification rates in shallow aquifers, Water Resour. Res., 47, W12522, doi:10.1029/2011WR010471, 2011.

Tesoriero, A. J., Saad, D. A., Buriow, K. R., Frick, E. A., Puckett, L. J., Barbash, J. E.: Linking groundwater age and chemistry data along flow-paths: implications for trends and transformations of nitrate and pesticides, J. Contam. Hydrol. 94, 139–155, 2007.

van der Raaij, R. and Beyer, M.: Use of CFCs and SF$_6$ as groundwater age tracers in New Zealand, J. Hydrol., accepted, 2014.

Weissman, G. S., Zhang, Y., LaBolle, E. M., and Fogg, G. E.: Dispersion of groundwater age in an alluvial aquifer, Water Resour. Res., 38, 16.1–16.8, 2002.

White, P. A. and Rutherford, K.: Groundwater catchment boundaries of Lake Rotorua, Institute of Geological & Nuclear Sciences Consultancy Report 2009/75LR, GNS Science, Lower Hutt, New Zealand, 2009.

White, P. A., Cameron, S. G., Kilgour, G., Mroczek, E., Bignall, G., Daughney, C., and Reeves, R. R.: Review of groundwater in the Lake Rotorua catchment. Institute of Geological & Nuclear Sciences Consultancy Report 2004/130, GNS Science, Lower Hutt, New Zealand, 2004.

White, P. A., Kilgour, G. N., Hong, T., Zemansky, G., and Wall, M.: Lake Rotorua groundwater and Lake Rotorua nutrients Phase 3 science programme technical report, Institute of Geological & Nuclear Sciences Consultancy Report 2007/220, GNS Science, Lower Hutt, New Zealand, 2007.

Wilson, C. J. N., Houghton, B. F., McWilliams, M. O., Lanphere, M. A., Weaver, S. D., and Briggs, R. M.: Volcanic and structural evolution of Taupo Volcanic Zone: A review, J. Volcanol. Geoth. Res., 68, 1–28, 1995.

Zoellmann, K., Kinzelbachach, W., and Fulda, C.: Environmental tracer transport (^{3}H and SF$_6$) in the saturated and unsaturated zones and its use in nitrate pollution management, J. Hydrol., 240, 187–205, 2001.

Zuber, A., Witczak, S., Rózaniski, K., Sìliwka, I., Opoka, M., Mochalski, P., Kuc, T., Karlikowska, J., Kania, J., Jackowicz-Korczynìski, M., and Dulinìski, M.: Groundwater dating with ^{3}H and SF$_6$ in relation to mixing patterns, transport modelling and hydrochemistry, Hydrol. Process., 19, 2247–2275, 2005

11

An efficient workflow to accurately compute groundwater recharge for the study of rainfall-triggered deep-seated landslides, application to the Séchilienne unstable slope (western Alps)

A. Vallet, C. Bertrand, O. Fabbri, and J. Mudry

UMR6249 – Chrono-Environnement, Université de Franche-Comté, 16 route de Gray, 25030 Besançon CEDEX, France

Correspondence to: A. Vallet (aurelien.vallet@univ-fcomte.fr)

Abstract. Pore water pressure build-up by recharge of underground hydrosystems is one of the main triggering factors of deep-seated landslides. In most deep-seated landslides, pore water pressure data are not available since piezometers, if any, have a very short lifespan because of slope movements. As a consequence, indirect parameters, such as the calculated recharge, are the only data which enable understanding landslide hydrodynamic behaviour. However, in landslide studies, methods and recharge-area parameters used to determine the groundwater recharge are rarely detailed. In this study, the groundwater recharge is estimated with a soil-water balance based on characterisation of evapotranspiration and parameters characterising the recharge area (soil available water capacity, runoff and vegetation coefficient). A workflow to compute daily groundwater recharge is developed. This workflow requires the records of precipitation, air temperature, relative humidity, solar radiation and wind speed within or close to the landslide area. The determination of the parameters of the recharge area is based on a spatial analysis requiring field observations and spatial data sets (digital elevation models, aerial photographs and geological maps). This study demonstrates that the performance of the correlation with landslide displacement velocity data is significantly improved using the recharge estimated with the proposed workflow. The coefficient of determination obtained with the recharge estimated with the proposed workflow is 78 % higher on average than that obtained with precipitation, and is 38 % higher on average than that obtained with recharge computed with a commonly used simplification in landslide studies (recharge = precipitation minus non-calibrated evapotranspiration method).

1 Introduction

Pore water pressure build-up by recharge of aquifers is one of the main triggering factors of destabilisation of deep-seated landslides (Noverraz et al., 1998; Van Asch et al., 1999; Guglielmi et al., 2005; Bogaard et al., 2007; Bonzanigo et al., 2007). In most deep-seated landslides, pore water pressure data are not available since piezometers, if any, have a very short lifespan because of slope movements. In addition, landslides show heterogeneous, anisotropic and discontinuous properties (Cappa et al., 2004; Binet et al., 2007a) and local measurements are rarely representative of the overall behaviour of the landslide aquifers. In the absence of piezometric measurements, the groundwater recharge is used as the most relevant parameter to characterise the pore water pressure of the landslide aquifers. Groundwater recharge (hereafter recharge), also referred to as deep percolation, is the part of the precipitation which recharges the saturated zones (aquifers).

Landslide studies involve a wide range of specialities (subsurface geophysics, structural geology, modelling, geotechnics, and geomechanics). Scientists or engineers in charge of landslides may not have the required hydrology knowledge to accurately estimate the recharge. In most cases, deep-seated landslide studies devoted to characterise the rainfall–destabilisation relationships do not take into account recharge with enough accuracy. In particular, some studies estimate the recharge without calibration of the evapotranspiration estimation methods and without soil-water balance (Canuti et al., 1985; Alfonsi, 1997; Hong et al., 2005; Binet et al., 2007b; Durville et al., 2009; Pisani et al., 2010; Prokešová et al., 2013). Lastly, several studies use precipitation data instead of the recharge (Rochet et al., 1994; Zêzere et al., 2005; Meric et al., 2006; Helmstetter and Garambois, 2010; Belle et al., 2014). These approaches can overestimate the groundwater recharge and can thus bias the characterisation of the relationship between rainfall and destabilisation. A more accurate estimation of the groundwater recharge signal can improve the accuracy of these studies. So far, no computation workflow has been proposed to estimate simply and accurately the recharge in the context of landslide studies.

Patwardhan et al. (1990) showed that the soil-water balance method is an accurate way to estimate groundwater recharge. Recharge computation with a soil-water balance depends mainly on the surface runoff, the soil available water capacity (SAWC) and the specific vegetation (so-called crop) evapotranspiration (ET_c, also referred to as potential evapotranspiration), itself being deduced from reference vegetation evapotranspiration (ET_0) with a vegetation coefficient (K_c). The Penman–Monteith method (Eq. A6 in Appendix A), hereafter referred to as the ET_0 standard equation or FAO-56 PM, developed in the paper FAO-56 (Food and Agriculture Organization of the United Nations) is considered by the scientific community as a global standard method to estimate ET_0 worldwide (Jensen et al., 1990; Allen et al., 1998). This method requires the knowledge of the air relative humidity, the air temperature, the wind speed and the solar radiation. However, most weather stations in landslide areas record only air temperature and rainfall. Unlike the FAO-56 PM method, methods based only on air temperature and solar radiation (R_S) allow for a simpler expression of ET_0 (Tabari et al., 2013). Besides, R_S can also be estimated only from air temperature (Almorox, 2011), thus allowing ET_0 to be obtained only from air temperature records. These reduced-set methods are developed under specific site conditions and must be calibrated in order to improve accuracy (Allen et al., 1994; Shahidian et al., 2012).

The objective of this study is to develop a parsimonious, yet robust, guideline workflow to calculate time series of groundwater recharge at the scale of the recharge area, time series that can subsequently be used as a deterministic variable in landslide studies. To maximise the accessibility to various user groups, we strive to develop an efficient method, balancing technical accuracy with operational simplicity. The proposed workflow is applied on the deep-seated Séchilienne landslide. To test its reliability, a correlation analysis is used to evaluate whether the calculated groundwater recharge is more strongly correlated with measured land mass displacement velocities than with precipitation or with recharge estimated with a common simplification in landslide studies (recharge = precipitation minus non-calibrated ET_0 (Canuti et al., 1985; Binet et al., 2007b; Pisani et al., 2010; Prokešová et al., 2013). The significance of the correlations is assessed with bootstrap tests. The proposed study aims at showing that an accurate estimation of the recharge can significantly improve the results of rainfall–displacement studies.

2 Method

2.1 General workflow

In the case of deep-seated landslides triggered by deep water-saturated zones, the impact of a multiday cumulative rainfall is far more significant than rainfall duration or intensity (Van Asch et al., 1999; Guzzetti et al., 2008). For these reasons, the workflow is developed to compute daily groundwater recharge. Similarly, this study is based on displacement recorded at a daily time step. For the sake of simplicity, the daily displacement, equivalent to a velocity measurement in millimetres per day, is hereafter referred to as displacement. The groundwater recharge is estimated with a soil-water balance based on characterisation of ET_0 and parameters characterising the recharge area (SAWC, runoff and K_c). The computation workflow (Fig. 1), hereafter referred to as LRIW (Landslide Recharge Input Workflow), includes four steps.

The estimation of the ET_0 requires the records of air temperature within the landslide area and relative humidity, solar radiation and wind speed within or close to the landslide area. In the case of a landslide-located weather station recording only the temperature, the first step (detailed in Sect. 2.2) consists of a regional calibration of ET_0 and R_S reduced-set equations (equations detailed in Appendix A). The calibrated methods then allow estimating evapotranspiration based only on temperature records. In the case of a landslide weather station recording the full set of parameters, the first step can be skipped and the FAO-56 PM method can then be used to estimate ET_0. The second step (detailed in Sect. 2.3) consists in estimating the recharge-area parameters (surface runoff, SAWC and K_c) using a GIS (geographic information systems) composite method requiring field observations and spatial data sets (digital elevation models (DEMs), aerial photographs and geological maps). The third step (detailed in Sect. 2.4) uses a soil-water balance to estimate the recharge with the estimated ET_0 and the estimation of the recharge-area parameters. The fourth step (detailed in Sect. 2.5) consists of a sensitivity analysis based on a recharge-displacement velocity correlation and is performed in order to refine the estimations of SAWC and runoff coefficient.

2.2 Step 1: regional calibration of ET_0 and R_S methods

ET_0 reduced-set and R_S temperature methods were initially developed for given regions or sites with their own climatic conditions and must be calibrated to take into account the weather conditions of the study site. Details about calibration can be found in the literature (Allen et al., 1994; Itenfisu et al., 2003; Lu et al., 2005; Alkaeed et al., 2006; Alexandris et al., 2008; Shahidian et al., 2012; Tabari et al., 2013).

The regional calibration method (step 1; Fig. 1) is performed using the records of nearby weather stations (here-

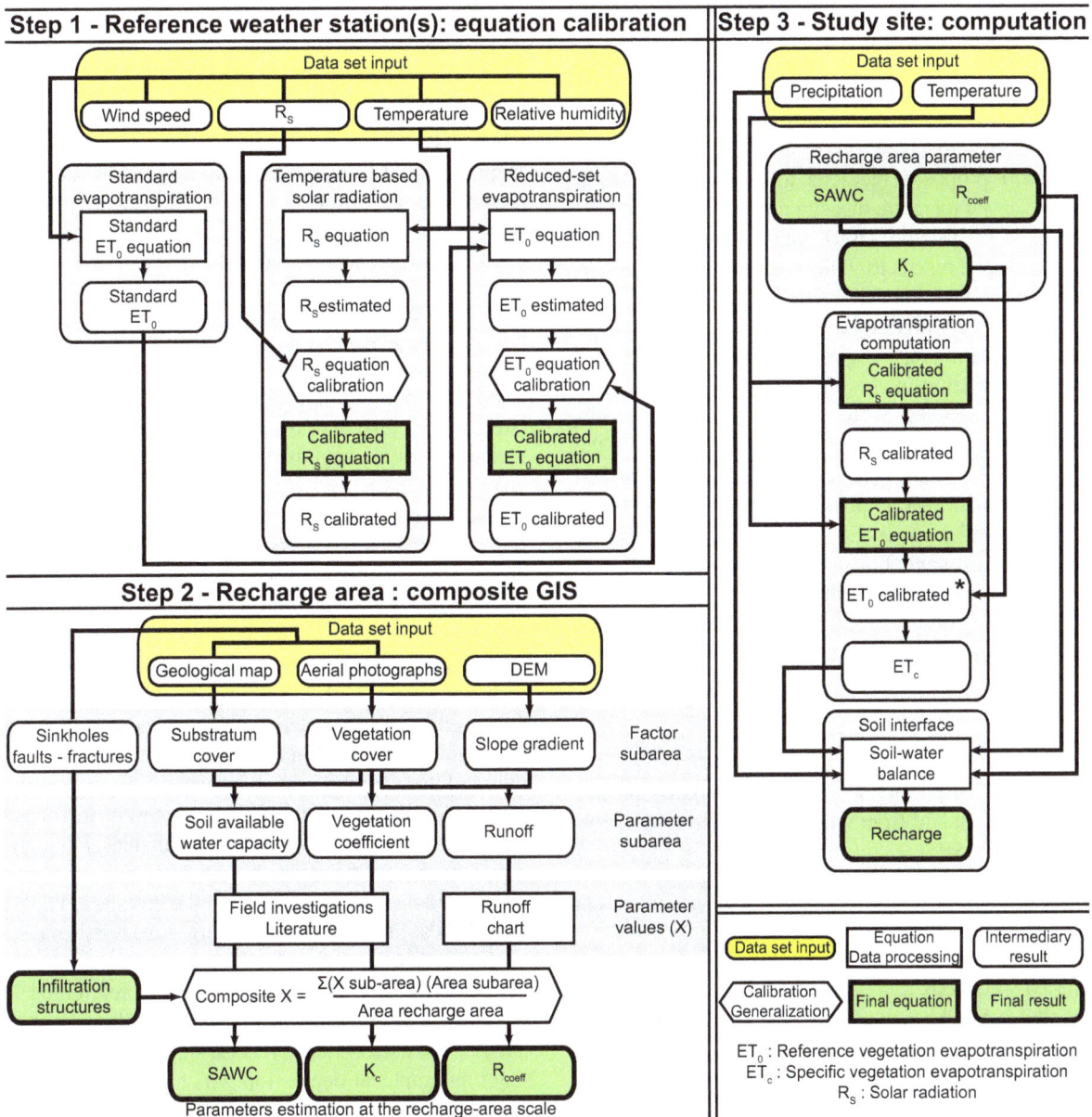

Figure 1. LRIW diagram. Step 1: calibration of standard ET_0 and R_S methods. Step 2: estimation of recharge-area parameters required for the soil-water balance (R_{coeff}, K_c and SAWC) and the infiltration structures. Step 3: computation of the recharge with the soil-water balance. * In the case of a landslide-located weather station recording the full set of parameters, the first step can be skipped and the ET_0 of step 3 can be estimated directly at the study site with the standard ET_0 method (FAO-56 PM method).

after referred to as reference weather stations) having similar climatic conditions as the study site and recording the required meteorological parameters. The calibration of R_S and ET_0 methods are performed for each reference weather station (local scale). The local adjustment coefficients of the reference stations are then averaged in order to define a regional calibration. The user has to maintain a balance between the number of selected reference stations and the necessity for these stations to be located in areas with climatic conditions similar to those of the study site. For sites with a sparse weather station network, one reference station can be sufficient for the calibration, provided that this station has the same weather conditions as those of the studied site.

The performance assessment of regional-scale calibrated methods is based on the comparison between observed measurements and calibrated estimates for R_S and between FAO-56 PM estimates and calibrated estimates for ET_0 for each reference weather station. Performance indicators are the coefficient of determination (R^2), the slope and the intercept from linear regression (independent variable: estimated parameter; dependant variable: reference parameter), and the root mean square error (RMSE).

2.2.1 Solar radiation methods

Bristow and Campbell (1984) and Hargreaves and Samani (1985) proposed methods to compute R_S based solely on the air temperature measurement (Eqs. A1 and A2 in Appendix A). Castellvi (2001) demonstrated that both methods show good results for daily frequencies. The coefficients of the Bristow–Campbell method have to be evaluated. The coefficients of the Hargreaves–Samani method have default values. However, Trajkovic (2007) showed that the regional calibration of the Hargreaves–Samani method is significantly improved by an adjustment of the coefficients rather than by a linear regression. Therefore, all the HS_{mod} R_s coefficients are adjusted. In this study, modified forms of the Bristow–Campbell method (Eq. A3) and Hargreaves–Samani method (Eq. A4) are used. For the R_S equations, the adjustment of the local calibration coefficients is non-linear. To adjust the calibration coefficients, a grid search iterative algorithm is used to maximise the R^2 value while minimising the RMSE at each reference weather station.

2.2.2 Evapotranspiration methods

ET_0 is the evapotranspiration from a reference grass surface and is used as a standard from which ET_c is deduced as follows (Allen et al., 1998):

$$ET_c = ET_0 \times K_c, \tag{1}$$

where K_c is the vegetation coefficient.

Several ET_0 methods using a reduced data set in comparison to the FAO-56 PM method have been developed worldwide. Only a few methods are commonly used. This is the case with the five ET_0 methods selected for this study, which have shown good performance when using daily to weekly frequencies (Trajkovic, 2005; Yoder et al., 2005; Alexandris et al., 2008; Shahidian et al., 2012; Tabari et al., 2013). The five selected ET_0 methods, namely the methods of Hargreaves and Samani (1985), Makkink (1957), Turc (1961), Priestley and Taylor (1972), and the Penman–Monteith reduced-set method (Allen et al., 1998), require records of R_S and temperature (Eqs. A7–A12 in Appendix A). As R_S can be estimated with a calibrated R_S temperature-based method, ET_0 can thus be obtained with temperature records only.

ET_0 is calculated using data collected at each reference weather station (independent ET_0 estimates). These calculations follow the FAO-56 PM method outlined in the FAO-56 document (Allen et al., 1998). These independent ET_0 estimates are then used as pseudo-standards for the purpose of calibrating the regional-scale ET_0 methods. A linear regression is performed for each of the evapotranspiration methods and for each reference weather station (Eq. 2). The slope a

and the intercept b of the best-fit regression line are used as local calibration coefficients.

$$ET_{0FAO-56\,PM} = aET_{0method} + b, \tag{2}$$

where $ET_{0FAO-56PM}$ is the ET_0 estimated with the standard method and $ET_{0method}$ is the ET_0 obtained by any of the five methods tested in this study. The linear regression method has been widely used to calibrate ET_0 methods (Allen et al., 1994; Trajkovic, 2005; Shahidian et al., 2012).

2.3 Step 2: estimation of the parameters of the recharge area

The estimation of the recharge with the soil-water balance (step 3; Sect. 2.4) requires the calculation, at the scale of the recharge area, of three parameters which are SAWC, runoff coefficient R_{coeff}, and K_c. These three parameters are controlled by one or several factors which are, in this study, the slope gradient, the geological nature of the substratum and the type of vegetation cover. Moreover, at the scale of the recharge area, the controlling factors are commonly heterogeneous and thus the recharge-area parameters cannot be readily computed. For each of the controlling factors, the recharge area is divided into subareas (hereafter referred to as factor subareas) characterised by homogenous factor properties. Factor subareas can be either continuous or discontinuous, and their number and shape can differ, depending of the spatial distribution of the factors. Relevant factor subareas are in turn used to define parameter subareas. For a given parameter subarea, the value of the parameter is estimated from either field measurements or from the literature. The parameter values at the scale of the recharge area are then calculated by taking into account the relative surface of the parameter subareas (step 2; Fig. 1). Lastly, if preferential infiltration structures (hereafter referred to as infiltration structures) such as sinkholes, cracks, reverse slope areas, bare ground or any topographical depression which can collect the surface runoff are present in the recharge area, the above-mentioned parameters have to be adjusted. For such areas, the SAWC and R_{coeff}, being very low, will be set at 0 in the calculations. Similarly, for such areas, ET_0 is negligible and therefore the surface of these areas is disregarded for the K_c computation. The parameter values are afterwards refined by a sensitivity analysis (step 4; Sect. 2.5) in order to find the optimal set of recharge-area parameters.

The K_c parameter takes into account four key characteristics (vegetation height, albedo, canopy resistance and evaporation from soil) that distinguish the vegetation type of a given subarea from the reference grass used to estimate ET_0 (Allen et al., 1998). The K_c subareas are defined according to the type of vegetation (e.g. meadows and forests) obtained from aerial photographs. The dominant vegetation species assigned to each vegetation type can be obtained from the literature (e.g. forest agency data) or from field observations. Since the K_c parameter depends on the stage of development of the vegetation, it varies from a minimum value during winter to a maximum value during summer. The minimum and maximum K_c values are estimated from the literature and are assigned respectively to 4 February (middle of winter) and 6 August (middle of summer) of each year. A daily linear interpolation is performed for K_c between these two dates (Verstraeten et al., 2005).

The SAWC parameter refers to the difference between a maximum water content above which all free water is drained through gravity (field capacity) and a minimum moisture content below which plant roots cannot extract any water (permanent wilting point). The SAWC is mainly affected by soil texture and thickness, both depending primarily on the geological substratum and the vegetation. The SAWC subareas are defined according to the type of vegetation (obtained from aerial photographs) and to the geological substratum (obtained from geological maps). SAWC values can be either calculated with pedotransfer functions (Bruand et al., 2004; Pachepsky and Rawls, 2004) from soil properties (type of horizon, texture and bulk density) and thickness or obtained directly from the literature. Soil properties and thickness can be obtained from the literature (e.g. pedological maps), from morphological description or laboratory measurements of auger hole cores.

The method used to estimate the surface runoff is similar to the commonly used "runoff rational method". The R_{coeff} parameter depends mainly on topography and vegetation. The R_{coeff} subareas are defined according to the vegetation (obtained from aerial photographs). An average slope gradient obtained from the DEM is assigned to each vegetation subarea. The R_{coeff} values can then be calculated from vegetation cover and slope gradient through the use of charts such as the Sautier chart (Musy and Higy, 2011).

Infiltration structures are first located through examination of aerial photographs (lineament analysis) and geological maps, and then inspected in the field.

2.4 Step 3: recharge computation with soil-water balance

The soil-water balance workflow used to estimate the recharge at a daily frequency is detailed in Fig. 2. All terms required for the soil-water balance estimation are expressed in water amount (millimetres), except for R_{coeff} expressed in percentage. The soil-water balance is based on ET_c, SAWC, K_c and R_{coeff}. The precipitation (P) is the amount of liquid (rain) or solid (snow) water which falls on the recharge area. The precipitation will be taken here as the sum of snowmelt and rainfall. A part of this water amount is intercepted by the vegetative canopy (interception; Fig. 2a). The remainder of precipitation reaches the ground surface and forms (i) the runoff (Rf), which is the water joining the surface drainage network and (ii) the infiltration (I) into the soil layer which supplies the SAWC. The remaining part of the precipitation not taken-up by evapotranspiration and runoff and not stored

in the SAWC is called the recharge (R). It corresponds to deep percolation and is the component of the precipitation which recharges the saturated zone (Fig. 2a).

The ET_c is a lumped parameter including potential transpiration, potential soil evaporation and canopy interception evaporation (Verstraeten et al., 2005). In the proposed computation diagram workflow (Fig. 2b) the interception component is therefore integrated in the ET_c component. The ET_c is the water evapotranspired without any other restrictions than the atmospheric demand (assuming unlimited soil water availability). However, field conditions do not always fulfil these requirements, particularly during low rainfall periods when water supplies are inadequate to support vegetation uptakes. The actual evapotranspiration (ET_a) corresponds to the actual amount of evapotranspired water.

Runoff takes place when the intensity of a precipitation event exceeds the soil infiltration capacity. The use of a daily measurement frequency for precipitation does not allow for an accurate estimation of rainfall intensity. Instead, a R_{coeff} is applied only for days when precipitation is greater than the average. Such days are considered as high intensity rainfall days. The R_{coeff} is applied only to excess precipitation, after the demands of evapotranspiration and SAWC are met, i.e. when SAWC is fulfilled (Fig. 2b).

2.5 Step 4: sensitivity analysis of the recharge-area parameters

In the landslide recharge area, recharge can be considered as spatially heterogeneous. Indeed, in fractured rocks, the groundwater flow is mainly driven by an anisotropic fracture network. The proportion of infiltrated water which flows toward the landslide aquifer can significantly differ between two zones of the recharge area. Nevertheless, the GIS composite method considers that any part of the recharge area has the same weight with respect to the groundwater which flows toward the landslide aquifer. This homogeneous recharge assumption can lead to biased estimations of the recharge-area parameters. However, uncertainties in the delimitation of the recharge area can also lead to biased estimations.

A sensitivity analysis evaluates the possible overestimation or underestimation of the set of recharge-area parameters. The infiltration-structure subareas are used as fitting factors (varying from 0 to 100 % of the recharge area surface) to adjust the estimation of the set of recharge-area parameters. A variation of the infiltration structure percentage corresponds to a variation of the contribution weight of the infiltration structures to the recharge of the landslide aquifer. Consequently, a variation of the infiltration structure percentage does not affect the relative proportion of the other subarea surfaces but only their contribution weights. The sensitivity analysis is based on the performance of a linear correlation between daily time series of recharge and displacement. The landslide displacement triggered by pore water pressure is therefore related to the hydrodynamic variations

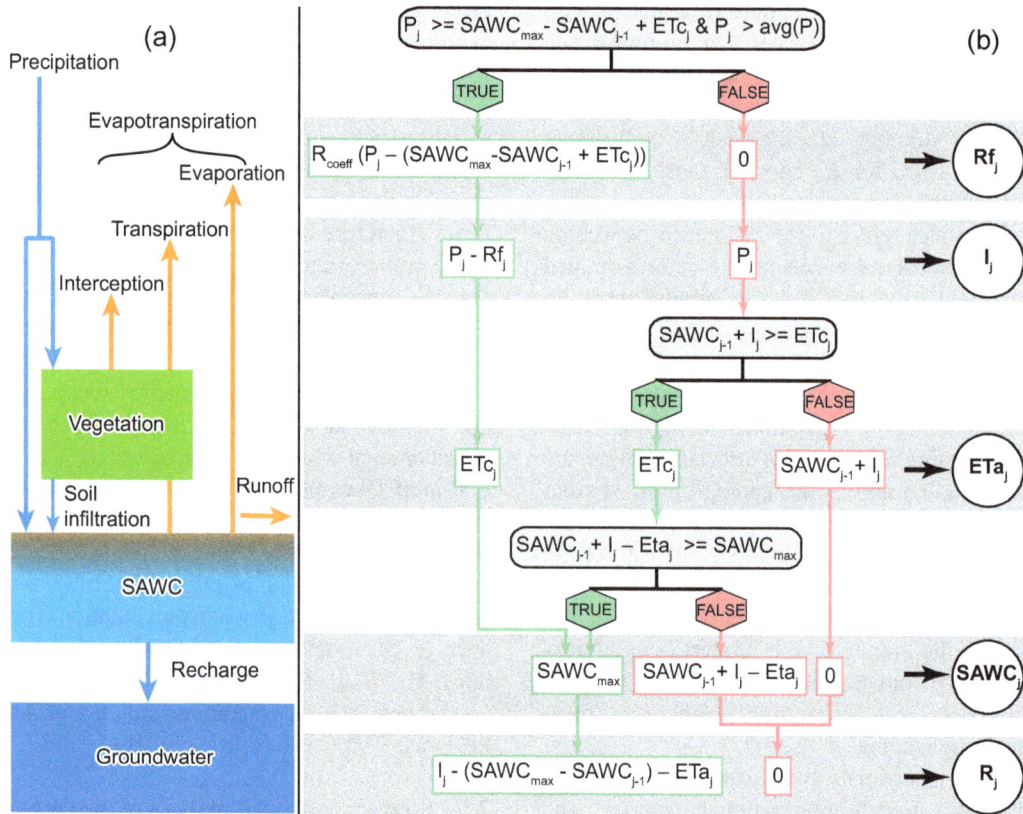

Figure 2. Soil-water balance: **(a)** soil-water balance conceptual representation and **(b)** soil-water balance diagram used for recharge computation on a daily frequency. SAWC: soil available water capacity; $SAWC_{max}$: SAWC threshold (possible maximum); P: precipitation (rainfall + snowmelt); avg (P): precipitation average of the entire record; I: part of precipitation which infiltrates the soil; Rf: surface runoff; R_{coeff}: runoff coefficient; ET_c: specific vegetation evapotranspiration; ET_a: actual vegetation evapotranspiration; R: recharge. Units: millimetres of water, except R_{coeff} in percentage. Subscript j is the computation day and subscript $j-1$ is the day before. TRUE and FALSE are the answers of the conditional inequality statements.

of the landslide aquifers. For this reason, the performance of the correlation between recharge and displacement informs whether the recharge-area parameters are satisfactorily estimated. The sensitivity analysis allows determining the optimal set of recharge-area parameters which maximise the performance of the correlation.

2.6 Correlation between water input and displacement

2.6.1 Antecedent cumulative sum

The correlation between water input and displacement requires measurements of landslide displacements at the same temporal frequency (daily frequency in this study) as the measurements of water input (precipitation or recharge). The groundwater hydrodynamic processes in aquifers are nonlinear. A former rainfall event displays less impact (though not negligible) than a recent one on the aquifer hydrodynamic fluctuations (Canuti et al., 1985; Crozier, 1986; Diodato et al., 2014). The daily precipitation/recharge time series cannot therefore be used without appropriate corrections. An an-

tecedent cumulative sum of precipitation/recharge weighted by a factor α is applied as a moving window to the daily precipitation/recharge time series (Eq. 3). The antecedent cumulative sum allows approximating the daily triggering impact of the aquifer (ATI) on the landslide destabilisation. In order to take into account the groundwater transit time, a β time-lag factor is introduced. This factor can shift the moving window from the target date t.

$$\text{ATI}_t = \sum_{i=t+\beta}^{t+\beta+n} \frac{W_i}{1 + \alpha(i - (t + \beta))}, \tag{3}$$

where ATI_t is the aquifer triggering impact (in mm) at the date t; β is the time shift of the moving window (in days); i is the ith day from the date t ($i = t + \beta$: start of the moving window and $i = t + \beta + n$: end of the moving window); n is the length of the moving window of the cumulative period (in days); W_i water input, i.e. precipitation or recharge at the ith day (in mm), and α is the weighting factor.

An iterative grid search algorithm is used to find the optimal set of parameters of the antecedent cumulative sum.

The optimal set of parameters is the set that maximises the correlation performance itself based on the R^2 indicator. The grid search algorithm investigates the following parameter ranges: n from 1 to 250 days (increment: 1 day), α from 0 to 0.5 (increment: 0.0001) and β from 1 to 10 days (increment: 1 day).

2.6.2 Significance of the water input-displacement correlation

The bootstrap method, which is an inference statistical re-sampling method, is used to estimate the confidence interval (CI) of estimated parameters and to perform statistical hypothesis tests (Chernick, 2008). The bootstrap method uses resampling with replacement and preserves the pair-wise relationship. However, for interdependent data (such as time series), the structure of the data set has to be preserved during the resampling. The moving block bootstrap is a variant of the bootstrap method. It divides data into blocks for which the structure is kept (Cordeiro and Neves, 2006). The moving block bootstrap method is performed with a 90-day block size (season) and 50 000 iterations for each run.

To estimate the significance of the linear regression, the lower bound of the confidence interval (LBCI) of R^2 is used at the level of confidence of 90 % (equivalent to a one-tailed test at the significance level of 5 %). An LBCI value greater than 0 means that the relationship is significant. Particular to statistical hypothesis tests is the definition of the tested null hypothesis which is often a default position opposite to the aim of the test, i.e. by stating that "there is no relationship between the two considered quantities". The null hypothesis is assumed to be true until it is rejected by statistical evidence in favour of the alternative opposite hypothesis. The recharge estimated with the LRIW workflow is hereafter called R_{LRIW}. The recharge estimated by subtracting a non-calibrated ET_0 from precipitation is hereafter called R_{PMNE}, PMNE standing for precipitation minus non-calibrated ET_0.

To estimate whether the R_{PMNE}–displacement correlation R^2 is significantly better than the precipitation–displacement correlation R^2 value, the Null Hypothesis 1 (NH1) is tested. The NH1 states that the R_{PMNE}–displacement correlation R^2 value is not significantly greater than the R^2 value obtained from precipitation. In other words, the NH1 statistic test is the difference between the R_{PMNE} R^2 value and the precipitation R^2 value, expected to be 0 if no difference. Similarly, the Null Hypothesis 2 (NH2) and the Null Hypothesis 3 (NH3) are tested. NH2 estimates whether the R_{LRIW}–displacement correlation R^2 is significantly better than the precipitation–displacement correlation R^2 value. NH3 estimates whether the R_{LRIW}–displacement correlation R^2 is significantly better than the R_{PMNE}–displacement correlation R^2 value.

To estimate whether the best precipitation-R_{LRIW}-displacement correlation R^2 value computed from the sensitivity analysis is significantly better than the other R^2 values obtained, the Null Hypothesis 4 (NH4) is tested. The NH4 states that the best R^2 value is not significantly greater than the ones obtained with all the remaining combinations. In other words, the NH4 statistic test is the difference between the best R^2 value and the R^2 values obtained with the remaining combinations, expected to be 0 if no difference.

For all null hypotheses, the decision of rejection is made by determining how much of the bootstrap distribution (among 50 000 iterations) falls below 0 by using the LBCI at the level of confidence of 90 %, equivalent to a one-tailed test at the significance level of 5 %. An LBCI value greater than 0 allows rejecting the null hypotheses.

3 Application to the Séchilienne landslide

3.1 Geological settings and rainfall triggering

The Séchilienne landslide is located in the French Alps on the right bank of the Romanche River, on the southern slope of the Mont Sec Massif (Fig. 3). The climate is mountainous with a mean annual precipitation of 1200 mm. The geological nature of the area is composed of vertical N–S foliated mica schists unconformably covered by carboniferous to Liassic sedimentary deposits along the massif ridge line above the unstable zone. Quaternary glacio-fluvial deposits are also present. The Séchilienne landslide is limited eastwards by a N–S fault scarp and northwards by a major head scarp of several hundred metres wide and tens of metres high below the Mont Sec. The slope is cut by a dense network of two sets of near-vertical open fractures trending N110–N120 and N70 (Le Roux et al., 2011).

The Séchilienne landslide is characterised by a deep progressive deformation controlled by the network of faults and fractures. A particularity of the Séchilienne landslide is the absence of a well-defined basal sliding surface. The landslide is affected by a deeply rooted (about 100–150 m) toppling movement of the 50–70° N slabs to the valley (accumulation zone) coupled with the sagging of the upper slope (depletion zone) beneath the Mont Sec (Vengeon, 1998; Durville et al., 2009; Lebrouc et al., 2013). A very active moving zone is distinguishable from the unstable slope where high displacement velocities can be 10 times higher than the rest of the landslide.

The landslide shows a higher hydraulic conductivity than the underlying stable bedrock (Vengeon, 1998; Meric et al., 2005; Le Roux et al., 2011), thus leading to a landslide perched aquifer (Guglielmi et al., 2002). The recharge of the landslide perched aquifer is essentially local, enhanced by the trenches and the counterscarps which tend to limit the runoff and to facilitate groundwater infiltration in the landslide area. However, the hydrochemical analyses of Guglielmi et al. (2002) show that the sedimentary deposits distributed above the landslide hold a perched aquifer which can recharge the landslide perched aquifer. The fractured metamorphic bedrock beneath the landslide contains a

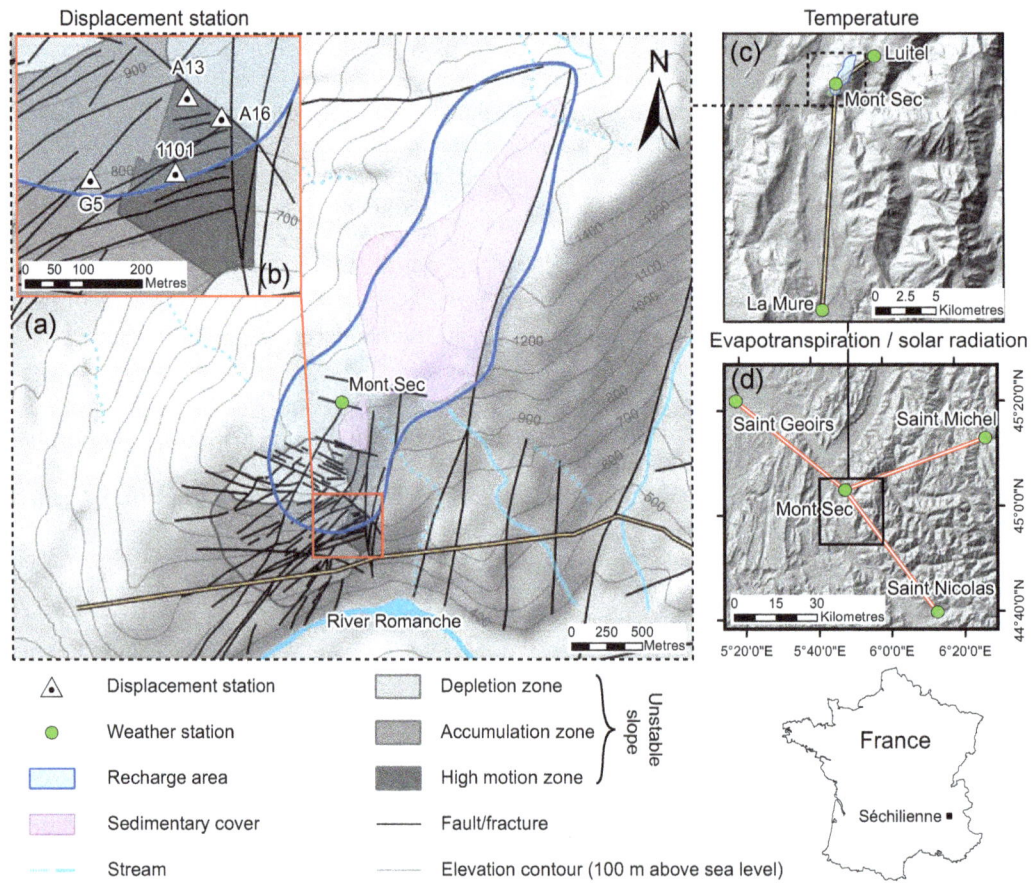

Figure 3. Location map of the Séchilienne landslide. (**a**) Map of the Séchilienne unstable slope and recharge area with the Mont Sec weather station. (**b**) Enlarged map of the most active area showing displacement stations. (**c**) Map showing the weather stations used for the temperature estimation at Mont Sec. (**d**) Map showing the weather stations used for evapotranspiration and solar radiation method calibration.

deep saturated zone at the base of the slope and an overlying vadose zone. The groundwater flow of the entire massif is mainly controlled by the network of fractures with high flow velocities (up to a few kilometres per day; Mudry and Etievant, 2007). The hydromechanical study of Cappa et al. (2014) shows that the deep aquifer can also trigger the Séchilienne landslide destabilisation as a result of stress transfer and frictional weakening. Thus, the Séchilienne landslide destabilisation is likely triggered by a two-layer hydrosystem: the landslide perched aquifer and the deep aquifer. The Séchilienne landslide behaviour is characterised by a good correlation between precipitations and displacement velocities (Rochet et al., 1994; Alfonsi, 1997; Durville et al., 2009; Chanut et al., 2013).The seasonal variations of the daily displacements are clearly linked to the seasonal variations of the recharge (high displacements during high flow periods and low displacements during low flow periods).

3.2   Method implementation

The recharge computation uses the daily rainfall recorded at the weather station located at Mont Sec, a few hundred metres above the top of the landslide (Table 1, Fig. 3). This station is equipped with rain and snow gauges and a temperature sensor. However, the temperature measurements at the Mont Sec station are considered unreliable because of a non-standard setting of the temperature sensor and numerous missing data. Consequently, the temperature at the Mont Sec station has to be estimated in order to estimate the evapotranspiration at the landslide site (see details about the computation in Appendix B).

Since the Mont Sec station does not record the full set of parameters (relative humidity, temperature, wind speed and solar radiation), a regional calibration of ET_0 and R_S reduced-set methods is required. Three weather stations located at less than 60 km from the studied site are used as reference weather stations: Grenoble-Saint-Geoirs, Saint-Jean-Saint-Nicolas and Saint-Michel-Maur (Table 1, Fig. 3). The Saint-Michel-Maur weather station does not measure R_S, which is estimated with the Angström formula (Eq. A5 in

Table 1. Summary of weather data sets with parameters used (X) at the various locations. Distance is measured from the Séchilienne landslide, R_S is the solar radiation, N is the sunshine duration, W is the wind speed, H is the humidity, T is the temperature and P is the precipitation depth.

Station name	Elevation (m a.s.l.)	Distance (km)	From	To	R_S	N	W	H	T	P	Number of days with data
Saint-Jean-Saint-Nicolas	1210	55	1 Jan 2004	1 Jan 2012	X	X	X	X	X		2876
Saint-Michel-Maur	698	54	1 Jan 2004	1 Jan 2012		X	X	X	X		2864
Grenoble-Saint-Geoirs	384	51	8 Jul 2009	1 Jan 2012	X	X	X	X	X		907
Chamrousse	1730	9	12 Sep 2002	1 Mar 2012			X	X			3261
La Mure	881	18	9 Sep 1992	1 Jan 2012					X		7517
Luitel	1277	4	6 Jul 2006	23 Jul 2012					X		2193
Mont Sec	1148	0.2	9 Sep 1992	1 Jan 2012						X	7517

Table 2. Statistics of the displacement records and results of the best linear correlation between precipitation/R_{LRIW} and displacement records for four displacement stations (1101, A13, A16 and G5). The displacement column indicates basic statistics of the displacement records: first quartile (Q_1), median and third quartile (Q_3). Cumulative period (n), shift factor (β) and weighting factor (α) are the terms of Eq. (3). P stands for precipitation, R_1 stands for R_{PMNE} and R_2 stands for R_{LRIW}.

Station	Displacement mm day^{-1}			Cumulative period (n)			Shift factor (β)			Weighting factor (α)			R^2		
	Q_1	median	Q_3	P	R_1	R_2	P	R_1	R_2	P	R_1	R_2	P	R_1	R_2
1101	1.75	2.50	3.84	42	54	68	2	2	2	0.071	0.065	0.091	0.28	0.35	0.50
A13	1.18	1.75	3.41	52	80	82	3	2	2	0.102	0.070	0.091	0.28	0.37	0.52
A16	1.94	2.98	4.39	64	71	76	2	2	2	0.163	0.125	0.168	0.34	0.44	0.59
G5	0.02	0.05	0.08	8	169	132	0	6	6	0.039	0.003	0.011	0.001	0.08	0.24

Appendix A) using sunshine duration data recorded at the station. The Angström formula empirical default coefficients are tuned with the two other weather stations ($a_S = 0.232$ and $b_S = 0.574$).

The delimitation of the recharge area of the two-layer hydrosystem (Fig. 3) of the Séchilienne landslide is based on the geological and hydrochemical studies of Vengeon (1998), Guglielmi et al. (2002) and Mudry and Etievant (2007). The recharge area is delimited by the spatial extent of the sedimentary cover of which the hosting perched aquifer recharges the two-layer hydrosystem. Groundwater flow of the entire Mont Sec Massif is controlled by faults and fractures. The N20 fault bordering the sedimentary cover to the east as well as the N–S fault zone bordering the landslide to the east are structures which delimitate the recharge area. The scarcity of information does not allow accurately defining the actual extent of the recharge area. The sensitivity analysis mentioned in Sect. 2.5 allows compensating for the possible biases introduced by this uncertainty. The following spatial data sets are used for the estimation of the parameters of the recharge area. The aerial photographs (0.5 m resolution) and a DEM of 25 m resolution are provided by the Institut National de l'Information Géographique et Forestière (IGN) and geological maps are provided by the French Geological Survey (BRGM).

The Séchilienne landslide is permanently monitored by a dense network of displacement stations managed by the CEREMA Lyon (Duranthon et al., 2003). In this study, one infrared station (1101) and three extensometer stations (A16, A13 and G5) are used. Stations 1101, A13 and A16 are representative of the most active zone (median displacements of 2.5, 1.75 and 2.98 mm day^{-1}, respectively), while G5 is located on a much less active zone (median displacement of 0.05 mm day^{-1}; Fig. 3, Table 2).

The sensitivity analysis is performed on the A16 extensometer on the period from 1 May 1994 to 1 January 2012, period during which both A16 extensometer and recharge data sets are available. The performance test of the LRIW workflow against precipitation and R_{PMNE} is performed on the four displacements station in the period from 1 January 2001 to 1 January 2012, period during which the four stations and recharge data sets are available. The R_{PMNE} is estimated with the non-calibrated Turc equation (Eq. A8) which is the most appropriate ET$_0$ reduced-set equation for the Séchilienne site. Indeed, the Turc equation was developed initially for the climate of France. The Turc equation requires the estimation of R_S which is performed with the non-calibrated Hargreaves–Samani equation (Eq. A2).

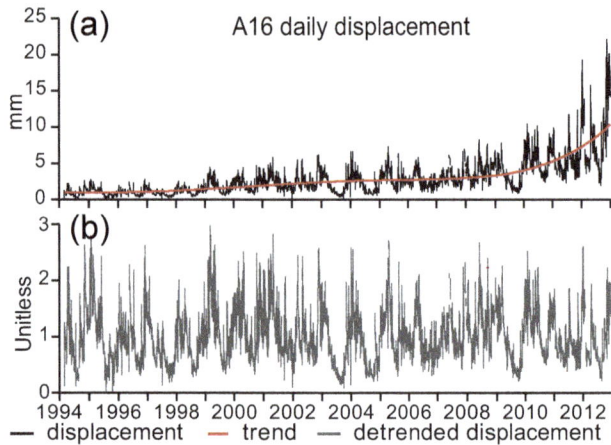

Figure 4. Trend removal of A16 extensometer displacement data. (a) A16 displacement data and the fourth-order polynomial curve fitting considered as the displacement trend; (b) A16 detrended data (unitless) corresponding to A16 displacement data for which the trend is removed by a multiplicative method.

3.3 Displacement data detrending

The long-term displacement monitoring shows that displacement rate and amplitude exponentially increased with time as illustrated by the records of extensometer A16 (Fig. 4a). The rainfall data series does not show any trend over the year, meaning that the displacement trend is independent of the recharge amount. Consequently, on the Séchilienne landslide, for the same amount of rainfall, the displacement rate and magnitude responses increase steadily with time. The observed trend is the consequence of a progressive weakening of the landslide due to long-term repetitive stresses. The accumulating deformation can be assimilated to long-term creep (Brückl, 2001; Bonzanigo et al., 2007) and can be explained by a decrease of the slope shear strength (Rutqvist and Stephansson, 2003). As shown by the detrended displacement, the Séchilienne landslide is constantly moving and shows large daily to seasonal variations which seem to be the landslide response to the precipitation trigger. Consequently, the precipitation-displacement correlation is performed on the detrended displacement.

The exponential trend is removed with the statistical multiplicative method ($y_t = T_t S_t I_t$) where the time series (y_t) is composed of three components (Madsen, 2007; Cowpertwait and Metcalfe, 2009; Aragon, 2011): trend (T_t), seasonal (S_t) and irregular (I_t). In this study, the irregular and seasonal components are both assumed to be linked to the rainfall triggering factor ($y_t = T_t D_t$ with $D_t = S_t I_t$). The trend is determined by curve fitting of a fourth-order polynomial (parametric detrending). The result is a detrended unitless time series (D_t) with both variance and mean trend removed. The time series decomposition process is illustrated with the A16 extensometer in Fig. 4.

Table 3. Calibration and performance of the five tested ET_0 methods, in relation to the FAO-56 PM ET_0 standard (Penman–Monteith method defined in the FAO-56 paper). All the ET_0 methods are detailed in Appendix A. a, b and R^2 are the results of linear regression between FAO-56 PM ET_0 and tested ET_0 methods. RMSE is the root mean square error.

Method	a	b	R^2	RMSE
HS ET_0	0.920	0.130	0.917	0.548
Turc ET_0	0.880	0.434	0.900	0.588
PS ET_0	0.352	0.365	0.919	0.533
M ET_0	1.107	−0.018	0.910	0.565
PM_{red} ET_0	0.994	0.013	0.932	0.505

4 Results of the recharge estimation with the LRIW method

4.1 Calibration of R_S and ET_0 methods

The two calibrated R_S methods show good results with respect to R_S measured at the reference weather stations. The BC_{mod} R_S method is selected as it shows a better performance ($R^2 = 0.864$; RMSE $= 1.567$) than the HS_{mod} R_S method ($R^2 = 0.847$; RMSE $= 1.625$). Equation (4) presents the calibrated BC R_S method with all the calibrated coefficients.

$$BC_{mod}R_S = 0.669Ra\left[1 - \exp\left(-0.010(\alpha\Delta T)^{2.053}\right)\right] + 1.733 \quad (4)$$

The cloud cover adjustment factor α is either equal to 0.79 (cloud impact) or to 1. All the equation terms are described in the Appendix A. The BC_{mod} R_S calibrated method is then used to compute R_S input data of the five ET_0 reduced-set methods.

Overall, all of the ET_0 methods tested show good results for regional calibration and are all suitable for the Séchilienne site (Table 3). Among the ET_0 methods tested, the PM_{red} ET_0 method shows the best performance ($R^2 = 0.932$; RMSE $= 0.505$) and requires only a low regional adjustment ($a = 0.994$ and $b = 0.013$). Therefore, the PM_{red} ET_0 method is selected to compute ET_0 for the Séchilienne site (hereafter referred to as $ET_{0Séch}$). Figure 5 displays the estimated $ET_{0Séch}$ versus the FAO-56 PM computation for each reference weather station.

Equation (5) is the final calibrated PM_{red} ET_0 method with all the calibrated coefficients. The input R_n term is deduced from the calibrated BC_{mod} R_S method (Eq. 4).

$$ET_{0\,Séch} = 0.994\frac{0.408\Delta\,(R_n - 0) + \gamma\frac{900}{T_{avg}+273}1.5(e_s - e_a)}{\Delta + \gamma(1 + 0.34\,1.5)} + 0.013 \quad (5)$$

4.2 Recharge-area parameters

Subareas are expressed in percentages of the whole recharge area (Table 4, Fig. 6). Two types of vegetation cover, pasture and forest, are defined using aerial photographs, with

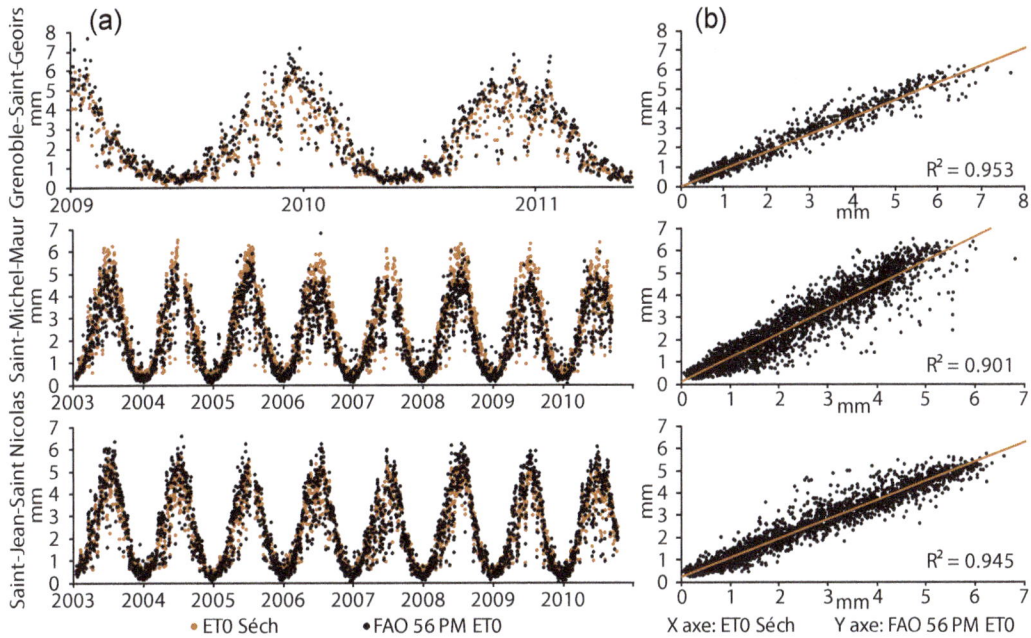

Figure 5. ET_0 regional calibration results at the three reference weather stations (Grenoble-Saint-Geoirs, Saint-Jean-Saint-Nicolas and Saint-Michel-Maur). **(a)** $ET_{0Séch}$ and FAO-56 PM ET_0 as a function of time. **(b)** Linear regression between $ET_{0Séch}$ (*x* axis) and FAO-56 PM ET_0 (*y* axis). $ET_{0Séch}$ stands for ET_0 computed with the combination of calibrated ET_0 Penman–Monteith reduced-set method and calibrated R_S modified Bristow–Campbell method.

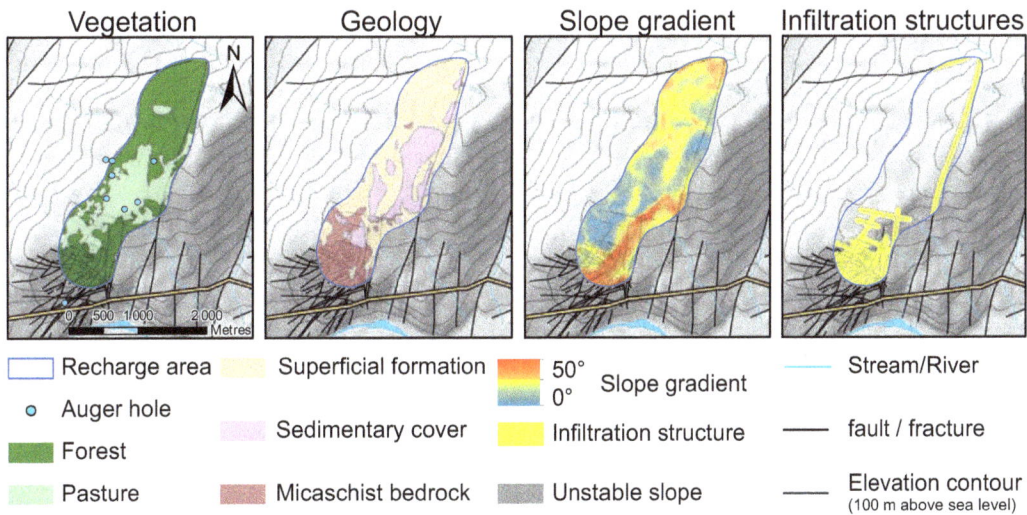

Figure 6. Factor subareas, auger holes and infiltration structures used for the estimation of recharge-area parameters.

proportions of 23 and 53 %, respectively. The Séchilienne forest is mainly composed of beeches (*Fagus sylvatica*) and conifers (*Picea excelsa*), which are associated occasionally with ashes (*Fraxinus*) and sweet chestnuts (*Castanea sativa*). Three main geology subareas, mica schist bedrock (15 %), sedimentary cover (20 %) and superficial formations (41 %), are defined through examination of the geological map and field investigations. Infiltration structures are centred on the major faults identified on the geological map, on lineaments deduced from aerial-photograph analysis and on geomorphological features (sinkholes, cracks, etc.). A 50 m wide influence zone is added to the identified objects, leading to an infiltration-structure subarea representing 24 % of the recharge area.

For K_c estimation, the proportion of beeches and conifers is assumed to be identical for the Séchilienne forest (each 50 % of forest subarea) and other species are ignored. K_c values are set to 0.71 and 0.97 for conifers, and to 0.78 and 0.9

Table 4. Estimation of K_c, SAWC and runoff for the recharge area of the Séchilienne landslide. Geology and vegetation are the subarea factors identified and expressed in relative proportion of the recharge area. The average slope gradient is the slope gradient for each identified vegetation subarea factor. K_c, R_{coeff} and SAWC columns are the estimated values for each subarea factor. K_c RA, SAWC RA and R_{coeff} RA columns are the contribution of each subarea parameter at the scale of the recharge area. The recharge area (bottom row) stands for the estimation at the scale of the recharge area.

Geology subarea (%)		Vegetation subarea (%)	Average slope gradient (°)	K_c min. max.	K_c RA min. max.	R_{coeff} (%)	R_{coeff} RA (%)	SAWC (mm)	SAWC RA (mm)
Mica schist	3	Pasture 23	14.0	0.85 1	0.256 0.301	22	5.1	173	5
Sedimentary	9							100	9
Superficial formations	11							112	12
Mica schist	12	Forest 53	20.6	0.745 0.935	0.521 0.654	15	7.7	254	30
Sedimentary	11							81	9
Superficial formations	30							133	41
Outcrop no soil	24	24	–	–	–	0	0	0	0
Recharge area	100	100	–	–	0.777 0.955	–	12.8	–	106

for beeches according to Verstraeten et al. (2005). Most pastures are anthropogenic and consist of grass. K_c values are set to 0.85 and 1 according to Allen et al. (1998). Infiltration structure subareas are not taken into account in the K_c estimation, so the relative proportions of pasture and forest become 30 and 70 %, respectively. The contribution of each subarea (column K_c RA, Table 4) allows determining the recharge area K_c values at the scale of the recharge area (0.777 to 0.955).

The combination of geology and vegetation subareas results in six types of SAWC subareas (Table 4). For each SAWC subarea, at least one auger hole was drilled. For each soil auger core, the soil texture, the stoniness and the organic-matter content are estimated by morphological description (Baize and Jabiol, 2011). Based on these estimations, the SAWC is then computed using the pedotransfer functions of Jamagne et al. (1977) and Bruand et al. (2004). The average estimation of SAWC at the recharge area scale is 106 ± 10 mm (rounded to 105 mm).

To estimate R_{coeff}, an average slope gradient is computed from slope gradient analysis of the DEM and is assigned to each vegetation subarea. Pasture and forest subareas show an average slope gradient of 14 and 20.6°, respectively. R_{coeff} values of 22 % for pasture and 15 % for forest are deduced from the Sautier chart (Musy and Higy, 2011). This chart was developed for Switzerland where environmental conditions are similar to the French Alps. A 12.8 % runoff coefficient is then estimated at the recharge area scale, according to the respective proportions of vegetation subareas (Table 4).

4.3 Sensitivity analysis of the parameters of the recharge area

Sensitivity analysis is performed for SAWC ranging from 0 (100 % of infiltration structures corresponding to precipitation) to 145 mm of SAWC (0 % infiltration structures +10 mm of SAWC uncertainties measurement) with increments of 10 mm. The coupled surface R_{coeff} ranges from 0 to 16.3 % (with increments of about 1 %). For each combination, recharge is computed according to the soil-water balance (step 3; Figs. 1, 2) with (i) the temperature estimated for the recharge area (Appendix B), (ii) the precipitation recorded at Mont Sec weather station, and (iii) the parameters of the recharge area.

All the best computations have a 1-day lag, with periods ranging from 56 to 104 days (Fig. 7a, Table 5). The best R^2 obtained from recharge is obtained with both the estimated recharge-area parameters (SAWC = 105 mm, $R^2 = 0.618$) and the recharge-area parameters for SAWC adjusted from 75 ($R^2 = 0.616$) to 115 mm ($R^2 = 0.617$; Fig. 7b, Table 5). One of the best correlation performances is obtained for the estimated recharge-area parameters. This shows that the delimitation of the recharge area properly reflects the actual field conditions. The best correlation performance is assumed to be obtained with the estimated recharge-area parameters for NH4, i.e. testing R^2 obtained with the estimated recharge-area set (SAWC = 105 mm) minus R^2 obtained with each of the other adjusted recharge-parameter sets of the sensitivity analysis (Table 5).

For all the recharge combinations tested, the LBCI values from bootstrap testing of NH2 are greater than 0, allowing for

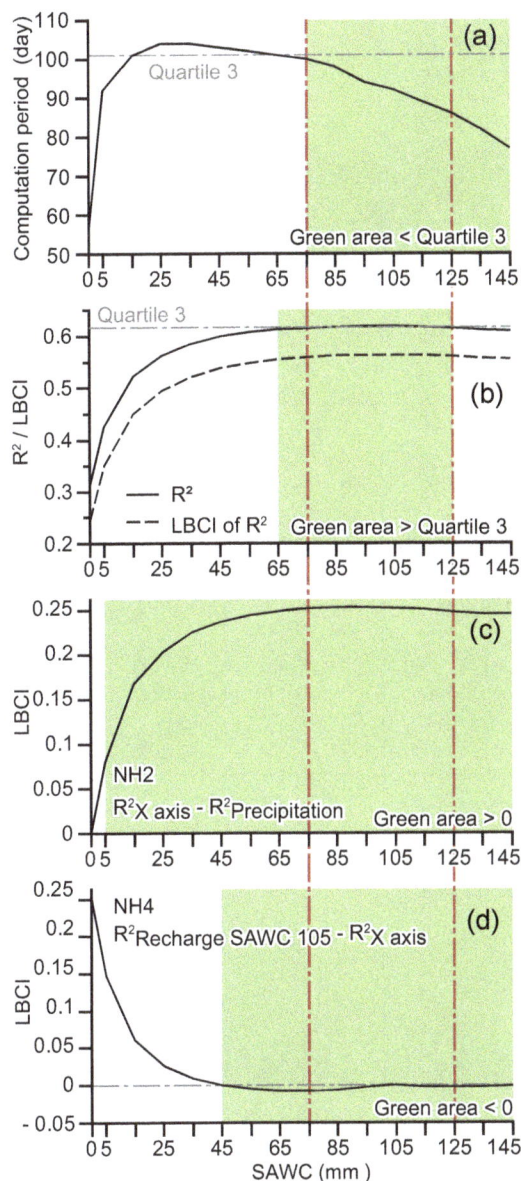

Figure 7. Results of the sensitivity analysis relative to SAWC for (a) the computation period, (b) the R^2 and the LBCI of R^2, (c) the LBCI of NH2 and (d) the LBCI of NH4.

the rejection of NH2 (Fig. 7c). In other words, it shows that the R^2 obtained with recharge is always significantly higher than the one computed with precipitation ($R^2 = 0.311$) even for a SAWC of 5 mm ($R^2 = 0.426$; Table 5). For the adjusted recharge-area parameter scenarios having SAWC values above 45 mm, the LBCI values from bootstrap testing of NH4 are lower than 0, not allowing for the rejection of NH4 (Table 5, Fig. 7d). In other words, it shows that the R^2 obtained with a SAWC of 105 mm is not significantly higher than the ones obtained from SAWC above 45 mm. Recharge–displacement correlations for SAWC values ranging from 75 (runoff = 9 %) to 115 mm (runoff = 13.9 %) show (i) a cumu-

lative period computation (n) below 101 days that is within the third quartile, (ii) an R^2 greater than 0.616 that is within the third quartile, (iii) LBCI values of NH2 greater than 0, and (iv) LBCI values of NH4 lower than 0 (Table 5, Fig. 7). These SAWC and runoff values seem to statistically reflect the recharge area properties of the landslide and are suggested for further work on the Séchilienne landslide.

4.4 Estimation of the recharge for the Séchilienne landslide

For the remaining part of this paper, R_{LRIW} is based on the estimated recharge-area parameters (infiltration structures = 24 %, SAWC = 105 mm, and $R_{coeff} = 12.8$ %). Indeed, among all solutions giving satisfying performances in the sensitivity analysis, these parameters arise from actual field data. R_{LRIW} is compared with the precipitation signal in Fig. 8.

The R_{LRIW} signal differs significantly from the precipitation signals, marked by a high seasonal contrast. This is especially true during summer when ET_c is important. Indeed, the first rainfall events after a dry period do not reach the aquifer until the SAWC is exceeded. Figure 9 shows the best correlation results for precipitation and R_{LRIW}, together with A16 detrended daily displacements. The cumulative recharge signal reproduces well the displacement acceleration and deceleration phases, and especially the dry summers where displacement dramatically dropped (summers 1997, 1998, 2003, 2004 and 2009; Fig. 9b). On the contrary, the cumulative precipitation signal is more contrasted and more noisy, and does not manage to reproduce several peaks (in width as well as in intensity) of the detrended displacement signal (winters 1997, 2000, 2004, 2005 and 2010). In addition, the cumulative precipitation signal shows a weak correlation with displacement deceleration phases (summers 1998, 1999, 2000 2006, 2009 and 2010).

5 Discussion

5.1 Relevance of the LRIW method

Figure 10 summarises the comparison of the performances between the precipitation, the R_{PMNE} and the R_{LRIW} based on the NH1, NH2 and NH3 tests for the four displacement stations. LBCI values from bootstrap testing of NH1 are lower than 0 for 1101, A13 and A16 stations and greater than 0 for the G5 station. NH1 cannot be rejected, meaning that the R^2 values obtained with R_{PMNE} are not significantly higher than those computed with precipitation, except for the G5 station. All LBCI values from bootstrap testing of NH2 and NH3 are greater than 0, allowing for the rejection of these two null hypotheses for the four stations (Fig. 10a). Rejection of NH2 shows that the R^2 values obtained with R_{LRIW} are significantly higher than those computed with precipitation. Similarly, rejection of NH3 shows that R^2 values

Table 5. Sensitivity analysis results of the best correlation between precipitation/R_{LRIW} and A16 extensometer detrended displacement. IS is for infiltration structures. SAWC is the soil available water capacity. LBCI is the lower bound of the confidence interval. R^2 row is the R^2 computed from recharge-area parameters indicated in each table row. Cumulative period (n), shift factor (β) and weighting factor (α) are the terms of Eq. (3). NH2 test: $R^2_{\text{row}} - R^2_{\text{precipitation}}$. NH4 test: $R^2_{\text{SAWC 105}} - R^2_{\text{row}}$.

SAWC mm	R_{coeff} %	IS %	Cumulative period (n) day	Shift factor (β) day	Weighting factor (α)	R^2	LBCI of R^2	LBCI of NH2	LBCI of NH4
0	0.0	100	56	1	0.1697	0.311	0.243	0	0.252
5	0.6	96	92	1	0.1362	0.426	0.350	0.080	0.148
15	1.8	89	101	1	0.1226	0.522	0.450	0.167	0.061
25	3.0	82	104	1	0.1259	0.563	0.494	0.203	0.027
35	4.2	75	104	1	0.1317	0.585	0.521	0.224	0.009
45	5.4	68	103	1	0.1374	0.599	0.538	0.236	0.000
55	6.6	61	102	1	0.143	0.608	0.548	0.244	−0.005
65	7.8	53	101	1	0.1484	0.613	0.555	0.249	−0.007
75	9.0	46	100	1	0.155	0.616	0.559	0.251	−0.007
85	10.3	39	98	1	0.1609	0.618	0.562	0.253	−0.006
95	11.5	32	94	1	0.1648	0.618	0.563	0.253	−0.003
105	12.8	24	92	1	0.1689	0.618	0.563	0.252	0
115	13.9	18	89	1	0.1727	0.617	0.562	0.251	−0.002
125	15.1	10	86	1	0.1745	0.614	0.560	0.248	−0.002
135	16.3	3	82	1	0.1746	0.611	0.556	0.245	−0.002
145	16.3	–	77	1	0.1731	0.609	0.555	0.245	−0.002

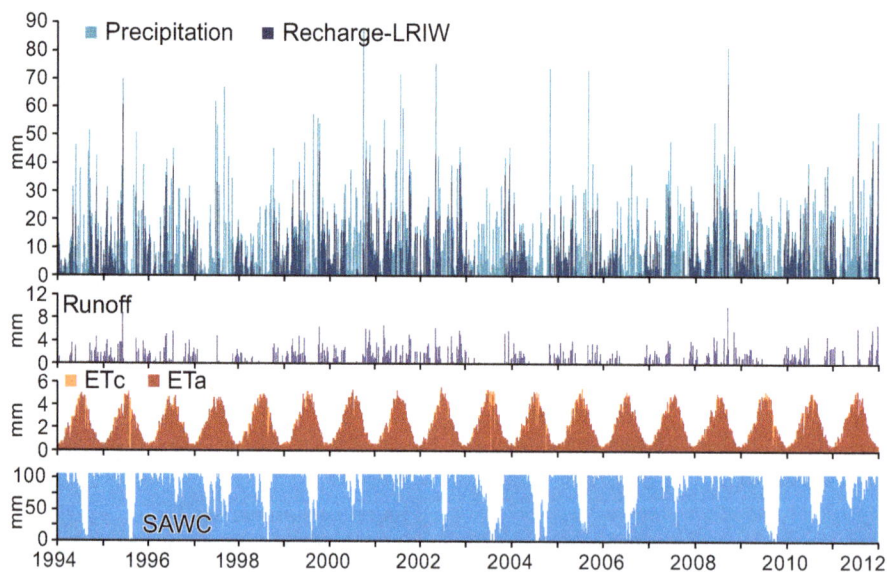

Figure 8. Recharge computation with the LRIW method at Séchilienne with an SAWC of 105 mm and a runoff coefficient of 12.8 %. ET_c: specific vegetation evapotranspiration; ET_a: actual vegetation evapotranspiration, SAWC: soil available water capacity.

obtained with R_{LRIW} are significantly higher than those computed with R_{PMNE}. R^2 values vary from 0.0006 to 0.343 for precipitation, from 0.076 to 0.444 for R_{PMNE} and from 0.243 to 0.586 for R_{LRIW}, for G5 and A16 extensometer, respectively (Table 2). On average, R_{PMNE} allows increasing the R^2 value by 29 % relative to precipitation, while R_{LRIW} allows increasing the R^2 by 78 % (Fig. 10b). The R^2 values

obtained with R_{LRIW} are 38 % higher on average than those obtained with R_{PMNE}.

These results are confirmed by the LBCI and by the observed values of the NH2 test which are always greater than those from the NH1 test as well as by the positive LBCI values of the NH3 test (Fig. 10). The correlation performance for the recharge estimated with the LRIW method signif-

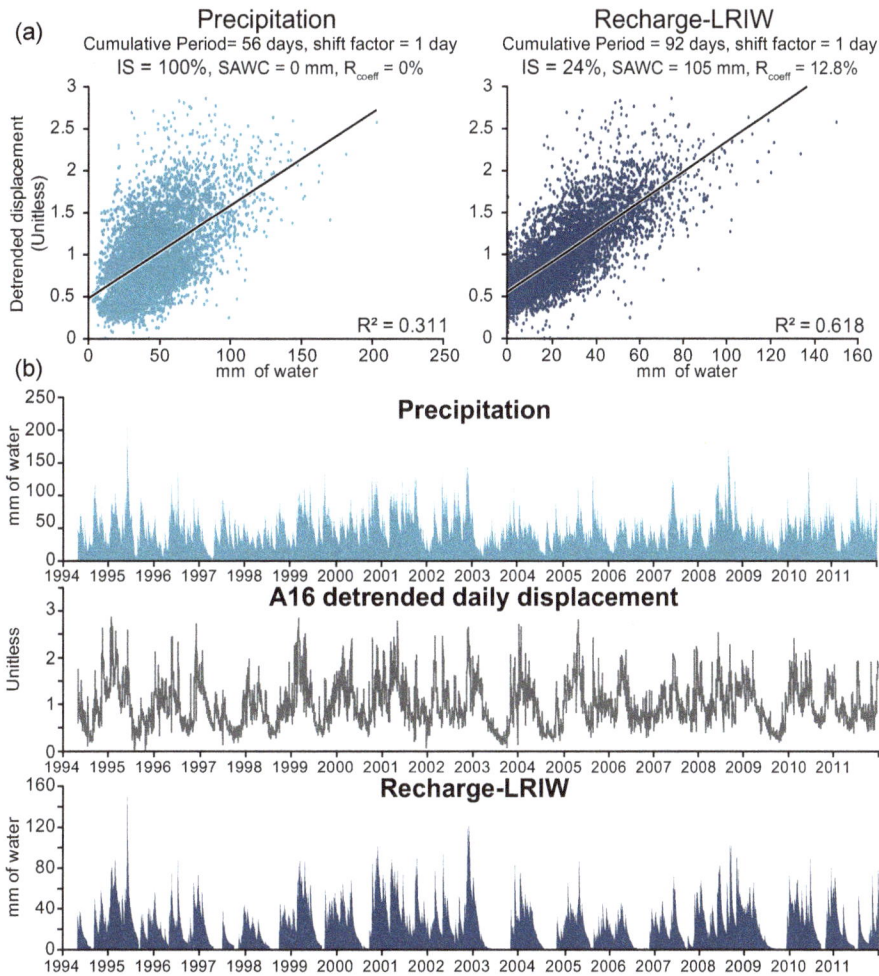

Figure 9. Best linear correlations for precipitation and recharge computed with the LRIW method. IS is for infiltration structures. SAWC is soil available water capacity. Cumulative period (n) and shift factor (β) are the terms of Eq. (3). (**a**) Linear regression between precipitation/R_{LRIW} and A16 detrended displacement. (**b**) Correlation between precipitation/R_{LRIW} and A16 detrended displacement as a function of time.

icantly exceeds the performance of the two other signals, making the LRIW method particularly appropriate to be used in landslide studies. A discussion about the benefit of this study for the understanding of the rainfall–displacement relationship in the case of the Séchilienne landslide can be found in Appendix C.

5.2 Applicability of the LRIW method to other landslides

Several studies have shown the relevance of the recharge signal for various landslide types: coastal landslides (Maquaire, 2000; Bogaard et al., 2013), unstable embankment slope landslides (Cartier and Pouget, 1987; Delmas et al., 1987; Matichard and Pouget, 1988) and deep-seated earth flow landslides (Malet et al., 2003; Godt et al., 2006). In addition, destabilisation of shallow landslides is known to be influenced by antecedent soil moisture and precipitation

(Brocca et al., 2012; Garel et al., 2012; Ponziani et al., 2012). Recharge, which implicitly combines antecedent soil moisture and precipitation, can be a significant parameter to consider.

Although the method proposed in this study has not yet been tested at other sites, there are several arguments which suggest its applicability elsewhere. First, the FAO Penman–Monteith method used in this study is considered worldwide as the evapotranspiration method standard (Allen et al., 1998; Shahidian et al., 2012). Several evapotranspiration methods were developed locally and many of them can be calibrated against reference methods in other contexts (Hargreaves and Allen, 2003; Yoder et al., 2005; Alkaeed et al., 2006; Igbadun et al., 2006; Trajkovic, 2007; Alexandris et al., 2008; López-Moreno et al., 2009; Sivaprakasam et al., 2011; Tabari and Talaee, 2011; Shahidian et al., 2012; Tabari et al., 2013). Otherwise, the Penman–Monteith or Hargreaves–Samani methods are recommended (Allen et al., 1998). Several solar radi-

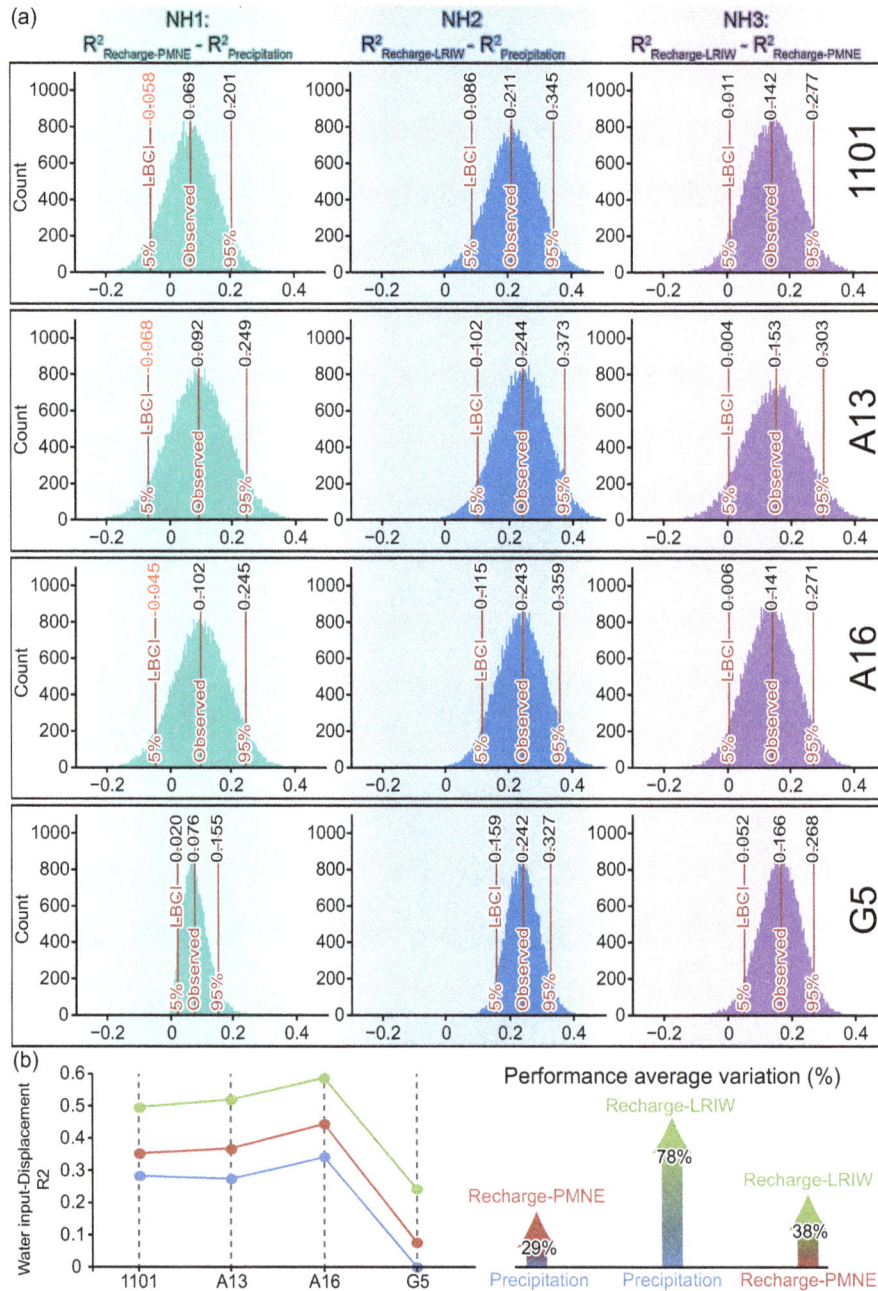

Figure 10. Performance of the LRIW workflow. **(a)** Bootstrap distribution of NH1, NH2 and NH3 tests for four displacement recording stations. **(b)** R^2 values for the four displacement recording stations obtained with the precipitation, recharge PMNE, and recharge LRIW. LBCI is the lower bound of the confidence interval. G5 station is disregarded in the the performance average variation calculation since the R^2 value obtained at G5 from precipitation is close to 0, therefore leading to a non-representative variation.

ation methods were developed and can be applied worldwide if locally calibrated, allowing for the estimation of evapotranspiration from temperature alone (Allen et al., 1998; Almorox, 2011). Recharge-area parameters can be estimated locally or with local or global literature reference values. The use of global values will increase recharge estimation uncertainties. However, the implementation of a sensitivity analysis allows refining the recharge-area parameters in order to

compensate for the lack of site-specific data. Pachepsky and Rawls (2004) developed pedotransfer functions to estimate SAWC for various regions of the world. R_{coeff} values from the widely used rational method can be applied, as well as most of the runoff coefficients from the literature (McCuen, 2005; Musy and Higy, 2011). In addition, pedotransfer functions can also be used for runoff estimation. Lastly, vegetation coefficients are available from local surveys (Gochis and

Cuenca, 2000; Verstraeten et al., 2005; Hou et al., 2010), but can also be found in the literature for many species (Allen et al., 1998).

6   Conclusion and perspectives

A method based on a soil-water balance, named LRIW, is developed to compute recharge on a daily interval, requiring the characterisation of evapotranspiration and parameters characterising the recharge area (soil available water capacity and runoff). A workflow is developed to compute daily groundwater recharge and requires the records of precipitation, air temperature, relative humidity, solar radiation and wind speed within or close to the landslide. The determination of the parameters of the recharge area is based on a spatial analysis requiring field observations and spatial data sets (digital elevation models, aerial photographs and geological maps). Once determined, the parameters are refined with a sensitivity analysis.

The method has been tested on the Séchilienne landslide. The tests demonstrate that the performance of the correlation with landslide displacement velocity data is significantly enhanced using the LRIW estimated recharge. The R^2 values obtained with the LRIW recharge are 78 % higher on average than those obtained with precipitation and are 38 % higher on average than those obtained with recharge computed with a commonly used simplification in several landslide studies (recharge = precipitation minus non-calibrated ET_0). The sensitivity analysis of the LRIW workflow appears to be an appropriate alternative to estimate or to refine soil-water balance parameters of the recharge area, especially in the case of insufficient field investigations or in the absence of the necessary spatial data set.

The LRIW workflow is developed to be as universal as possible in order to be applied to other landslides. The workflow is developed in order to be sufficiently simple to guide any non-hydrogeology specialist who intends to estimate the recharge signal in the case of rainfall–landslide displacement studies. Within this scope, a software is planned to be developed in the near future in order to provide a user-friendly tool for recharge estimation. In addition, the LRIW workflow also enables the reconstruction of retrospective time series for sites recently equipped with weather stations designed to measure a full set of parameters. A further step will have to account for the spatial and temporal variabilities of precipitation and recharge area properties, thus providing a better estimation of the recharge. In addition, taking recharge into account can assist in determining a warning rainfall threshold for the deep-seated slope movements.

Appendix A: Equations for evapotranspiration and solar radiation methods

A1 Equation parameter terms for all equations

R_a	extraterrestrial solar radiation ($\mathrm{MJ\,m^{-2}\,day^{-1}}$)
R_S	solar radiation ($\mathrm{MJ\,m^{-2}\,day^{-1}}$)
R_n	net solar radiation ($\mathrm{MJ\,m^{-2}\,day^{-1}}$)
N	maximum possible duration of sunshine (h)
n	actual daily duration of sunshine (h)
T_{avg}	average air temperature at 2 m height (°C)
T_{min}	minimum air temperature at 2 m height (°C)
T_{max}	maximum air temperature at 2 m height (°C)
G	soil heat flux density ($\mathrm{MJ\,m^{-2}\,day^{-1}}$)
γ	psychometric constant ($\mathrm{kPa\,°C^{-1}}$)
u_2	wind speed at 2 m height ($\mathrm{m\,s^{-1}}$)
e_s	mean saturation vapour pressure (kPa)
e_a	actual vapour pressure (kPa)
e^o	saturation vapour pressure at the air temperature T (kPa)
Δ	slope of vapour pressure curve ($\mathrm{kPa\,°C^{-1}}$]
RH	relative humidity (%)
α	cloud cover adjustment factor (unitless)

The procedure for calculating these equation terms are given in the FAO-56 guidelines for computing crop water requirements (Allen et al., 1998).

A2 Solar radiation (R_S)

The solar radiation BC R_S is obtained from the Bristow–Campbell method (Bristow and Campbell, 1984):

$$BC R_S = A_{BC} Ra \left[1 - \exp\left(-B_{BC}(\alpha \Delta T_{BC})^{C_{BC}} \right) \right]$$
$$\text{with } \Delta T_{BC} = T_{\max(j)} - \frac{T_{\min(j)} + T_{\min(j+1)}}{2}. \quad \text{(A1)}$$

The solar radiation HS R_S obtained from the Hargreaves–Samani method (Hargreaves and Samani, 1985):

$$HS R_S = A_{HS} Ra (\Delta T_{HS})^{B_{HS}}$$
$$\text{with } \Delta T_{HS} = T_{\max(j)} - T_{\min(j)}, \quad \text{(A2)}$$

where j is for the current target day and $j+1$ is for the following day; A_{BC}, B_{BC}, and C_{BC} are the Bristow–Campbell empirical coefficients (no default values) and A_{HS} and B_{HS} are the Hargreaves–Samani empirical coefficients ($A_{HS} = 0.16$ and $B_{HS} = 0.5$).

In this study, the modified forms of the R_S equations of Bristow–Campbell and Hargreaves–Samani are implemented: (i) a constant is added to take into account the possibility of a R_S estimation shift, (ii) the ΔT from the Bristow–Campbell method is used in both equations, and (iii) a cloud cover adjustment factor α is applied to ΔT since, for cloudy

conditions, ΔT can produce an estimate larger than the incoming solar radiation (Bristow and Campbell, 1984).

Bristow–Campbell modified equation (BC$_{mod}$ R_S):

$$BC_{mod} R_S = A_{BC} Ra \left[1 - \exp\left(-B_{BC}(\alpha \Delta T)^{C_{BC}} \right) \right] + D_{BC}. \quad \text{(A3)}$$

Hargreaves–Samani modified equation (HS$_{mod}$ R_S):

$$HS_{mod} R_S = A_{HS} Ra (\alpha \Delta T)^{B_{HS}} + C_{HS} \quad \text{(A4)}$$

with $\Delta T = T_{\max(j)} - \frac{T_{\min(j)} + T_{\min(j+1)}}{2}$, where j is for the current day and $j+1$ is for the following day; A_{BC}, B_{BC}, C_{BC}, and D_{BC} are the Bristow–Campbell regional calibration coefficients and A_{HS}, B_{HS}, and C_{HS} are the Hargreaves–Samani regional calibration coefficients.

The α coefficient is applied for the two first rain-event days since, for a rain period longer than two days, the value of the R_S estimated from ΔT and the actual R_S value become almost identical. If ΔT on the day before a rain event (ΔT_{j-1}) is less than ΔT_{j-2} by more than 2 °C, the coefficient α is also applied assuming that cloud cover was already significantly present. For the remaining days, α is not applied ($\alpha = 1$). A 2 °C threshold and a 2 day period are used (Bristow and Campbell, 1984). In this study, the calibration of α is based on the principle that if this adjustment is not relevant, a calibrated α coefficient would be equal to 1 (no effect).

R_S can also be calculated with the Angström formula using sunshine duration data recorded at a weather station (FAO-56 guidelines, Allen et al., 1998):

$$R_S = \left(a_s + b_s \frac{n}{N} \right) R_a \quad \text{(A5)}$$

where: $a_s + b_s$ is the fraction of extraterrestrial solar radiation reaching the Earth surface on clear days (default values, $a_s = 0.25$ and $b_s = 0.5$).

A3 Reference vegetation evapotranspiration (ET₀)

The reference vegetation evapotranspiration FAO-56 PM ET$_0$ obtained from the Penman–Monteith method modified form from the FAO paper number 56 (Allen et al., 1998) is

$$\text{FAO-56 PM ET}_0 = \frac{0.408\Delta\,(R_n - G) + \gamma \frac{900}{T_{avg}+273} u_2 (e_s - e_a)}{\Delta + \gamma\,(1 + 0.34 u_2)}. \quad \text{(A6)}$$

The reference vegetation evapotranspiration HS ET$_0$ obtained from the Hargreaves–Samani method (Hargreaves and Samani, 1985) is

$$\text{HS ET}_0 = 0.0135\ 0.408 R_S \left(T_{avg} + 17.8 \right). \quad \text{(A7)}$$

The 0.408 unit conversion factor is added to the original formula in order to compute ET$_0$ in millimetres per day with R_S in megajoules per square metre per day.

The reference vegetation evapotranspiration Turc ET_0 obtained from the Turc method (Turc, 1961) is

for RH > 50%, Turc ET_0

$$= 0.01333 \frac{T_{avg}}{T_{avg} + 15} (23.9001 R_S + 50), \quad \text{(A8)}$$

for RH < 50%, Turc ET_0

$$= 0.01333 \frac{T_{avg}}{T_{avg} + 15} (23.9001 R_S + 50)$$
$$\left(1 + \frac{50 - RH}{70}\right). \quad \text{(A9)}$$

For the Séchilienne landslide, Eq. (A8) is preferred to Eq. (A9) because of an average greater than 50% relative humidity (RH) of the nearby mountain weather stations (Chamrousse, 70%; Saint-Michel-Maur, 66%; Saint-Jean-Saint-Nicolas, 66%).

The reference vegetation evapotranspiration PT ET_0 obtained from the Priestley–Taylor method (Priestley and Taylor, 1972) is

$$PT\ ET_0 = 1.26 \frac{\Delta}{\Delta + \gamma} (R_n - G). \quad \text{(A10)}$$

The reference vegetation evapotranspiration M ET_0 obtained from the Makkink method (Makkink, 1957) is

$$M\ ET_0 = 0.61 \frac{\Delta}{(\Delta + \gamma)} \frac{R_S}{2.45} - 0.012. \quad \text{(A11)}$$

The Penman–Monteith reduced-set method which allows calculating the reference vegetation evapotranspiration $PM_{red}\ ET_0$ is identical to the PM FAO-56 method (Eq. A6), but humidity and wind speed are estimated according to FAO-56 guidelines (Allen et al., 1998). The actual vapour pressure is estimated with Eq. (A12):

$$e_a = e^0 (T_{min}) = 0.611 \exp\left(\frac{17.27 T_{min}}{T_{min} + 237.3}\right). \quad \text{(A12)}$$

In the case of the Séchilienne landslide, the wind speed is fixed at $1.5\ m\,s^{-1}$ at a 2 m height ($2\ m\,s^{-1}$ by default), which is the daily average of the nearby mountain weather stations (Chamrousse, $2.33\ m\,s^{-1}$; Saint-Michel-Maur, $0.95\ m\,s^{-1}$; Saint-Jean-Saint-Nicolas, $1.26\ m\,s^{-1}$).

A4   Practical information

The ET_0 methods used in this study were developed for irrigation scheduling, for which the scope of application involves positive temperatures (plant water supply during the spring–summer growing period). However, in mountainous sites, winter temperatures are often below 0 °C, and ET_0 empirical methods can compute negative ET_0 values. Negative

ET_0-computed values do not have any physical meaning and are therefore set to 0 for this study.

The Priestley–Taylor and Penman–Monteith ET_0 methods use net solar radiation (R_n) instead of R_S, which can be deduced from R_S following the FAO-56 guideline (Allen et al., 1998).

ET_0 reduce-set methods do not take into account the wind speed variations. By removing saturated air from the boundary layer, wind increases evapotranspiration (Shahidian et al., 2012). Several studies show the influence of the wind speed on ET_0 method performance and therefore on calibration (Itenfisu et al., 2003; Trajkovic, 2005; Trajkovic and Stojnic, 2007). For this study, the days with average wind speed above the 95th percentile of the data set (extreme values) are disregarded in the calibration.

Appendix B: Temperature estimation at the Mont Sec weather station

B1   Method

The temperatures at the Mont Sec weather station are estimated with the characterisation of the local air temperature gradient using two surrounding weather stations recording the temperatures at a daily rate (Luitel et La Mure weather stations). Once the local air temperature gradient is characterised, one of the stations is used to estimate the Mont Sec temperatures.

The decrease in air density with elevation leads to a decrease in air temperature known as the lapse rate (Jacobson, 2005). A commonly used value of this rate is $-6.5\ °C\ 1000\ m^{-1}$. The air temperature can thus be related to elevation. In order to compute a local air temperature gradient, two weather stations surrounding the Séchilienne site are used: Luitel and La Mure (Table 1, Fig. 3). The Luitel station is located on the Séchilienne massif whereas the La Mure station is located about 18 km from the landslide. Both stations have weather conditions similar to the Séchilienne recharge area. Although the temperature estimation from the Luitel station would probably be more accurate, in order to maximise common interval lengths of temperatures with displacement records from 1994 to 2012, the La Mure station with records from 1992 to 2012 is preferred to estimate temperatures at Mont Sec.

The local air temperature gradient in relation to elevation is defined by Eq. (B1). The La Mure station minimum and maximum temperatures are used to estimate the temperatures at Luitel in relation to elevation, over their common recording period. A linear regression between temperatures measured at La Mure and Luitel is performed to determine the a and b coefficients. The b coefficient, which combines the lapse rate (λ) and the elevation difference, is then divided by the elevation difference of the two stations used for the calibration.

$$T_{(\text{Station})} = aT_{(\text{Mure})} + b = aT_{(\text{Mure})} + \lambda\text{Diff}_{\text{elevation}}$$

$$\text{with Diff}_{\text{elevation}} = \text{Elevation}_{\text{Mure}} - \text{Elevation}_{\text{Station}}, \quad (B1)$$

where a and b are regional calibration coefficients; T is the temperature minimum or maximum ($°C$); λ is the temperature lapse rate ($°C\,m^{-1}$); $\text{Diff}_{\text{elevation}}$ is the difference of elevation between two weather stations (m); elevation refers to the weather station elevation (m a.s.l.) and station to the target station (Luitel for calibration, Mont Sec for computation).

B2 Results

The recording period used for temperature calibration is from 6 July 2006 to 23 July 2012 (2193 records). This is a common data interval for the two weather stations used (La Mure and Luitel). The estimation of the local air temperature gradient shows a very good performance with R^2 equal to 0.895 (LBCI = 0.839) and 0.916 (LBCI = 0.862), and RMSE equal to 2.12 and 2.48, respectively, for minimum and maximum daily temperature calibrations. Equations (B2) and (B3) are used to estimate temperatures at Mont Sec with temperatures measured at La Mure. Rather than taking the elevation of the Mont Sec weather station (1147 m), the average elevation of recharge area (1200 m) is used, resulting in a difference of elevation with La Mure of 319 m. The estimated local air temperature gradient is 0.7 °C per 100 m of elevation; the average of the λ of the two following equations:

$$T_{\min(\text{Mont Sec})} = 0.911 T_{\min(\text{Mure})} - 0.0056 \times 319\,, \quad (B2)$$

$$T_{\max(\text{Mont Sec})} = 0.928 T_{\max(\text{Mure})} - 0.0087 \times 319\,. \quad (B3)$$

Appendix C: Rainfall–displacement relationship in the case of the Séchilienne landslide

The rainfall–displacement relationship is hereafter discussed for the precipitation and the R_{LRIW} signals. Although the R^2 values are significantly variable from one station to another, the 5th and 95th percentiles and the observed value of the NH2 test are rather constant for the four displacement stations (respectively about 0.116, 0.351 and 0.235; Fig. 10a). These results show that the improvement of the correlation performance by using recharge rather than precipitation has the same order of magnitude for the four stations, whereas R^2 values vary considerably between the four stations. This may be explained by the fact that groundwater hydrodynamics probably triggers the entire Séchilienne landslide while the displacement velocity response depends on the damage level of the rock at the location of the displacement station. This interpretation is supported by the variability of the cumulative period, the shift factor, the weighting factor and the R^2 value, especially between G5 and the three other stations (Table 2).

The cumulative period and the shift factor deduced from the antecedent cumulative sum allow determining the response time of the Séchilienne landslide to rainfall events. Displacement stations located in the high motion zone show homogenous time delays with shift factors of 2–3 days. The average cumulative periods beyond which precipitation or R_{LRIW} have no longer any influence on the landslide destabilisation are estimated at about 50 days for precipitation and 75 days for R_{LRIW}. Station G5 shows significantly different time delays and cumulative periods, whatever the precipitation or R_{LRIW} data used. This difference can be explained by the low signal-to-noise ratio which makes the correlations difficult to interpret.

Concerning the A16 extensometer, regarding precipitation, R^2 is better for the recent short testing interval (0.343) than for the former long interval of the sensitivity analysis (0.311). Conversely, regarding the recharge, R^2 is better for the former long interval (0.618) than for the recent short testing interval (0.586). This could be the consequence of a degradation of the near-surface rock mechanical properties of the Séchilienne landslide (as suggested by the displacement trend; Fig. 4), which makes the landslide more sensitive to precipitation events in the recent period.

Lastly, the best correlations from the sensitivity analysis suggest that infiltration structures could gather a large proportion of the flow (up to 68 % for SAWC = 45 mm; NH4 LBCI < 0) with respect to their recharge surface area (24 %; Table 5). If so, fractures can play an important role in the groundwater drainage from the massif towards the landslide aquifers.

Acknowledgements. This research was funded by the SLAMS program (Séchilienne Land movement: Multidisciplinary Studies) of the National Research Agency (ANR). The meteorological data were provided by Météo France, LTHE, ONF and CEREMA Lyon. Aerial photographs and the digital elevation model were provided by IGN. Displacement data were supplied by CEREMA Lyon. The authors acknowledge the support of Jean-Pierre Duranthon and Marie-Aurélie Chanut from the CEREMA Lyon and Jean-Paul Laurent from the LTHE. Appreciation is also given to Eric Lucot for his kind advice on soil log interpretation, to Patrick Giraudoux for his support implementing bootstrap tests and to Pete Milmo for English and technical proof reading. Lastly, the authors thank the editor and the three anonymous referees for their insightful comments.

Edited by: N. Romano

References

Alexandris, S., Stricevic, R., and Petkovic, S.: Comparative analysis of reference evapotranspiration from the surface of rainfed grass in central Serbia, calculated with six empirical methods against the Penman–Monteith formula, European Water, 21/22, 17–28, 2008.

Alfonsi, P.: Relation entre les paramètres hydrologiques et la vitesse dans les glissements de terrains, Exemples de La Clapière et de Séchilienne, Revue Française de Géotechnique, 79, 3–12, 1997.

Alkaeed, O. A., Flores, C., Jinno, K., and Tsutsumi, A.: Comparison of Several Reference Evapotranspiration Methods for Itoshima Peninsula Area, Fukuoka, Japan, Memoirs of the Faculty of Engineering, Kyushu University, 66, 1–14, 2006.

Allen, R. E., Pereira, L. S., Raes, D., and Smith, M.: Crop evapotranspiration: guidelines for computing crop water requirements, FAO Irrigation and drainage paper 56, Food and Agriculture Organization of the United Nations, Rome, 1998.

Allen, R. G., Smith, M., Pereira, L. S., and Perrier, A.: An Update for the Definition of Reference Evapotranspiration, ICID Bull. Int. Commis. Irrig. Drain., 43, 1–34, 1994.

Almorox, J.: Estimating global solar radiation from common meteorological data in Aranjuez, Spain, Turk. J. Phys., 35, 53–64, 2011.

Aragon, Y.: Séries temporelles avec R: Méthodes et cas, Springer Editions, Paris, 2011.

Baize, D. and Jabiol, B.: Guide pour la description des sols, 2nd Edn., Ed. Quae, Versailles, 2011.

Belle, P., Aunay, B., Bernardie, S., Grandjean, G., Ladouche, B., Mazué, R., and Join, J.-L.: The application of an innovative inverse model for understanding and predicting landslide movements (Salazie cirque landslides, Reunion Island), Landslides, 11, 343–355, doi:10.1007/s10346-013-0393-5, 2014.

Binet, S., Guglielmi, Y., Bertrand, C., and Mudry, J.: Unstable rock slope hydrogeology: insights from the large-scale study of western Argentera-Mercantour hillslopes (South-East France), Bulletin de la Société Géologique de France, 178, 159–168, doi:10.2113/gssgfbull.178.2.159, 2007a.

Binet, S., Mudry, J., Scavia, C., Campus, S., Bertrand, C., and Guglielmi, Y.: In situ characterization of flows in

a fractured unstable slope, Geomorphology, 86, 193–203, doi:10.1016/j.geomorph.2006.08.013, 2007b.

Bogaard, T., Guglielmi, Y., Marc, V., Emblanch, C., Bertrand, C., and Mudry, J.: Hydrogeochemistry in landslide research: a review, Bulletin de la Société Géologique de France, 178, 113–126, doi:10.2113/gssgfbull.178.2.113, 2007.

Bogaard, T., Maharjan, L. D., Maquaire, O., Lissak, C., and Malet, J.-P.: Identification of Hydro-Meteorological Triggers for Villerville Coastal Landslide, in: Landslide Science and Practice, edited by: Margottini, C., Canuti, P., and Sassa, K., Springer, Berlin, Heidelberg, 141–145, available at: http://link.springer.com.biblioplanets.gate.inist.fr/chapter/10.1007/978-3-642-31427-8_18 (last access: 12 June 2014), 2013.

Bonzanigo, L., Eberhardt, E., and Loew, S.: Long-term investigation of a deep-seated creeping landslide in crystalline rock, Part I. Geological and hydromechanical factors controlling the Campo Vallemaggia landslide, Can. Geotech. J., 44, 1157–1180, doi:10.1139/T07-043, 2007.

Bristow, K. L. and Campbell, G. S.: On the relationship between incoming solar radiation and daily maximum and minimum temperature, Agr. Forest Meteorol., 31, 159–166, doi:10.1016/0168-1923(84)90017-0, 1984.

Brocca, L., Ponziani, F., Moramarco, T., Melone, F., Berni, N., and Wagner, W.: Improving Landslide Forecasting Using ASCAT-Derived Soil Moisture Data: A Case Study of the Torgiovannetto Landslide in Central Italy, Remote Sensing, 4, 1232–1244, doi:10.3390/rs4051232, 2012.

Bruand, A., Duval, O., and Cousin, I.: Estimation des propriétés de rétention en eau des sols à partir de la base de données SOLHYDRO: Une première proposition combinant le type d'horizon, sa texture et sa densité apparente, Etude et Gestion des Sols, 11, 323–334, 2004.

Brückl, E. P.: Cause-Effect Models of Large Landslides, Nat. Hazards, 23, 291–314, doi:10.1023/A:1011160810423, 2001.

Canuti, P., Focardi, P., and Garzonio, C.: Correlation between rainfall and landslides, Bull. Eng. Geol. Environ., 32, 49–54, doi:10.1007/BF02594765, 1985.

Cappa, F., Guglielmi, Y., Soukatchoff, V. M., Mudry, J., Bertrand, C., and Charmoille, A.: Hydromechanical modeling of a large moving rock slope inferred from slope levelling coupled to spring long-term hydrochemical monitoring: example of the La Clapière landslide (Southern Alps, France), J. Hydrol., 291, 67–90, doi:10.1016/j.jhydrol.2003.12.013, 2004.

Cappa, F., Guglielmi, Y., Viseur, S., and Garambois, S.: Deep fluids can facilitate rupture of slow-moving giant landslides as a result of stress transfer and frictional weakening, Geophys. Res. Lett., 41, 61–66, doi:10.1002/2013GL058566, 2014.

Cartier, G. and Pouget, P.: Corrélation entre la pluviométrie et les déplacements de pentes instables, in: 9th European conference on soil mechanics and foundation engineering, vol. 1, CRC Press, Dublin, available at: http://www.crcpress.com/product/isbn/9789061917229 (last access: 2 April 2014), 1987.

Castellvi, F.: A new simple method for estimating monthly and daily solar radiation. Performance and comparison with other methods at Lleida (NE Spain); a semiarid climate, Theor. Appl. Climatol., 69, 231–238, doi:10.1007/s007040170028, 2001.

Chanut, M.-A., Vallet, A., Dubois, L., and Duranthon, J.-P.: Mouvement de versant de Séchilienne: relations entre déplacements de surface et précipitations – analyse statistique, in: Journées Aléa Gravitaire 2013, Grenoble, France, 2013.

Chernick, M. R.: Bootstrap methods: A guide for practitioners and researchers, Wiley-Interscience, Hoboken, NJ, 2008.

Cordeiro, C. and Neves, M.: The Bootstrap methodology in time series forecasting, in: COMPSTAT 2006 – Proceedings in Computational Statistics, Springer, Rome, Italy, 1067–1073, available at: http://www.springer.com.biblioplanets.gate.inist.fr/statistics/computational+statistics/book/978-3-7908-1708-9 (last access: 12 June 2014), 2006.

Cowpertwait, P. S. P. and Metcalfe, A.: Introductory Time Series with R, Édition: 2009, Springer-Verlag New York Inc., Dordrecht, New York, 2009.

Crozier, M. J.: Landslides: Causes, consequences et environment, Croom Helm, London, Dover, NH, 1986.

Delmas, P., Cartier, G., and Pouget, G.: Méthodes d'analyse des risques liés aux glissements de terrain, Bulletin Liaison Laboratoire Ponts et Chaussées, 150/151, 29–38, 1987.

Diodato, N., Guerriero, L., Fiorillo, F., Esposito, L., Revellino, P., Grelle, G., and Guadagno, F. M.: Predicting Monthly Spring Discharges Using a Simple Statistical Model, Water Resour. Manage., 28, 969–978, doi:10.1007/s11269-014-0527-0, 2014.

Duranthon, J.-P., Effendiaz, L., Memier, M., and Previtali, I.: Apport des méthodes topographiques et topométriques au suivi du versant rocheux instable des ruines de Séchilienne, Association Francaise de Topographie, 94, 31–38, 2003.

Durville, J.-L., Kasperki, J., and Duranthon, J.-P.: The Séchilienne landslide: monitoring and kinematics, in: First Italian Workshop on Landslides, vol. 1, Napoli, Italia, 174–180, 2009.

Garel, E., Marc, V., Ruy, S., Cognard-Plancq, A.-L., Klotz, S., Emblanch, C., and Simler, R.: Large scale rainfall simulation to investigate infiltration processes in a small landslide under dry initial conditions: the Draix hillslope experiment, Hydrol. Process., 26, 2171–2186, 2012.

Gochis, D. and Cuenca, R.: Plant Water Use and Crop Curves for Hybrid Poplars, J. Irrig. Drain. Eng., 126, 206–214, doi:10.1061/(ASCE)0733-9437(2000)126:4(206), 2000.

Godt, J. W., Baum, R. L., and Chleborad, A. F.: Rainfall characteristics for shallow landsliding in Seattle, Washington, USA, Earth Surf. Proc. Land., 31, 97–110, doi:10.1002/esp.1237, 2006.

Guglielmi, Y., Vengeon, J. M., Bertrand, C., Mudry, J., Follacci, J. P.. and Giraud, A.: Hydrogeochemistry: an investigation tool to evaluate infiltration into large moving rock masses (case study of La Clapière and Séchilienne alpine landslides), Bull. Eng. Geol. Environ., 61, 311–324, 2002.

Guglielmi, Y., Cappa, F., and Binet, S.: Coupling between hydrogeology and deformation of mountainous rock slopes: Insights from La Clapière area (southern Alps, France), Comptes Rendus Geoscience, 337, 1154–1163, doi:10.1016/j.crte.2005.04.016, 2005.

Guzzetti, F., Peruccacci, S., Rossi, M., and Stark, C. P.: The rainfall intensity–duration control of shallow landslides and debris flows: an update, Landslides, 5, 3–17, doi:10.1007/s10346-007-0112-1, 2008.

Hargreaves, G. H. and Allen, R. G.: History and Evaluation of Hargreaves Evapotranspiration Equation, J. Irrig. Drain. Eng., 129, 53–63, doi:10.1061/(ASCE)0733-9437(2003)129:1(53), 2003.

Hargreaves, G. and Samani, Z.: Reference Crop Evapotranspiration from Temperature, Appl. Eng. Agr., 1, 96–99, 1985.

Helmstetter, A. and Garambois, S.: Seismic monitoring of Séchilienne rockslide (French Alps): Analysis of seismic signals and their correlation with rainfalls, J. Geophys. Res., 115, F03016, doi:10.1029/2009JF001532, 2010.

Hong, Y., Hiura, H., Shino, K., Sassa, K., Suemine, A., Fukuoka, H., and Wang, G.: The influence of intense rainfall on the activity of large-scale crystalline schist landslides in Shikoku Island, Japan, Landslides, 2, 97–105, doi:10.1007/s10346-004-0043-z, 2005.

Hou, L. G., Xiao, H. L., Si, J. H., Xiao, S. C., Zhou, M. X., and Yang, Y. G.: Evapotranspiration and crop coefficient of Populus euphratica Oliv forest during the growing season in the extreme arid region northwest China, Agr. Water Manage., 97, 351–356, doi:10.1016/j.agwat.2009.09.022, 2010.

Igbadun, H., Mahoo, H., Tarimo, A., and Salim, B.: Performance of Two Temperature-Based Reference Evapotranspiration Models in the Mkoji Sub-Catchment in Tanzania, Agricultural Engineering International: the CIGR Ejournal, VIII, aailable from: http://ecommons.library.cornell.edu/handle/1813/10573 (last access: 15 April 2014), 2006.

Itenfisu, D., Elliott, R. L., Allen, R. G., and Walter, I. A.: Comparison of Reference Evapotranspiration Calculations as Part of the ASCE Standardization Effort, J. Irrig. Drain. Eng., 129, 440–448, doi:10.1061/(ASCE)0733-9437(2003)129:6(440), 2003.

Jacobson, P. M. Z.: Fundamentals of Atmospheric Modeling, Édition: 2, Cambridge University Press, Cambridge, UK, New York, 2005.

Jamagne, M., Bétrémieux, R., Bégon, J. C., and Mori, A.: Quelques données sur la variabilité dans le milieu naturel de la réserve en eau des sols, Bulletin Technique d'Information du Ministère de l'Agriculture, 324–325, 627–641, 1977.

Jensen, M. E., Burman, R. D., and Allen, R. G.: Evapotranspiration and irrigation water requirements: a manual, American Society of Civil Engineers, New York, 1990.

Lebrouc, V., Schwartz, S., Baillet, L., Jongmans, D., and Gamond, J. F.: Modeling permafrost extension in a rock slope since the Last Glacial Maximum: Application to the large Séchilienne landslide (French Alps), Geomorphology, 198, 189–200, doi:10.1016/j.geomorph.2013.06.001, 2013.

Le Roux, O., Jongmans, D., Kasperski, J., Schwartz, S., Potherat, P., Lebrouc, V., Lagabrielle, R., and Meric, O.: Deep geophysical investigation of the large Séchilienne landslide (Western Alps, France) and calibration with geological data, Eng. Geol., 120, 18–31, doi:10.1016/j.enggeo.2011.03.004, 2011.

López-Moreno, J. I., Hess, T. M., and White, S. M.: Estimation of reference evapotranspiration in a mountainous mediterranean site using the Penman–Monteith equation with limited meteorological data, Pirineos, 164, 7–31, 2009.

Lu, J., Sun, G., McNulty, S. G., and Amatya, D. M.: A comparison of six potential evapotranspiration methods for regional use in the Southeastern United States, J. Am. Water Resour. Assoc., 41, 621–633, doi:10.1111/j.1752-1688.2005.tb03759.x, 2005.

Madsen, H.: Time Series Analysis, 1st Edn., Chapman and Hall/CRC, Boca Raton, 2007.

Makkink, G.: Testing the Penman formula by means of lysimeters, J. Inst. Water Eng., 11, 277–288, 1957.

Malet, J. P., Maquaire, O., and Vanash, T. W.: Hydrological behaviour of earthflows developed in clay-shales: investigation, concept and modelling, The Occurrence and Mechanisms of

Flows in Natural Slopes and Earthfills, Patron Editore, Bologna, 175–193, 2003.

Maquaire, O.: Effects of Groundwater on the Villerville-Cricqueboeuf Landslides, Sixteen Year Survey (Calvados, France), in: 8th Landslides International symposium, Cardiff, 1005–1010, 2000.

Matichard, Y. and Pouget, P.: Pluviométrie et comportement de versants instables, in Landslides: proceedings of the fifth International Symposium on Landslides, Lausanne, Switzerland, 725–730, 1988.

McCuen, R. H.: Hydrologic analysis and design, Pearson Prentice Hall, Upper Saddle River, NJ, 2005.

Meric, O., Garambois, S., Jongmans, D., Wathelet, M., Chatelain, J. L., and Vengeon, J. M.: Application of geophysical methods for the investigation of the large gravitational mass movement of Séchilienne, France, Can. Geotech. J., 42, 1105–1115, doi:10.1139/t05-034, 2005.

Meric, O., Garambois, S., and Orengo, Y.: Large Gravitational Movement Monitoring Using a Spontaneous Potential Network, in: Proc. 19th Annual Symposium on the Application of Geophysics to Engineering and Environmental Problems, EEGS, Seattle, USA, 202–209, 2006.

Mudry, J. and Etievant, K.: Synthèse hydrogéologique du versant instable des Ruines de Séchilienne, Unpublished report, UMR Chrono-Environnement, University of Franche-Comté, Franche-Comté, 2007.

Musy, A. and Higy, C.: Hydrology: A Science of Nature, English Edn., CRC Press, Science Publishers, Boca Raton, FL, Enfield, NH, 2011.

Noverraz, F., Bonnard, C., Dupraz, H., and Huguenin, L.: Grands glissements de terrain et climat, VERSINCLIM – Comportement passé, présent et futur des grands versants instables subactifs en fonction de l'évolution climatique, et évolution en continu des mouvements en profondeur, Rapport final PNR31 (Programme National de Recherche), vdf Hochschulverlag AG an der ETH Zürich, Zürich, Switzerland, 1998.

Pachepsky, Y. and Rawls, W. J.: Development of pedotransfer functions in soil hydrology, Elsevier, Amsterdam, New York, 2004.

Patwardhan, A., Nieber, J., and Johns, E.: Effective Rainfall Estimation Methods, J. Irrig. Drain. Eng., 116, 182–193, doi:10.1061/(ASCE)0733-9437(1990)116:2(182), 1990.

Pisani, G., Castelli, M., and Scavia, C.: Hydrogeological model and hydraulic behaviour of a large landslide in the Italian Western Alps, Nat. Hazards Earth Syst. Sci., 10, 2391–2406, doi:10.5194/nhess-10-2391-2010, 2010.

Ponziani, F., Pandolfo, C., Stelluti, M., Berni, N., Brocca, L., and Moramarco, T.: Assessment of rainfall thresholds and soil moisture modeling for operational hydrogeological risk prevention in the Umbria region (central Italy), Landslides, 9, 229–237, doi:10.1007/s10346-011-0287-3, 2012.

Priestley, C. H. B. and Taylor, R. J.: On the Assessment of Surface Heat Flux and Evaporation Using Large-Scale Parameters, Mon. Weather Rev., 100, 81–92, doi:10.1175/1520-0493(1972)100<0081:OTAOSH>2.3.CO;2, 1972.

Prokešová, R., Medved'ová, A., Tábořík, P., and Snopková, Z.: Towards hydrological triggering mechanisms of large deep-seated landslides, Landslides, 10, 239–254, doi:10.1007/s10346-012-0330-z, 2013.

Rochet, L., Giraud, A., Antoine, P., and Évrard, H.: La déformation du versant sud du Mont-Sec dans le secteur des ruines de Séchilienne (Isère), Bull. Int. Assoc. Eng. Geol., 50, 75–87, doi:10.1007/BF02594959, 1994.

Rutqvist, J. and Stephansson, O.: The role of hydromechanical coupling in fractured rock engineering, Hydrogeol. J., 11, 7–40, doi:10.1007/s10040-002-0241-5, 2003.

Shahidian, S., Serralheiro, R., Serrano, J., Teixeira, J., Haie, N., and Santos, F.: Hargreaves and Other Reduced-Set Methods for Calculating Evapotranspiration, in: Evapotranspiration – Remote Sensing and Modeling, edited by: Irmak, A., InTech, Rijeka, Croatia, 60–80, 2012.

Sivaprakasam, S., Murugappan, A., and Mohan, S.: Modified Hargreaves equation for estimation of ETo in a Hot and Humid Location in Tamilnadu State, India, Int. J. Eng. Sci. Technol., 3, 592–600, 2011.

Tabari, H. and Talaee, P. H.: Local Calibration of the Hargreaves and Priestley-Taylor Equations for Estimating Reference Evapotranspiration in Arid and Cold Climates of Iran Based on the Penman-Monteith Model, J. Hydrol. Eng., 16, 837–845, doi:10.1061/(ASCE)HE.1943-5584.0000366, 2011.

Tabari, H., Grismer, M. E., and Trajkovic, S.: Comparative analysis of 31 reference evapotranspiration methods under humid conditions, Irrig. Science, 31, 107–117, doi:10.1007/s00271-011-0295-z, 2013.

Trajkovic, S.: Temperature-Based Approaches for Estimating Reference Evapotranspiration, J. Irrig. Drain. Eng., 131, 316–323, doi:10.1061/(ASCE)0733-9437(2005)131:4(316), 2005.

Trajkovic, S.: Hargreaves versus Penman–Monteith under Humid Conditions, J. Irrig. Drain. Eng., 133, 38–42, doi:10.1061/(ASCE)0733-9437(2007)133:1(38), 2007.

Trajkovic, S. and Stojnic, V.: Effect of wind speed on accuracy of Turc method in a humid climate, Facta universitatis – series: Architecture and Civil Engineering, 5, 107–113, doi:10.2298/FUACE0702107T, 2007.

Turc, L.: Evaluation des besoins en eau d'irrigation, évapotranspiration potentielle, formule simplifiée et mise à jour, Annales Agronomiques, 12, 13–49, 1961.

Van Asch, T. W. J., Buma, J., and van Beek, L. P.: A view on some hydrological triggering systems in landslides, Geomorphology, 30, 25–32, doi:10.1016/S0169-555X(99)00042-2, 1999.

Vengeon, J. M.: Déformation et rupture des versants en terrain métamorphique anisotrope: Apport de l'étude des Ruines de Séchilienne, PhD thesis, Université Joseph Fourier I, 3 November 1998, Grenoble, 1998.

Verstraeten, W. W., Muys, B., Feyen, J., Veroustraete, F., Minnaert, M., Meiresonne, L., and De Schrijver, A.: Comparative analysis of the actual evapotranspiration of Flemish forest and cropland, using the soil water balance model WAVE, Hydrol. Earth Syst. Sci., 9, 225–241, doi:10.5194/hess-9-225-2005, 2005.

Yoder, R. E., Odhiambo, L. O., and Wright, W. C.: Evaluation of methods for estimating daily reference crop evapotranspiration at a site in the humid Southeast United States, Appl. Eng. Agr., 21, 197–202, 2005.

Zêzere, J. L., Trigo, R. M., and Trigo, I. F.: Shallow and deep landslides induced by rainfall in the Lisbon region (Portugal): assessment of relationships with the North Atlantic Oscillation, Nat. Hazards Earth Syst. Sci., 5, 331–344, doi:10.5194/nhess-5-331-2005, 2005.

12

Climate change impacts on the seasonality and generation processes of floods – projections and uncertainties for catchments with mixed snowmelt/rainfall regimes

K. Vormoor[1]**, D. Lawrence**[2]**, M. Heistermann**[1]**, and A. Bronstert**[1]

[1]Institute of Earth and Environmental Science, University of Potsdam, Postdam, Germany
[2]Norwegian Water Resources and Energy Directorate (NVE), Oslo, Norway

Correspondence to: K. Vormoor (kvormoor@uni-potsdam.de)

Abstract. Climate change is likely to impact the seasonality and generation processes of floods in the Nordic countries, which has direct implications for flood risk assessment, design flood estimation, and hydropower production management. Using a multi-model/multi-parameter approach to simulate daily discharge for a reference (1961–1990) and a future (2071–2099) period, we analysed the projected changes in flood seasonality and generation processes in six catchments with mixed snowmelt/rainfall regimes under the current climate in Norway. The multi-model/multi-parameter ensemble consists of (i) eight combinations of global and regional climate models, (ii) two methods for adjusting the climate model output to the catchment scale, and (iii) one conceptual hydrological model with 25 calibrated parameter sets. Results indicate that autumn/winter events become more frequent in all catchments considered, which leads to an intensification of the current autumn/winter flood regime for the coastal catchments, a reduction of the dominance of spring/summer flood regimes in a high-mountain catchment, and a possible systematic shift in the current flood regimes from spring/summer to autumn/winter in the two catchments located in northern and south-eastern Norway. The changes in flood regimes result from increasing event magnitudes or frequencies, or a combination of both during autumn and winter. Changes towards more dominant autumn/winter events correspond to an increasing relevance of rainfall as a flood generating process (FGP) which is most pronounced in those catchments with the largest shifts in flood seasonality. Here, rainfall replaces snowmelt as the dominant FGP primarily due to increasing temperature. We further analysed

the ensemble components in contributing to overall uncertainty in the projected changes and found that the climate projections and the methods for downscaling or bias correction tend to be the largest contributors. The relative role of hydrological parameter uncertainty, however, is highest for those catchments showing the largest changes in flood seasonality, which confirms the lack of robustness in hydrological model parameterization for simulations under transient hydrometeorological conditions.

1 Introduction

The hydrological cycle is likely to intensify due to climate change (IPCC, 2007; Seneviratne et al., 2012), and a recent study indicates that global warming has caused more intense precipitation over the last century on the global scale (Benestad, 2013). These changes will, in turn, have direct implications for flood risk. For mountainous and Nordic regions, changes in the ratio of rainfall and snowfall due to temperature rise are of special interest since they have direct implications for flood seasonality and for the dominant processes generating flood discharge.

A coherent picture of observed positive annual and winter streamflow trends for the Nordic countries (Stahl et al., 2010; Wilson et al., 2010) has been linked to a pattern of generally increasing mean and extreme precipitation (Bhend and von Storch, 2007; Dyrrdal et al., 2012). Regarding flood seasonality, neither significant trends towards higher autumn floods as a result of increasing autumn rainfall, nor systematic trends in spring flood magnitudes are yet detected (Wilson et al., 2010). The same study found, however, a strong trend towards earlier spring floods at many stations. This is likely due to the observed increase in mean annual temperature during the last century, which has been reported to be 0.8 °C, with the strongest decadal temperature rise during the spring season (Hanssen-Bauer et al., 2009).

Climate projections for Norway for the end of the 21st century indicate increasing temperatures (2.3–4.6 °C) and precipitation (5–30 %) with the largest temperature increase during winter in northern Norway, and the largest precipitation increase during autumn and winter on the west coast (Hanssen-Bauer et al., 2009). Extreme precipitation is also likely to increase for all seasons across the whole of Norway (Beniston et al., 2007; Hanssen-Bauer et al., 2009; Seneviratne et al., 2012), although such projections are highly uncertain (Fowler and Ekström, 2009). Changes in temperature and precipitation regimes will have direct implications for the snow regime in Norway. For mountainous areas and in northern Norway where mean winter temperature is a few degrees below 0, snow depth is observed to have increased in recent decades (Dyrrdal et al., 2013) and climate projections suggest further increases until 2050 (Hanssen-Bauer et al., 2009). In other parts of Norway snow depths are projected to decrease. Towards the end of the 21st century, a decrease in snow depths and a shorter snow season are projected for the whole of the country due to temperature rise.

For the Nordic countries, several previous studies have investigated the hydrological impacts of climate change (e.g. Andréasson and Bergström, 2004; Roald, 2006; Beldring et al., 2008; Veijalainen et al., 2010; Lawrence and Hisdal, 2011; Lawrence and Haddeland, 2011). For Norway, Lawrence and Hisdal (2011) studied the changes in flood frequency in 115 Norwegian catchments and found coherent regional patterns of directional change in flood magnitudes under a future climate: the magnitudes of the 200-year flood, for example, is likely to increase in catchments in western and much of coastal Norway where flood generation is dominated by autumn/winter rainfall, while magnitudes are expected to decrease in the snowmelt-dominated catchments in inland areas and parts of northern Norway. This regional pattern reflects systematic changes in climate forcing, which lead to changes in hydrological flooding in terms of both seasonal prevalence and generation process (rainfall vs. snowmelt). There are, however, many catchments which are transitional between rainfall-dominated vs. snowmelt-dominated flood regimes, and interpretation of the likely direction of change in the magnitude of future floods is more difficult. In addition, such catchments may be subject to a shift in the flood season under a future climate. Considering the uncertainty in the projections for future (extreme) precipitation and subsequent flooding conditions (Bronstert et al., 2007), Blöschl et al. (2011) argue that seasonal change in the distribution of floods is the key to understanding climate change impacts on flooding rather than changes in flood magnitudes and fre-

quencies. Changes in the underlying flood generating processes (FGPs) are correspondingly important for interpreting the direction (i.e. increase vs. decrease) of climate change impacts on future floods. Therefore, we aim to study in detail the changing role of rainfall and snowmelt under future climate scenarios to aid in understanding flood regime changes in catchments which already show mixed snowmelt/rainfall flood regimes in today's climate.

For practical purposes, changes in flood seasonality have implications for future flood risk assessments, design flood estimations, and hydropower production management. In Norway, where hydropower represents about 96 % of the total electricity production, flood seasonality impacts reservoir management and accordingly hydropower production. In addition, design flood estimates for dam safety require that the season for the highest flood risk is assessed (e.g. Midttømme et al., 2011) and changes in the dominant flood season under a future climate have significant implications for these assessments. Despite the relevance of this issue, there has not yet been a detailed investigation of climate change impacts on future flood seasonality and the process-related factors contributing to those changes in Norway.

In this study, we investigate the impact of climate change on flood seasonality and the related FGPs in six Norwegian catchments representing different geographical and climatological conditions. The catchments were selected such that both rainfall and snowmelt sometimes play a role in the generation of high-flow events under the current climate; we investigate how the balance between these two flood generating factors changes. We apply a multi-model/multi-parameter ensemble to develop a range of hydrological projections which allows us to consider some of the uncertainties associated with such an analysis (e.g. Hall et al., 2014). The multi-model/multi-parameter ensemble used here consists of eight combinations of global and regional climate models (GCM/RCM combinations), two methods for locally adjusting the climate model output data to the catchment scale, and hydrological modelling implemented with the HBV model based on 25 calibrated parameter sets. Our particular research questions are (1) how might the existing patterns of flood seasonality change under a future climate? (2) How are shifts in seasonality related to changes in the magnitude vs. changes in the frequency of events? (3) Are changes in flood seasonality associated with changes in the dominant FGPs? (4) What is the relative importance of the different ensemble components in contributing to the overall variance as a measure of the uncertainty in the projected changes?

2   Study area

2.1   Climate and runoff regimes in Norway

Climatological gradients driven by latitude, topography and location relative to the coastal zone control the spatial pattern of temperature and precipitation regimes in Norway. The mean annual temperature varies from 7.7 °C at the south-western coast to about −3 °C in the inland areas of northern Norway and the high-altitude areas in central Norway (Hanssen-Bauer et al., 2009). Mean annual precipitation varies from about 300 mm in north-eastern and central Norway to more than 3500 mm in western Norway (Hanssen-Bauer et al., 2009). Seasonally, western Norway receives the largest precipitation volumes during the autumn and winter months, while the more inland region in the east receives these during the summer.

Mean annual runoff generally reflects the pattern of mean annual precipitation, and runoff coefficients tend to be high due to low evapotranspiration. However, due to differences in the temperature regime, snowpack volumes and the snow season vary considerably across the country, which leads to differences in the regional importance of snowmelt as a runoff generation process. Hence, two basic patterns in runoff regimes can be distinguished in Norway: (i) regions in inland and northernmost Norway with prominent high flows during spring and summer predominantly due to snowmelt, and (ii) regions in western Norway and in coastal regions with prominent high flows during autumn and winter predominantly due to rainfall. There are, though, numerous variations reflecting local climate as well as transitional, mixed, regimes. In addition, catchments with sources in high mountain areas can experience peak flows in late summer, due to glacier melt. A comprehensive classification of runoff regimes based on the seasonal occurrence of monthly high and low flows is given by Tollan (1975) and reviewed in Gottschalk et al. (1979). This classification defines five types of flood regimes for the Nordic countries and give detailed distinctions between possible combinations of high-flow and low-flow periods. However, in order to develop a broad picture of flood seasonality, it is most useful to apply the simple distinction between two high-flow seasons (spring/summer vs. autumn/winter) and to distinguish rainfall vs. snowmelt as the most fundamental flood generation processes.

2.2 Study catchments

Changes in flood seasonality and the FGPs were investigated in six catchments distributed across Norway: Krinsvatn, Fustvatn, Øvrevatn, Junkerdalselv, Atnasjø, and Kråkfoss (Fig. 1). These catchments represent some of the variability in climate conditions across the country. The focus in this work, however, is on catchments which already exhibit some tendency for both snowmelt- and rainfall-dominated flood regimes. Therefore, a full range of climatic conditions is not represented nor are some regions (e.g. western and southern coastal Norway) included in this analysis. In addition, the sample includes only catchments of moderate size which are suitable for hydrological modelling with a daily time step.

The catchments considered are largely unaffected by damming or regulation (Petterson, 2004), and anthropogenic land use (changes) can be neglected since land use constitutes only between 0 and 1 % of land cover in all catchments except Kråkfoss (11 %). The catchments are included in the benchmark data set for climate change studies for Norway and are classified as suitable for daily analyses of flood discharge (Fleig et al., 2013). The six catchments are mesoscale catchments and vary in size from 207 km^2 (Krinsvatn) to 526 km^2 (Fustvatn). Further catchment characteristics including elevation, land cover, as well as mean annual precipitation and runoff are given in Table 1. Figure 1 displays flood roses to illustrate the magnitudes of the annual maximum floods (AMFs) from observed daily series by their Julian date of occurrence. These plots indicate the flood seasonality for the six catchments for the period 1961–1990 (except for Kråkfoss where the observed time series begins in 1966).

Although Krinsvatn and Fustvatn have the lowest elevations amongst the catchments, they receive a considerably higher annual precipitation (2291 and 3788 mm, respectively) due to their coastal locations. Correspondingly, the catchments have large average annual runoff values, and both the majority of and the largest AMFs occur during late autumn and winter, representing rainfall-dominated flood generation. However, both catchments are also subject to snowmelt floods, as indicated by the comparatively smaller events occurring during spring.

Øvrevatn, Junkerdalselv and Atnasjø show the highest median elevation and elevation ranges, but differ considerably with respect to annual precipitation and runoff volumes. Øvrevatn and Atnasjø, though being the highest catchments within this comparison, receive considerably less precipitation (832 and 840 mm, respectively) due to their rain shadow locations. Junkerdalselv, being located further inland near the Swedish border, is not directly influenced by rain shadow effects and has annual precipitation and runoff volumes that are about 3 times larger than at Øvrevatn and Atnasjø. Because of the temperature regime, all three catchments receive a large portion of the annual precipitation as snow so that the majority of and the largest AMFs occur during spring and summer (May–July, Fig. 1), with snowmelt as the dominant FGP.

Kråkfoss, located inland of the Oslofjord, is the southernmost catchment within this study and has a slightly different flood regime. There is no definite seasonal prevalence for the AMFs; one-half of the events occur during spring and summer, the other half during autumn and early winter. The magnitude of the autumn events tends to be slightly larger than those occurring during spring. Snowmelt plays a definite role in the early events in the spring/summer period, whilst the events during autumn are triggered by rainfall. In addition, it is important to note that for catchments dominated by snowmelt floods, the largest events almost always represent a combination of snowmelt and heavy rainfall. Similarly,

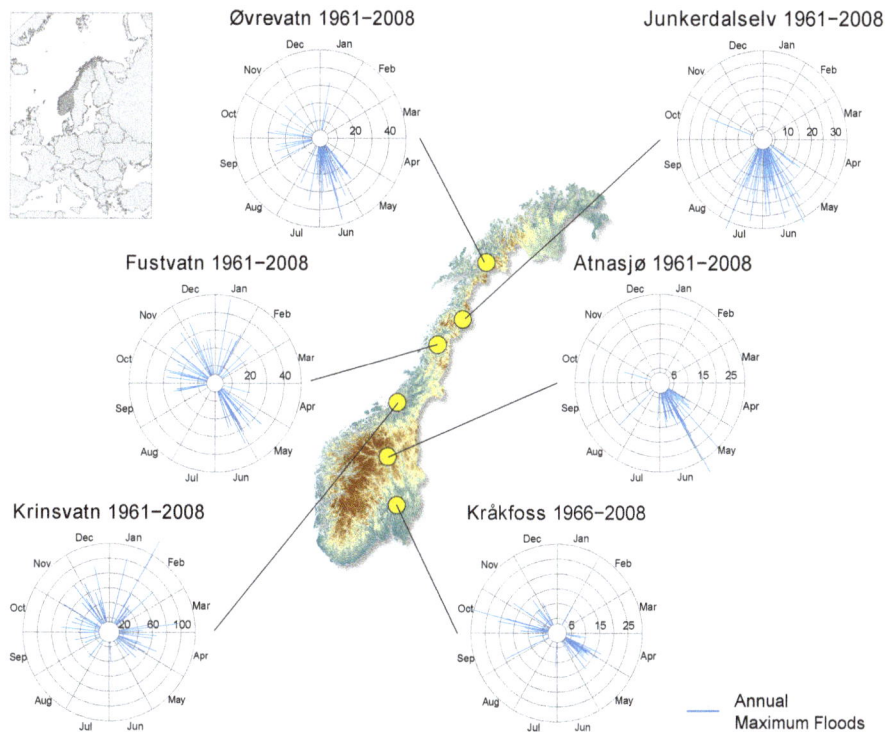

Figure 1. The location of the six study catchments and their current flood regime demonstrated by flood roses indicating the magnitude and timing of observed annual maximum floods. Values are given as specific discharge (mm day^{-1}). Note that secondary annual flood peaks can occur during contrasting seasons.

Table 1. Characteristics of the six study catchments.

Catchment property	Krinsvatn	Fustvatn	Øvrevatn	Junkerdalselv	Atnasjø	Kråkfoss
Area (km^2)	207	526	525	420	463	433
Median elevation (m a.s.l.)	349	436	841	835	1205	445
Elevation range (m a.s.l.)	87–629	39–812	145–1636	117–1703	701–2169	105–803
Average annual P (mm)	2291	3788	832	3031	840	2092
Average annual Q (mm)	1992	3017	564	2722	672	1798
Land cover, % lake	8	6	10	0	2	4
Land cover, % glacier	0	< 1	4	1	< 1	0
Land cover, % forest	20	38	23	25	20	76
Land cover, % marsh and bog	9	5	1	1	2	5
Land cover, % sparse vegetation above treeline	57	37	57	63	69	0
Anthropogenic land use (%)	0.4	0.0	0.7	0.5	0.4	11.2

most of the catchments dominated by rainfall-induced flooding have periods in which a transient snow cover also may contribute to runoff during rainfall. Therefore, for this study it is useful to define a third FGP ("rainfall + snowmelt"), which occurs to varying degrees in all six catchments considered.

The dominant land cover types in the six catchments are either exposed (crystalline) bedrock with sparse vegetation above tree line (Atnasjø, 69 %; Junkerdalselv, 63 %; Krinsvatn, 57 %; Øvrevatn, 57 %) or boreal forest (Kråkfoss, 76 %; Fustvatn, 38 %). Soils in all catchments are rather thin and poorly developed, and large, regional groundwater storage in aquifers is virtually non-existent due to the crystalline bedrock. However, in most catchments, surface water in the form of lakes, marshes and bogs can lead to water retention and, in some cases, significant attenuation of flood peaks.

3   Data and methods

3.1   Modelling strategy

The analyses of changes in flood seasonality and their associated FGPs are based on a multi-model/multi-parameter ensemble approach consisting of (i) eight GCM/RCM combinations, (ii) two methods for adjusting the temperature and precipitation outputs of the climate models at the catchment scale, and (iii) the HBV hydrological model with 25 different parameter sets for considering hydrological parameter uncertainty. It has become good practice to include more than one model for each member within the model chain to derive a range of possible projections and to allow drawing conclusions about the uncertainty that is associated by such approaches. We have only used one hydrological model in our ensemble setup; this is supported by Velázquez et al. (2013), who conclude that the use of multiple hydrological models in climate impact studies is important for the study of low flows and means, but not for high flows, as various lumped and distributed models lead to very similar results. Moreover, the HBV model has been widely applied in the Nordic countries since it represents a suitable conceptual representation of the dominant runoff generating processes and does not impose excessive data requirements. The following subsections describe the individual components of the ensemble in more detail.

3.2   Climate projections

The climate projections for precipitation and temperature chosen for the hydrological simulations are based on eight GCM/RCM combinations (Table 2) from the EU FP6 EN-SEMBLES project (van der Linden and Mitchell, 2009). The spatial resolution of all RCMs considered is 0.22° (approximately 25 km), and projections of daily values are available for the period 1950–2099. Within this study, two periods are compared: a reference period (1961–1990) for which the

Table 2. The GCM/RCM combinations from ENSEMBLES used for the hydrological projections. The full names of the institute abbreviations are SMHI – Swedish Meteorological and Hydrological Institute, met.no – the Norwegian Meteorological Institute, KNMI – The Royal Netherlands Meteorological Institute, MPI – Max Planck Institute for Meteorology (Germany), ICTP – International Centre for Theoretical Physics (Italy), METEO-HC – The Met Office Hadley Centre (UK).

Global climate model (GCM)	Regional climate model (RCM)	Institute
BCM	RCA HIRHAM	SMHI met.no
ECHAM5	RACMO REMO RegCM	KNMI MPI ICTP
HadCM3Q0	HadRM3Q0	METEO-HC
HadCM3Q3	HadRM3Q3	METEO-HC
HadCM3Q16	HadRM3Q16	METEO-HC

GCM/RCM combinations are driven by the IPCC-AR4 scenario C20, and a future period (2071–2099) for which the climate model combinations are driven by the SRES A1B scenario, which represents intermediate greenhouse gas emissions until the end of the 21st century (IPCC, 2000, 2007). We only focus on the far future period since the change signals are more pronounced by this time. We selected the eight RCMs from ENSEMBLES that are nested within as many different GCMs as possible to minimize the interdependency between the climate model outputs used (Sunyer et al., 2013).

3.3   Local adjustment methods (LAMs)

It is widely acknowledged that the RCM outputs for the variables of interest (in our case precipitation and temperature) are biased due to limited process description, biased fluxes at the RCM margins and insufficient spatial resolution relative to the catchment scale (Engen-Skaugen et al., 2007). Therefore, data post-processing is necessary to bridge the gap between the large-scale climate model and the local hydrological processes (e.g. Maraun et al. 2010; Chen et al., 2011). Considerable progress has been made during recent years regarding the development and improvement of such methods and Hanssen-Bauer et al. (2005), Fowler et al. (2007), Maraun et al. (2010), and Teutschbein and Seibert (2012) give comprehensive reviews on available approaches.

Amongst the LAMs, a useful distinction can be made between statistical downscaling and bias correction methods. In this study two different LAMs were applied: (i) empirical quantile mapping (Boé et al., 2007; Gudmundsson et al., 2012) representing a bias correction method, and (ii) expanded downscaling (Bürger, 1996; Bürger et al., 2009) which is a type of statistical downscaling.

3.3.1 Empirical quantile mapping (EQM)

EQM is a bias correction method that seeks a transfer function (h) to adjust RCM data so that it is in better agreement with observations. By adjusting the quantiles of the biased RCMs (x_m) to those of the locally observed data (x_o), the bias-corrected distribution of x_m should match the distribution of x_o, such that

$$x_o = h(x_m) = F_o^{-1}(F_m(x_m)),\qquad(1)$$

where F_m is the empirical cumulative distribution function (eCDF) of x_m, and F_o^{-1} is the inverse eCDF (the quantile function) corresponding to x_o. Based on the assumption that the shortcomings of the climate model are the same for the reference and future periods (van Roosmalen et al., 2011) and that the transfer function is stationary in time (Maraun et al., 2010), the function is applied to bias-correct projections from RCMs for both the reference and future periods.

For Norway, Gudmundsson et al. (2012) found that non-parametric transfer methods (as EQM) performed best for the bias correction of precipitation compared to parametric and distribution-derived transformations. For temperature, we found the same ranking though the differences are not as large as for precipitation. Therefore, EQM was considered as a suitable LAM for the correction of daily precipitation and temperature values for this study. The method was implemented as an add-on package (qmap; Gudmundsson, 2014) for the statistical programming environment R (R Core Team, 2012). Bias correction was performed on daily values for the full year, without distinguishing seasons, following work of Piani et al. (2009) which illustrated that the correction without seasonal subsampling performs remarkably well.

3.3.2 Expanded downscaling (XDS)

XDS is a statistical downscaling approach and, as such, it maps large-scale atmospheric fields (the predictors – x) to local data (the predictands – y). XDS has been applied for various purposes, e.g. for early flood warning (Bürger et al., 2009), downscaling extreme precipitation projections (Dobler et al., 2013), and hydrological impact studies (Dobler et al., 2012a).

At its core, XDS is based on multiple linear regression (MLR) which leads to minimizing the least square errors. The drawback of MLR, however, is that local climate variability will be smoothed significantly, which has strong implications for the simulation of extremes. To overcome this limitation, XDS adds an additional condition for retaining local co-variability between the variables:

$$\text{XDS} = \arg\min_Q \|x\mathbf{Q} - y\|, \text{ subjected to } \mathbf{Q}'x'x\mathbf{Q} = y'y,\qquad(2)$$

such that XDS is the solution of the error-minimizing matrix $\mathbf{Q}(x\,\mathbf{Q} - y)$ which is found amongst those that preserve the local covariance ($\mathbf{Q}'\,x'\,x\,\mathbf{Q} = y'\,y$). This approach is supposed to improve the estimation of extreme events, at the cost of a larger mean error as compared to conventional MLR.

For the present study, we used humidity, wind fields, temperature, and precipitation characteristics as predictor fields. XDS was calibrated on the RCM atmospheric fields driven by the ECMWF ERA-40 reanalysis (Uppala et al., 2005) for the period 1961–1980, and then applied to downscale the RCM outputs for the reference and future scenarios.

3.4 The HBV model

The analysis of climate change impacts at the catchment scale is based on daily streamflow simulated by the lumped, conceptual HBV model (Bergström, 1976, 1995), forced by the locally adjusted RCM data. In this study we apply the "Nordic" version of the model (Sælthun, 1996), which incorporates a snow module with 10 equal area height zones, such that snow accumulation and melting has a semi-distributed structure. For each equal area height zone, snow accumulation and melting is calculated individually, and the mean is finally used to represent the snow dynamics for each catchment. The principal advantage of the HBV model relative to more physically based models are that it only requires precipitation and temperature as climatological input. These are given as catchment mean values for the catchment centroid. Input data for precipitation and temperature are modified for the snow routine by three parameters defining the precipitation altitude gradient, and the temperature gradients for dry and wet days, respectively.

The HBV model was calibrated for each catchment using daily-averaged discharge data. Excepting Kråkfoss, where observed data are only available since 1966, the entire reference period (1961–1990) was used for model calibration. The use of such a long calibration period increases the chance that all relevant processes are covered (Merz et al., 2009). The model calibration uses the dynamically dimensioned search (Tolson and Shoemaker, 2007) (DDS) which is a global optimization algorithm for the calibration of multi-parameter models. A modified version of the Nash–Sutcliffe efficiency (NSE) was used as the objective function so as to focus on matching the high-flow events

$$\text{NSE}_w = 1 - \frac{\sum\limits_{i=1}^{n} Q_{obs}(Q_{sim} - Q_{obs})^2}{\sum\limits_{i=1}^{n} Q_{obs}(Q_{sim} - \overline{Q_{obs}})^2},\qquad(3)$$

where Q_{obs} represents the observed discharges and Q_{sim} represents the modelled discharges. The squared differences in the numerator and denominator are weighted by the observed discharge. A mismatch between high observed and simulated discharges is, therefore, penalized proportionally to the observed discharge value.

To account for parameter uncertainty, 25 best-fit parameter sets were identified and included for the hydrological simulations. Fifteen free parameters were subjected to the calibration by DDS, which was setup to 1200 model calls. The best-performing parameter set was taken directly from the DDS calibration. The remaining 24 parameter sets were identified by a subsequent Monte Carlo simulation with another 1200 model calls using a narrowed range in the parameter values which was defined by the range of parameter values of the 36 (3 %) best parameter sets identified by DDS. In that way, the effects of interdependency between the parameter sets are minimized.

3.5 Change analysis

The extreme events of the daily streamflow simulations were extracted using a peak over threshold (POT) approach, which leads to a more comprehensive selection of events (in terms of timing and flood processes) compared the block maximum method (i.e. AMF) (Lang et al., 1999). The threshold was set to the 98.5 streamflow percentile for both the control and future periods. Independency of events was achieved by enforcing that (i) only one event can occur within twice the normal flood duration (which is catchment specific) and (ii) that only the largest event will be considered if more than one peak is identified within that time period. The normal flood duration has been derived, for each of the six catchments considered, by a simple experiment using the HBV model. Each catchment was artificially drained to baseflow conditions before twice the amount of annual rainfall was added to completely saturate the catchment again. Concentration and recession time to baseflow were estimated from the resulting hydrographs. The normal flood duration for the catchment was then defined as the sum of the concentration and recession times.

3.5.1 Changes in flood seasonality

Detected POT events were divided into two seasons reflecting the basic flood regimes described in Sect. 2.1: (i) the spring/summer period from March to August, which is associated with snowmelt as an important FGP under the current climate, and (ii) the autumn/winter period from September to February, which is associated with rainfall as the most important FGP. To quantify the seasonality of flood events, we define a seasonality index S_D:

$$S_D = \frac{\mathrm{POT}_{Sep-Feb}}{\mathrm{POT}_{all}} - \frac{\mathrm{POT}_{Mar-Aug}}{\mathrm{POT}_{all}}, \tag{4}$$

where the first term describes the ratio between the flood peaks ($m^3 \, s^{-1}$) of the POT events occurring within the period September–February over all POT events, and the second

term describes the ratio between the POT events occurring within March–August over all POT events. The index ranges from -1 to $+1$: negative numbers indicate dominant events during spring/summer while positive numbers indicate dominant events during autumn/winter. S_D was estimated for each ensemble member for both the reference and the future periods. The difference in S_D between the future and the reference periods is an indicator of changes in flood seasonality. In addition, the magnitudes and frequencies of the detected spring/summer and autumn/winter events were analysed for the reference and the future periods. The changes in magnitudes and relative frequencies of the events within each season aid in explaining changes in flood seasonality.

3.5.2 Changes in FGPs

Each POT event was analysed to determine the dominant contribution to flood discharge. This contribution has been inferred from the runoff components simulated by the HBV model. A simple water balance approach was used to classify the events into floods generated by (i) "rainfall", (ii) "rainfall + snowmelt" and (iii) "snowmelt". The classification is based on the relative contribution of the volumes of rainfall and snowmelt to the flood event discharge: an event was classified as rainfall if the contribution of rainfall was larger than two-thirds, and classified as snowmelt if rainfall contribution was smaller than one-third. Other events were classified as rainfall + snowmelt. Note that there exist more detailed approaches for classifying types of flood processes, including the use of various process indicators (e.g. flood timing, storm duration, rainfall depth, snowmelt, catchment states), as suggested by Merz and Blöschl (2003). The classification proposed here, however, is very easy to apply and fully suitable for our analyses, given the broad distinction between rainfall and snowmelt flood generation that we are using in this work. In addition, the required runoff components can be readily extracted from the output of the HBV model.

Events were identified using a tool implemented in the R add-on package seriesdist (https://bitbucket.org/heisterm/seriesdist), which enables the detection of both flood peaks and their event-specific flood duration. In order to also account for the antecedent conditions in the catchment, the detected flood duration time of the core event was extended by adding the catchment-specific recession time (found in the definition of the normal flood duration) before the onset of the flood. The classification approach was then applied to the extended flood duration time such that all relevant contributions to the peak flow are considered.

Two statistics were applied to identify changes in the FGPs: (1) the ratios of rainfall-, rainfall + snowmelt- and snowmelt-generated events relative to all events for all ensemble realizations were estimated for the reference and future periods. The change in the ratios indicates the changes in the prevalence of the different FGPs. (2) Circular kernel density functions and the circular mean Julian date of occurrence of the rainfall-, rainfall + snowmelt- and snowmelt-generated events were calculated for both periods to illustrate changes in the annual distribution and mean timing of the events. The circular mean Julian dates of occurrence for the events with respect to each FGP are converted to mean radians ($\overline{\Theta}$) estimated from the Julian date of occurrence D for each event i:

$$\Theta_i = \frac{D2\pi}{365}, \tag{5}$$

where the Julian date $D = 1$ is for 1 January and $D = 365$ for 31 December. The $\overline{x}$ and $\overline{y}$ coordinates for the mean date as an angular value are derived from the sample of n events for each FGP group:

$$\overline{x} = \frac{1}{n} \sum_{i=1}^{n} \cos \Theta_i, \tag{6}$$

$$\overline{y} = \frac{1}{n} \sum_{i=1}^{n} \sin \Theta_i, \tag{7}$$

$$\overline{\Theta} = \tan^{-1}\left(\frac{\overline{y}}{\overline{x}}\right). \tag{8}$$

This approach was introduced by Bayliss and Jones (1993) and Burn (1997), and has been recently applied by Parajka et al. (2010) and Köplin et al. (2014). Note that these authors also estimate the variability of the date of occurrence. In this study, this is illustrated using the circular kernel density functions.

3.6 Sources of uncertainty

The range of all ensemble realizations provides a measure of the overall uncertainty represented by the ensemble, given that each projection is assumed to be equally likely. Similar to Déqué et al. (2007, 2011), the mean variance $\overline{\sigma^2}_{ensemble}$ (as a measure of uncertainty) of the entire ensemble is here defined as the additive mean variances from the ensemble components:

$$\overline{\sigma^2}_{ensemble} = \overline{\sigma^2}_{GCM/RCM} + \overline{\sigma^2}_{LAM} + \overline{\sigma^2}_{HP}. \tag{9}$$

We exemplify the computation of mean variances from the ensemble components for the hydrological model parameterization ($\overline{\sigma^2}_{HP}$): for each combination i out of n possible combinations of GCM/RCMs and LAMs, we compute the variance $\sigma_{HP,i}$ subject to 25 parameter sets of the hydrological model. Then, we compute $\overline{\sigma^2}_{HP}$ as the mean over all $\sigma_{HP,i...n} \cdot \overline{\sigma^2}_{GCM/RCM}$ and $\overline{\sigma^2}_{LAM}$ are computed accordingly.

This approach was used to identify the fractional uncertainty emerging from the different sources within the model chain for three variables: (i) the change in the index S_D, (ii) the change in the median magnitude of the POT events, and (iii) the change in the fraction of snowmelt- over rainfall-generated events.

4 Results and discussion

4.1 Model and ensemble validation

The performance of the HBV model is validated using the 25 best-fit parameter sets to estimate POT events during the reference period. These are compared with the distribution of observed POT events for the same period. In this case, the HBV simulations are based on observed meteorological data. Furthermore, we evaluated the ability of the entire ensemble (i.e. including all GCM/RCM combinations, LAMs, and hydrological parameter sets) to match the observed POT events for the reference period. A further comparison was made with HBV simulations based on the raw RCM data and the adjusted RCM data to assess the potential benefit of the adjustment procedures. The distribution of the POT events for each of these options is illustrated in Fig. 2.

The results indicate that the HBV model using the 25 best-fit parameter sets with observed climate data reproduces the observed POT events reasonably well for almost all of the catchments. For Junkerdalselv, the underestimation of the distribution of observed POT events is considerably larger than in other catchments. Junkerdalselv also has the lowest NSE_w value (0.77), which is due to systematic underestimation of flood peaks by the calibrated model. The NSE_w value for the other five catchments varies from 0.83 (Fustvatn) to 0.91 (Atnasjø).

As expected, the absolute range and the interquartile range of the POT event distribution from the full ensemble are larger. This mainly results from the large range introduced by the locally adjusted climate projections (see the fourth and fifth box in each plot). In four catchments the quartiles match the observed distribution fairly well (Krinsvatn, Øvrevatn, Atnasjø, Kråkfoss). The largest discrepancies occur for Fustvatn and Junkerdalselv. In both cases, the mismatch of the ensemble reflects the overestimation (Fustvatn) and underestimation (Junkerdalselv) resulting from the different LAMs. Nevertheless, the observed distributions of POT events are always captured by the full range of the ensemble and the data locally adjusted by EQM and XDS. The performance of the ensemble in reproducing the observed POT events is the only indicator we have of how reliable the ensemble is for future projections. For Fustvatn and Junkerdalselv, this implies a lower degree of reliability as compared with the remaining catchments.

Figure 2 also underlines the benefit of locally adjusting raw RCM data for hydrological simulations. The simulations iii–v are based on only one best-fit HBV parameter set assuring that the ranges in the distribution of the events are solely based on the range of the input data. The large ranges in the distribution of the simulations based on RCM raw data are narrowed considerably after adjustment at Øvrevatn, Junkerdalselv, Atnasjø and Kråkfoss by both LAMs. Moreover, the LAMs are able to correct the large discrepancies in the POT event distributions for the observed vs. the simulated

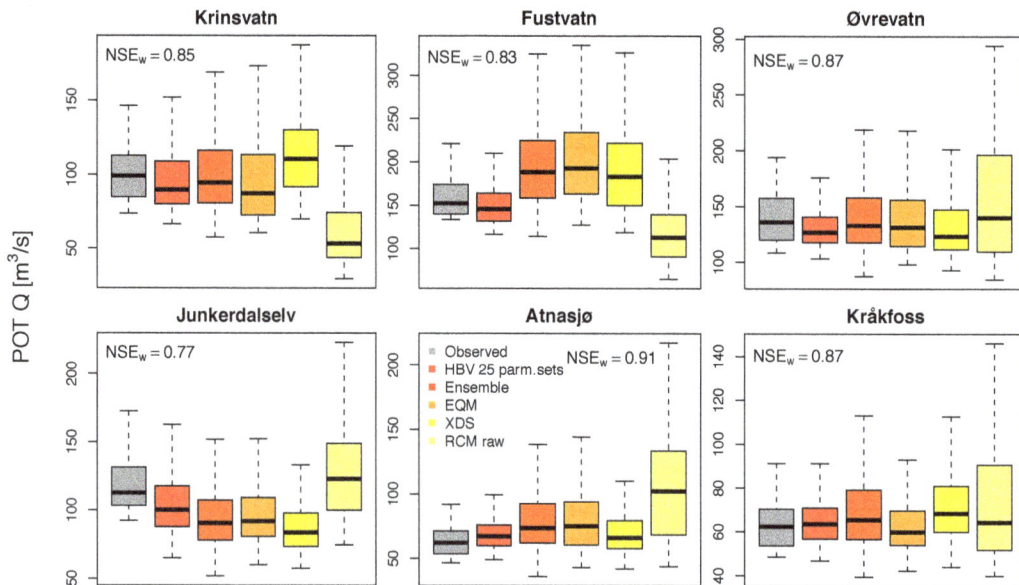

Figure 2. The distributions of POT events for the reference period from observed (grey) and simulated streamflow series generated by the calibrated HBV model using (from left to right) (i) observed climate data with the 25 best-fit parameter sets, (ii) the entire ensemble (i.e. all GCM/RCM combinations, LAMs, and hydrological parameter sets), (iii) the data locally adjusted by EQM, (iv) the data locally adjusted by XDS, and (v) the raw RCM data. For the simulations (iii–v) only one best-fit HBV parameter set is considered. The NSE_w values given for each catchment represent the goodness of fit of the HBV model for the entire series (not only POT events) using the best parameter set identified by the calibration. Note that the ordinate's point is not 0 and differs between the single plots.

series for Krinsvatn, Fustvatn and Atnasjø. For Fustvatn, the benefit of the local adjustment is least since the underestimation of the RCM raw data is only corrected to an overestimation of almost the same magnitude and range. It is not possible to conclude which of the two LAMs is better suited for high-flow estimations, neither in general nor for specific catchments.

4.2 Changes in the temperature and precipitation regime

Figure 3 summarizes the interquartile ranges of the projected changes in mean temperature and precipitation sums for the spring/summer and autumn/winter seasons after local adjustment by EQM and XDS for the six study catchments.

Increasing median temperatures from 2.9 °C (Krinsvatn, spring/summer) to 4.8 °C (Øvrevatn, autumn/winter) are projected for both seasons and all catchments considered. The temperature projections indicate a larger warming in autumn/winter than in spring/summer in all catchments, which agrees with Engen-Skaugen et al. (2007) and Hanssen-Bauer et al. (2003, 2009). Moreover, the largest warming is found for the northernmost catchments (Øvrevatn and Junkerdalselv) both for the spring/summer and autumn/winter periods. Generally, the results reflect findings from previous studies indicating an increasing warming signal with larger distances in latitudinal and longitudinal directions (Engen-Skaugen et al., 2007; Hanssen-Bauer et al., 2003). With the exception of

Kråkfoss, the interquartile ranges for the spring/summer season are higher as compared to the autumn/winter season for all catchments.

Regarding precipitation, the medians show increasing precipitation sums for both seasons and all catchments considered. The increase in spring/summer precipitation tends to be larger than autumn/winter precipitation at Krinsvatn, Fustvatn, Øvrevatn and Junkerdalselv. For Atnasjø and Kråkfoss the increase in precipitation during autumn/winter is projected to be larger than during spring and summer. The increase in autumn/winter precipitation in these two catchments is the largest projected change in precipitation ($> +30\%$) found within this study. Despite the positive median values, the ensemble does not consistently show positive changes in the projections. The first quartile for the changes in autumn/winter precipitation indicates decreasing precipitation sums for Krinsvatn, Fustvatn, Øvrevatn and Junkerdalselv. For Atnasjø and Kråkfoss the first quartile of the distribution indicates decreasing spring/summer precipitation sums. Generally, the results for these six catchments correspond to the regional differences in seasonal precipitation change previously presented in Hanssen-Bauer et al. (2009).

4.3 Changes in flood seasonality

Figure 4 summarizes the results for the index S_D for the reference and future period for the six study catchments. The boxplots represent the full ensemble.

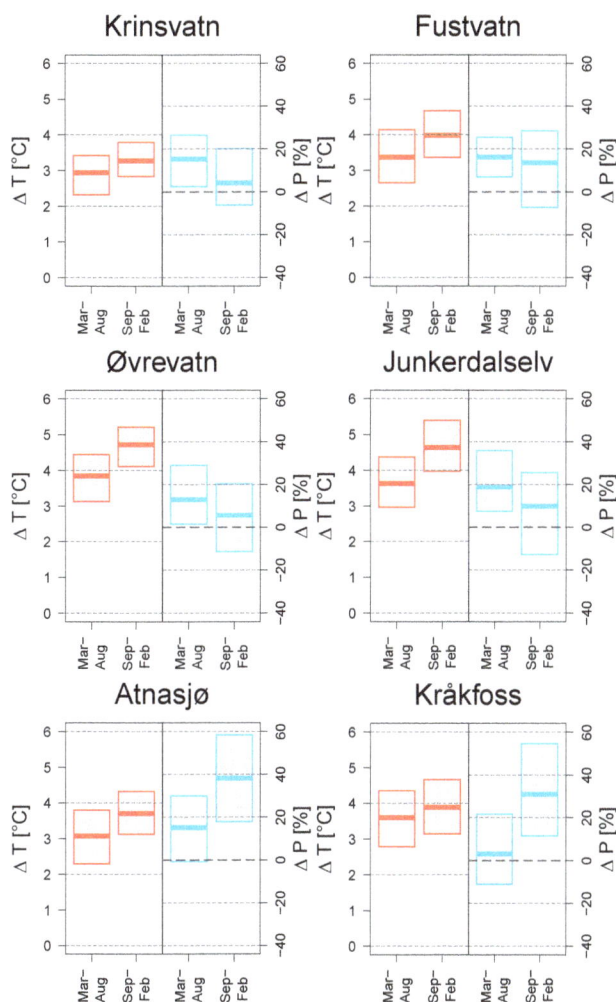

Figure 3. The interquartile ranges of the projected changes from the reference (1961–1990) to the future period (2071–2099) in mean temperature (left panel) and precipitation sums (right panel) for the spring/summer and autumn/winter seasons as they are locally adjusted by EQM and XDS for the six study catchments.

For the reference period, the S_D quartiles for the coastal catchments Krinsvatn and Fustvatn show positive values, indicating a dominance of autumn/winter POT events under the current climate. For Øvrevatn and Junkerdalselv in the north, as well as for Atnasjø and Kråkfoss in central and south-eastern Norway, the S_D quartiles indicate dominant POT events during spring/summer. The dominance of spring/summer events is largest for Atnasjø, but Junkerdalselv also shows a distinct spring/summer pattern with negative S_D values for all ensemble realizations. For Kråkfoss, this dominance is least pronounced. The observed flood seasonality (indicated by the green bars) is matched reasonably well in five of the six catchments (Krinsvatn, Fustvatn, Øvrevatn, Junkerdalselv, Atnasjø), with Fustvatn and Øvrevatn having the best matches. For Kråkfoss, however, the S_D values for the majority of the ensemble realizations are

rather low suggesting that the dominance of spring/summer events is exaggerated to some degree by the model simulations. Simulated S_D values based on observed meteorological input data and the 25 best-fit parameter sets were, however, found to be very similar to the S_D value based on observed runoff (not shown). Thus, the overestimation of spring/summer events at Kråkfoss is a consequence of climate input data, rather than the hydrological modelling.

For the future period, the S_D values are higher for all catchments. That means that the importance of autumn/winter events is projected to increase in all catchments considered. The lowest impact is found for Atnasjø where the dominance of spring/summer events persists into the future. However, for Øvrevatn and Kråkfoss considerably higher S_D values indicate a possible seasonal shift in the flood regimes since S_D becomes positive for almost the entire interquartile of all ensemble realizations. Changes towards dominant autumn/winter events are also indicated for some ensemble members for Junkerdalselv. However, the first and second quartiles still show negative S_D values.

The ranges in the projections given by the boxplots illustrate the uncertainty associated with the ensemble. For the reference period, this is highest for Fustvatn and Kråkfoss. For the future period, the highest ranges are found for Øvrevatn, Junkerdalselv and Kråkfoss, which show the largest change in flood seasonality. Note that the projected changes in seasonality are significant (with 95 % confidence) for all catchments, as none of the notches of the boxplots for the reference and future periods are overlapping.

4.4 Changes in the magnitude vs. the frequency of events

After having detected changes in flood seasonality, the question arises as to whether these result from changes in flood magnitude vs. frequency in the two respective seasons. Figure 5 summarizes the POT events for all ensemble realizations according to their associated magnitudes and number of occurrences for the two seasons.

For the coastal catchments, Krinsvatn and Fustvatn, Fig. 5 shows that both the relative number and the magnitude of POT events increase in autumn/winter during the future period. For spring/summer the magnitude also increases but the frequency decreases (i.e. the blue boxes show smaller widths). Together, this explains the intensification of the seasonality index S_D towards autumn/winter events. The seasonal shift towards autumn/winter events is even more pronounced for the northernmost catchments, Øvrevatn and Junkerdalselv (Fig. 4). Figure 5 indicates that this shift is mostly due to changes in the frequency (increasing in autumn/winter, decreasing in spring/summer) while the mean magnitudes are decreasing in both seasons. Note, however, that the observed seasonal POT magnitudes are not well reproduced by the ensemble for Junkerdalselv. For the high-altitude catchment in central Norway, Atnasjø, Fig. 4 indi-

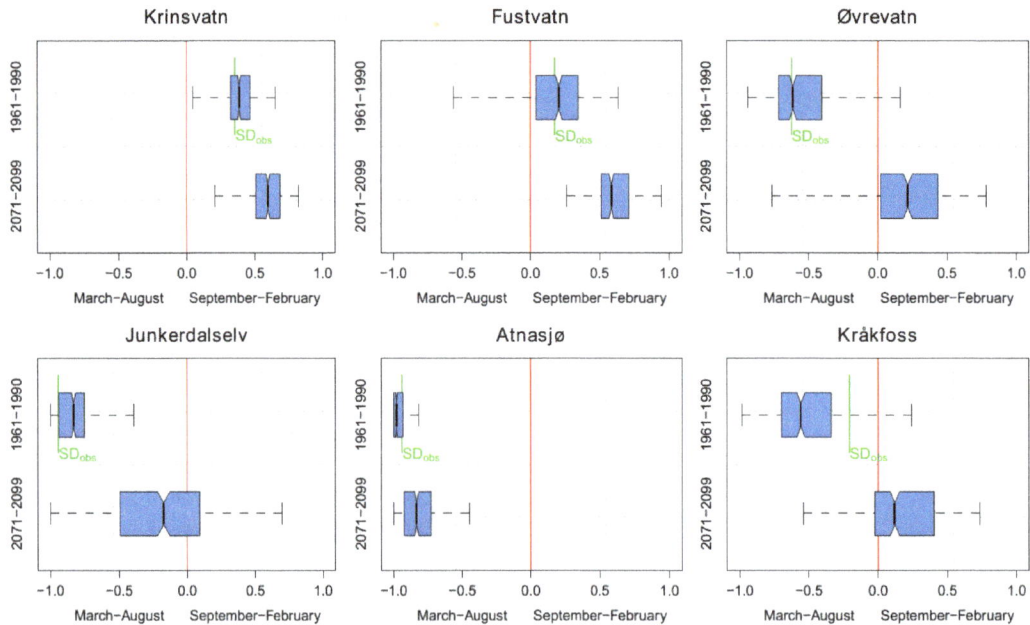

Figure 4. Boxplots showing the seasonality index S_D for all ensemble realizations for the reference and future periods. The boxes show the interquartile of the values; the whiskers show the full range of the projections. The green bars in the upper panel of each plot ($S_{D_{obs}}$) indicate the observed seasonality index S_D.

cates that spring/summer events are very dominant in both the current and future climates. Figure 5 establishes that this dominance reflects the frequency of the events in the POT series, and not necessarily the magnitude. Future flood magnitudes increase slightly for both seasons, while frequencies increase particularly in the autumn/winter period. This is responsible for the slight shift in seasonal index in Fig. 4. Finally, Fig. 5 also illustrates that the large seasonal shift for Kråkfoss is caused by both frequencies (decrease in spring/summer, increase in autumn/winter) and magnitudes. Future flood magnitudes increase in both seasons, but the increase in autumn/winter is considerably larger.

Note that the discrepancies between the observed and simulated POT magnitudes for the reference period (i.e. Fig. 2) are also reflected in the seasonal values in Fig. 5. The large discrepancy at Junkerdalselv (underestimation) and Atnasjø (overestimation) for the autumn/winter period is due to the limited number of observed events during these months. The correspondence between the observed and simulated seasonal median POT event magnitudes at Kråkfoss is comparatively better than for the seasonality index S_D (Fig. 4). Since the distribution of the POT event magnitudes are very similar both for the spring/summer and the autumn/winter seasons, the bias of S_D towards spring/summer results from an overestimation of the event frequency for this season.

The median changes in the POT event magnitudes from the references to the future period are significant (with 95 % confidence) for all catchments, as none of the illustrated notches for the respective period is overlapping.

4.5 Changes in FGPs

In the previous sections, we established that autumn/winter events will become more dominant in the future. This is consistent over all investigated catchments, although there are differences with respect to their underlying causes (i.e. changes in frequency, magnitude, or both). In general, we would expect that an increasing dominance of autumn/winter events corresponds to an increasing importance of rainfall as a FGP. Figure 6 shows how the percentage of different flood generating processes will change between the reference and the future periods.

Rainfall becomes the dominant FGP in the future period in all investigated catchments. For the coastal catchments, Krinsvatn and Fustvatn, where rainfall already dominates flood generation in the current climate, it will become even more important in the future. Snowmelt-generated floods, which play only a minor role in these catchments during the reference period, will be non-existent by the end of the 21st century. In the remaining four catchments, rainfall replaces snowmelt as the dominant FGP. The largest increases in the importance of rainfall are projected for the northernmost catchments, Øvrevatn and Junkerdalselv, and the southeastern catchment, Kråkfoss, where the changes in flood seasonality are also highest. This confirms that changes in flood seasonality are closely connected to changes in the FGPs and supports the conclusion of Köplin et al. (2014), who also found that the most pronounced changes in flood seasonality under a future climate will occur in catchments which are snowmelt-dominated during the current climate.

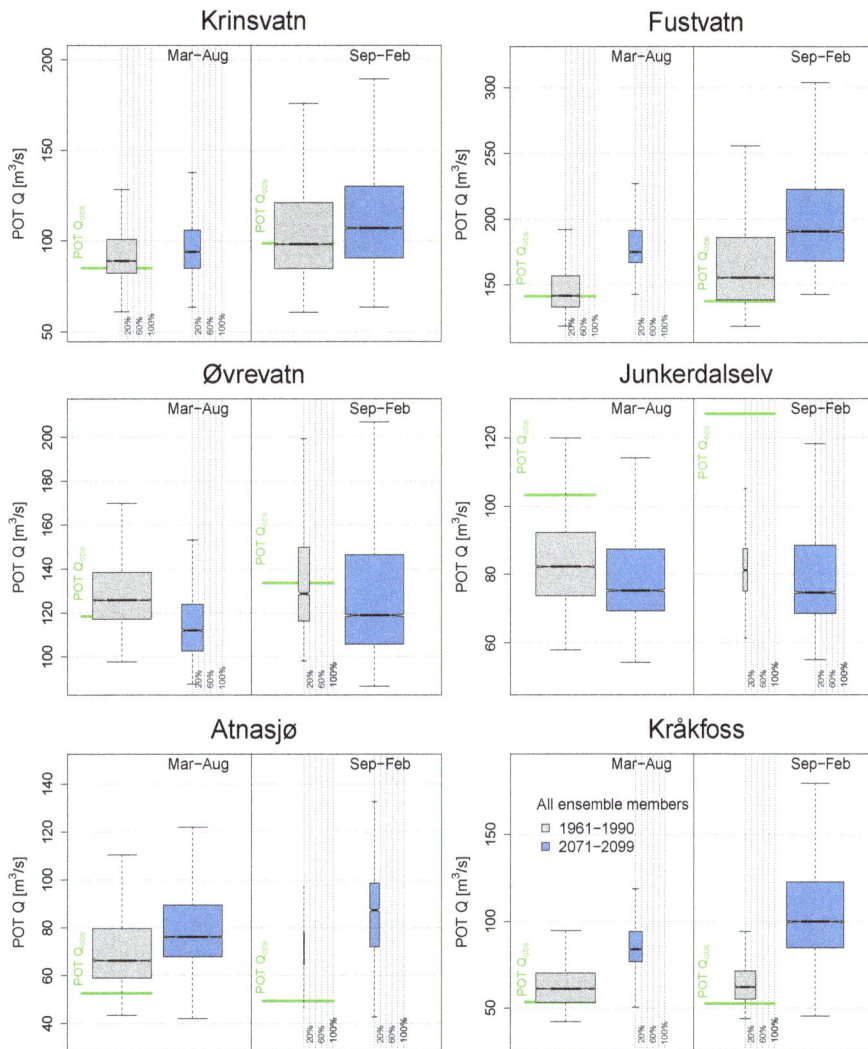

Figure 5. Boxplots showing the median and interquartile magnitudes of the simulated POT events from all ensemble realizations for the reference (grey boxes) and future periods (blue boxes), separated with respect to the two basic flood seasons in Norway (spring/summer – left panels; autumn/winter – right panels). The whisker range corresponds to twice the interquartile range. The green bars (POT$_{obs}$) indicate the median magnitudes of observed POT events. The width of the boxes illustrates the seasonal distribution in the frequency of the POT events: Per catchment and period, the smaller boxes are scaled compared to the larger boxes representing the dominant flood season in terms of flood frequency.

Figure 7 shows the circular kernel density functions of the events for each FGP and illustrates the relationship between the changes in the FGPs and the median magnitude of the events as a function of their mean Julian date of occurrence.

Snowmelt-associated events are connected with an earlier timing and POT events of a decreased magnitude in catchments where this FGP continues to be relevant in the generation of high flows. Higher mean temperatures in the future period (Fig. 3) lead to an earlier onset of the annual snowmelt season. For the catchments which continue to have peak discharges derived from snowmelt in the future period, the circular mean Julian dates of occurrence of the snowmelt-generated events is estimated to be 14–26 days earlier com-

pared to the reference period: Øvrevatn (26 days), Junkerdalselv (21 days), Atnasjø (14 days), Kråkfoss (22 days). This agrees with similar findings from streamflow observations and projections for the Nordic countries (Beldring et al., 2008; Stahl et al., 2010; Wilson et al., 2010) and for other parts of Europe and the world (e.g. Déry et al., 2009; Kormann et al., 2014; Renner and Bernhofer, 2011; Stewart et al., 2005; Hall et al., 2014, and reference list therein). With the exception of Kråkfoss, the mean magnitude of snowmelt-generated POT events will decrease in all catchments where snowmelt has an influence on flooding in the future period. This is because of smaller snowpack volumes due to shorter

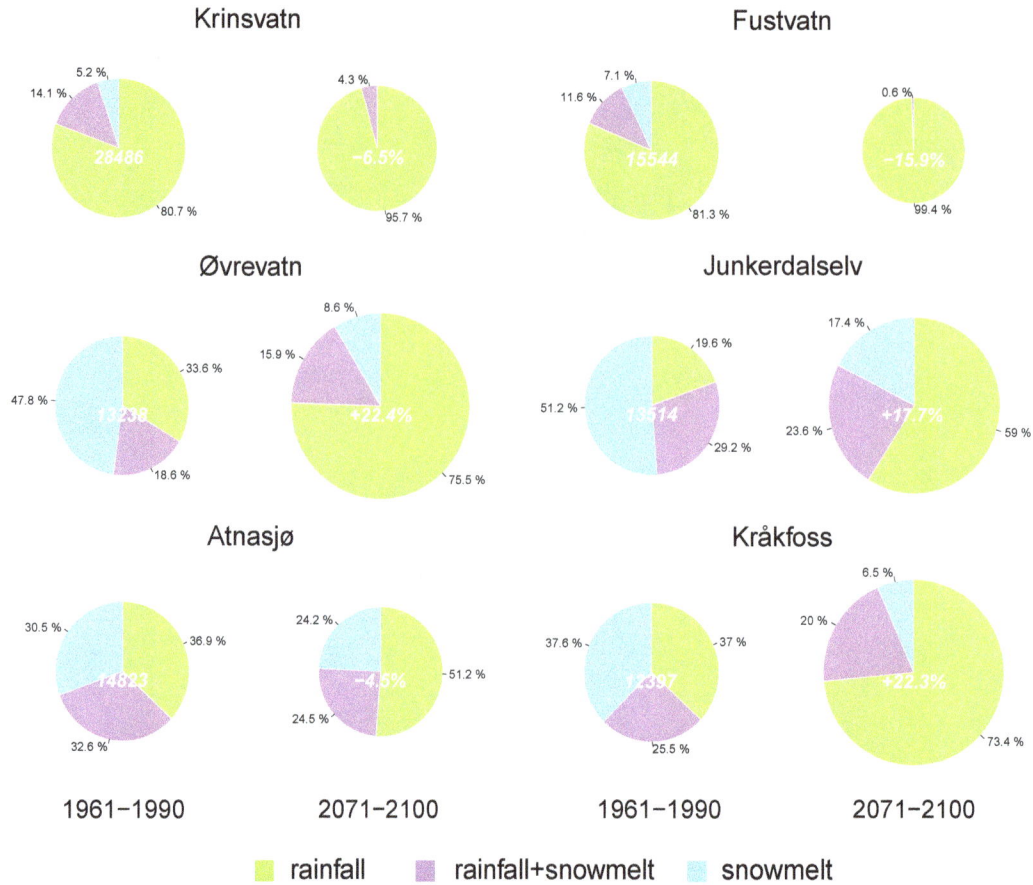

Figure 6. Percentage of POT events according to their FGPs in relation to the total number of events for the reference (left pie charts) and future periods (right pie charts) derived by all ensemble realizations. The diameter of the pie charts for the future period indicates the direction of change in the total number of events. Total numbers of events for the reference period and the percentage of change in the number of events for the future period are given by the white numbers within the pie charts.

and warmer winters in the future period (Vikhamar Schuler et al., 2006).

Rainfall-generated events tend to occur later within the year across all catchments. The later mean timing of rainfall-generated events highlights the increasing importance of winter rainfall floods in the future period. This corresponds to projected changes in the temperature and precipitation regimes (Fig. 3), which lead to a shorter snow season and reduced snow storage, and to an increasing relevance of episodes with intermittent rainfall and winter snowmelt due to higher winter temperatures (Hanssen-Bauer et al., 2009). For Øvrevatn, Junkerdalselv and Kråkfoss, this suggests that winter precipitation is no longer primarily received as snowfall such that the contribution of snowmelt to runoff is considerably less in the future period. Thus, the strongest changes in flood seasonality are observed for these three catchments which is in line with Arnell (1999), who concludes that the most significant changes in flow regimes occur where snowfall becomes less important due to higher temperatures. The effects of increased evaporation during late summer due to higher temperatures may also amplify the

later mean timing of rainfall-generated events, as soil moisture deficits may have a more pronounced role in attenuating heavy rainfalls during the autumn period.

The mean magnitudes of rainfall-generated events are projected to increase at Fustvatn, Atnasjø and Kråkfoss which explains the increasing POT-event magnitudes during autumn and winter in these catchments as shown in Fig. 5. The increasing magnitudes of autumn/winter events at Krinsvatn (Fig. 5) result from an earlier circular mean timing of the rainfall + snowmelt-generated events in the future period (from March to February) rather than from larger rainfall-generated POT-event magnitudes during autumn and winter (Fig. 7). The circular density functions show that rainfall has an influence on flooding throughout the year, particularly during autumn and winter for both the reference and future periods. Prominent seasonal peaks of rainfall-generated events during the reference period, as observed for Kråkfoss (October–November), will be smoothed in the future period. Thus, rainfall becomes more relevant for spring and summer, as well as winter events in the future period.

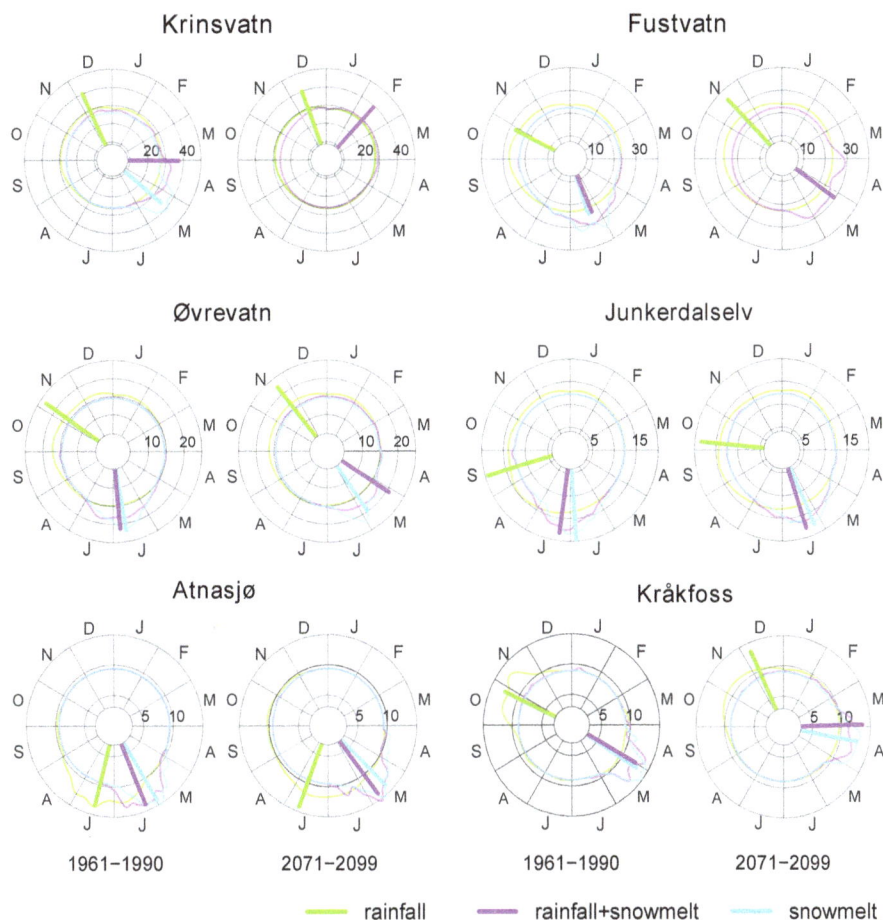

Figure 7. Circular plots showing (i) the circular kernel density function of the simulated POT events according to their FGPs (normalized; no units), and (ii) the median POT-event magnitude (mm day^{-1}) as bars according to their circular mean Julian date of occurrence and their FGPs for the reference and future periods.

The results also illustrate that changes in flood seasonality cannot be directly inferred from seasonal changes in precipitation and temperature. Hydrological modelling is required to highlight the changing role of snow storage and its effect on flood generation.

4.6 Contribution of ensemble components to uncertainty

Figure 8 shows the fractional variance from the different sources of the ensemble as they contribute to the total variance regarding the changes in the index S_D, the POT magnitudes and the FGPs presented in Figs. 4–7.

First of all, the GCM/RCM combinations and the LAMs are the dominant sources of uncertainty for all catchments and variables considered. Hydrological model parameterization tends to be the smallest contributor to overall uncertainty, which is in line with earlier studies (e.g. Wilby and Harris, 2006; Kay et al., 2008; Prudhomme and Davies, 2008; Dobler et al., 2012b). Note, however, that there are exceptions where the variance due to the hydrological model

parameterization is as high as that due to the LAMs or the climate projections (i.e. Junkerdalselv, second and third columns). Focusing on the target variables, hydrological parameter uncertainty tends to be less important for changes in the seasonality index S_D as compared with changes in the POT-event magnitudes and the dominant FGPs.

A possible pattern becomes apparent regarding the relative role of hydrological parameter uncertainty, which seems to be closely connected to the changes in flood seasonality and FGPs. Hydrological parameter uncertainty is rather high in those catchments for which a considerable change in their flood seasonality and the FGPs is expected (Øvrevatn, Junkerdalselv, Kråkfoss). This is probably due to changes in the dominant flood generation mechanisms. It is likely that the parameter sets, which are calibrated for the climate conditions in the entire reference period, are not sufficiently stable given the likely changes in hydroclimatological and runoff generation processes under future conditions. This highlights the difficulties associated with transferring model parameters in time under non-stationary conditions (Brigode et al.,

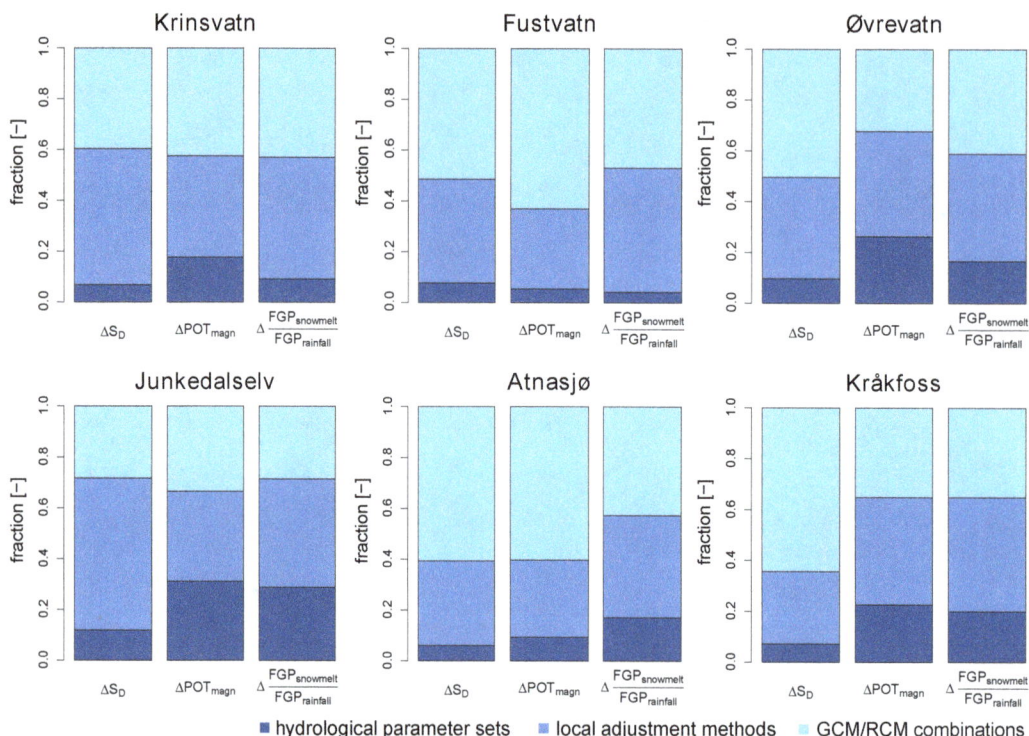

Figure 8. The fractions of total variance (–) as a measure for uncertainty, explained by (i) the GCM/RCM combinations (light blue), (ii) the local adjustment methods (medium blue), and (iii) the hydrological parameterization (dark blue) with respect to three target variables: (1) change in the seasonality index S_D, (2) change in the mean POT event magnitude, and (3) change in the ratio of snowmelt- and rainfall-generated POT events.

2013; Merz et al., 2011). The choice of the reference period (1961(1966)–1990) may imply that we have detected quite stable parameters for that period since the pronounced warming during the recent years are not included in the calibration. These parameters, however, may be even less representative for the future conditions. Merz et al. (2011) have calibrated a similar version of the HBV model for six consecutive 5-year periods for a comprehensive set of catchments in Austria. They found notable time trends in the calibrated parameters representing snow dynamics and soil moisture processes which lead to considerable biases especially in high flows. For our results, that implies that the hydrological model parameter uncertainty limits the reliability of the estimated changes in the proportion of rainfall and snowmelt and their effects on flood seasonality and FGPs.

One option for dealing with that issue are differential split sample tests (Klemeš, 1986). These are usually used to evaluate parameter sets which are optimized for contrasting conditions. Seibert (2003) calibrated the HBV model in four Swedish catchments on years with lower runoff peaks and tested the calibrated parameters for years with higher peaks, finding a decrease in model performance. Coron et al. (2012) introduced generalized split-sample tests, which systematically test all possible combinations of calibration-validation periods using a 10-year moving window over the observation time period. They also pointed out a lack of robustness in hydrological model parameters tested in climate conditions which differ to those used for model calibration. Similar schemes need to be adapted for seasonality purposes, i.e. identifying contrasting periods in terms of seasonal flood prevalence and dominant FGPs. Differential split sample testing can then indicate parameter robustness when applied under contrasting seasonality conditions. They may also indicate parameter sets which are suitable for runoff simulations under future conditions. This approach, however, presupposes that relevant changes can already be detected in the observation data and that contrasting periods are long enough for a sufficient model calibration.

5 Conclusions

Using a multi-model/multi-parameter ensemble approach, the impacts of climate change on flood seasonality and their underlying flood generating processes (FGPs) have been investigated in six catchments representing different hydroclimatological regions in Norway.

The results indicate that the HBV model, including the use of 25 best-fit parameter sets, is able to reproduce observed distributions of flood events reasonably well for five out of six study catchments for the reference period. Small discrepancies between the event distributions simulated by the locally adjusted climate projection data and the observed event distributions slightly reduce the reliability of the ensemble setup for two catchments (Fustvatn, Junkerdalselv). For the remaining four catchments the ensemble reproduces the observed flood-event distributions fairly well. The benefit of post-processing the RCM raw data has also been demonstrated. However, no distinct ranking emerged regarding the performance of the two LAMs applied.

Reconsidering our research questions, the following conclusions can be drawn:

– *How might the existing patterns of flood seasonality change under a future climate?* Autumn/winter floods become more important in all the catchments considered. For the two coastal catchments that suggests an intensification of the current autumn/winter flood regime. For the high-mountain catchment, Atnasjø, in central Norway, the dominance of spring/summer floods will be slightly reduced. For the northernmost catchments, Øvrevatn and Junkerdalselv, as well as for the south-eastern catchment, Kråkfoss, the increase in autumn/winter floods is largest and may lead to a systematical shift in the current flood regimes from spring/summer to autumn/winter.

– *How are the shifts in seasonality related to changes in the magnitude vs. changes in the frequency of events?* Changes in flood seasonality from spring/summer towards autumn/winter are the result of increasing event magnitudes or frequencies, or a combination of both, during the autumn and winter months. Changes in seasonal frequency, however, are more relevant than changes in seasonal magnitude since two of the catchments with the strongest changes in flood seasonality (Øvrevatn and Junkerdalselv) show decreasing flood magnitudes but large shifts in the seasonal frequency of events.

– *Are changes in flood seasonality associated with changes in the FGPs?* The change towards more autumn/winter events corresponds to an increasing relevance of rainfall as a FGP. Rainfall becomes more dominant where it has already been dominant and it replaces snowmelt as the dominant FGP in the remaining catchments. The largest increases in the relative role of rainfall correspond with the largest shifts in flood seasonality (Øvrevatn, Junkerdalselv, Kråkfoss). In these catchments, less snow accumulation and shorter snow seasons due to increased winter temperatures lead to a considerable decrease in the frequency and magnitude of snowmelt-generated events. Additionally, rainfall-generated events occur more often and also later within the autumn/winter period. Thus, the largest changes in the FGPs are closely connected with temperature effects which determine the relative role of snowmelt vs. rainfall. This has a major influence on the seasonal distribution of floods.

– *What is the relative importance of the different ensemble components in contributing to the overall variance as a measure of the uncertainty in the projected changes?* For changes in flood seasonality the ensemble range is largest in those catchments for which the largest seasonal changes are projected. The climate projections (i.e. the GCM/RCM combinations) or the LAMs tend to be the largest contributor to the total variance. However, the relative role of the hydrological model parameterization compared to the other two contributors is highest for those catchments showing the most pronounced seasonal changes. This is consistent with an earlier study of climate change impacts in four Norwegian catchments (Lawrence and Haddeland, 2011), and confirms the lack of robustness in HBV parameterizations for simulations with transient hydroclimatological conditions which lead to changes in the flood regime. It further stresses the need for alternative calibration approaches which improve the transferability of hydrological model parameters under non-stationary conditions.

Although the catchments analysed within this study represent a large variety of climate conditions in Norway, the sample size is too small to allow for robust regional conclusions on changes in the seasonality and generation processes of floods. The results presented here can only indicate possible responses to climate change in terms of flood seasonality and FGPs for catchments with similar hydroclimatological regimes and physical conditions. For robust regional conclusions, the proposed methodology needs to be applied for a larger sample of catchments. Alternatively, a grid-based modelling approach covering the whole country could also be used, although such results must be interpreted with care in areas lacking data for model calibration.

Acknowledgements. The first author acknowledges the Helmoltz graduate research school GeoSim for funding the PhD studentship and NVE for funding study visits to Norway. The second author acknowledges support from NVE for the internally funded project Climate change and future floods. The regional climate model simulations stem from the EU FP6 project ENSEMBLES, whose support is gratefully acknowledged. Gerd Bürger (University of Potsdam) is thanked for his great support on downscaling the RCM data by XDS. The Potsdam Graduate School (PoGS) is acknowledged for supporting the service charge cover for this open access publication. Daniel Viviroli and two anonymous referees are thanked for their comments on an earlier version of this manuscript.

Edited by: R. Merz

References

Andréasson, J. and Bergström, S.: Hydrological change-climate change impact simulations for Sweden, AMBIO A J. Hum. Environ., 33, 228–234, doi:10.1579/0044-7447-33.4.228, 2004.

Arnell, N. W.: The effect of climate change on hydrological regimes in Europe: a continental perspective, Global Environ. Change, 9, 5–23, doi:10.1016/S0959-3780(98)00015-6, 1999.

Bayliss, A. C. and Jones, R. C.: Peaks-Over-Threshold Flood Database: Summary Statistics and Seasonality, Wallingford, UK, 1993.

Beldring, S., Engen-Skaugen, T., Førland, E. J., and Roald, L. A.: Climate change impacts on hydrological processes in Norway based on two methods for transferring regional climate model results to meteorological station sites, Tellus A, 60, 439–450, doi:10.1111/j.1600-0870.2008.00306.x, 2008.

Benestad, R. E.: Association between trends in daily rainfall percentiles and the global mean temperature, J. Geophys. Res.-Atmos., 118, 10802–10810, doi:10.1002/jgrd.50814, 2013.

Beniston, M., Stephenson, D. B., Christensen, O. B., Ferro, C. A. T., Frei, C., Goyette, S., Halsnaes, K., Holt, T., Jylhä, K., Koffi, B., Palutikof, J., Schöll, R., Semmler, T., and Woth, K.: Future extreme events in European climate: an exploration of regional climate model projections, Climatic Change, 81, 71–95, doi:10.1007/s10584-006-9226-z, 2007.

Bergström, S.: Development and application of a conceptual runoff model for Scandinavian catchments, Report No. 7 RHO, Swedish Meteorological and Hydrological Institute – SMHI, Nörrköping, 1976.

Bergström, S.: The HBV model, in: Computer Models of Watershed Hydrology, edited by: Singh, V. P., Water Resources Publications, Highlands Ranch, CO, 443–476, 1995.

Bhend, J. and von Storch, H.: Consistency of observed winter precipitation trends in northern Europe with regional climate change projections, Clim. Dynam., 31, 17–28, doi:10.1007/s00382-007-0335-9, 2007.

Blöschl, G., Viglione, A., Merz, R., Parajka, J., Salinas, J. L., and Schöner, W.: Auswirkungen des Klimawandels auf Hochwasser und Niederwasser, Österreichische Wasser- und Abfallwirtschaft, 63, 21–30, doi:10.1007/s00506-010-0269-z, 2011.

Boé, J., Terray, L., Habets, F., and Martin, E.: Statistical and dynamical downscaling of the Seine basin climate for hydro-meteorological studies, Int. J. Climatol., 27, 1643–1655, doi:10.1002/joc.1602, 2007.

Brigode, P., Oudin, L., and Perrin, C.: Hydrological model parameter instability: A source of additional uncertainty in estimating the hydrological impacts of climate change?, J. Hydrol., 476, 410–425, doi:10.1016/j.jhydrol.2012.11.012, 2013.

Bronstert, A., Kolokotronis, V., Schwandt, D., and Straub, H.: Comparison and evaluation of regional climate scenarios for hydrological impact analysis?: General scheme and application example, Int. J. Climatol., 1594, 1579–1594, doi:10.1002/joc.1621, 2007.

Bürger, G.: Expanded downscaling for generating local weather scenarios, Clim. Res., 7, 111–128, doi:10.3354/cr007111, 1996.

Bürger, G., Reusser, D., and Kneis, D.: Early flood warnings from empirical (expanded) downscaling of the full ECMWF Ensemble Prediction System, Water Resour. Res., 45, W10443, doi:10.1029/2009WR007779, 2009.

Burn, D. H.: Catchment similarity for regional flood frequency analysis using seasonality measures, J. Hydrol., 202, 212–230, doi:10.1016/S0022-1694(97)00068-1, 1997.

Chen, J., Brissette, F. P., and Leconte, R.: Uncertainty of downscaling method in quantifying the impact of climate change on hydrology, J. Hydrol., 401, 190–202, doi:10.1016/j.jhydrol.2011.02.020, 2011.

Coron, L., Andréassian, V., Perrin, C., Lerat, J., Vaze, J., Bourqui, M., and Hendrickx, F.: Crash testing hydrological models in contrasted climate conditions: An experiment on 216 Australian catchments, Water Resour. Res., 48, W05552, doi:10.1029/2011WR011721, 2012.

Déqué, M., Rowell, D. P., Lüthi, D., Giorgi, F., Christensen, J. H., Rockel, B., Jacob, D., Kjellström, E., Castro, M., and Hurk, B.: An intercomparison of regional climate simulations for Europe: assessing uncertainties in model projections, Climatic Change, 81, 53–70, doi:10.1007/s10584-006-9228-x, 2007.

Déqué, M., Somot, S., Sanchez-Gomez, E., Goodess, C. M., Jacob, D., Lenderink, G., and Christensen, O. B.: The spread amongst ENSEMBLES regional scenarios: regional climate models, driving general circulation models and interannual variability, Clim. Dynam., 38, 951–964, doi:10.1007/s00382-011-1053-x, 2011.

Déry, S. J., Stahl, K., Moore, R. D., Whitfield, P. H., Menounos, B., and Burford, J. E.: Detection of runoff timing changes in pluvial, nival, and glacial rivers of western Canada, Water Resour. Res., 45, W04426, doi:10.1029/2008WR006975, 2009.

Dobler, C., Bürger, G., and Stötter, J.: Assessment of climate change impacts on flood hazard potential in the Alpine Lech watershed, J. Hydrol., 460–461, 29–39, doi:10.1016/j.jhydrol.2012.06.027, 2012a.

Dobler, C., Hagemann, S., Wilby, R. L., and Stötter, J.: Quantifying different sources of uncertainty in hydrological projections in an Alpine watershed, Hydrol. Earth Syst. Sci., 16, 4343–4360, doi:10.5194/hess-16-4343-2012, 2012b.

Dobler, C., Bürger, G., and Stötter, J.: Simulating future precipitation extremes in a complex Alpine catchment, Nat. Hazards Earth Syst. Sci., 13, 263–277, doi:10.5194/nhess-13-263-2013, 2013.

Dyrrdal, A. V., Isaksen, K., Hygen, H., and Meyer, N.: Changes in meteorological variables that can trigger natural hazards in Norway, Clim. Res., 55, 153–165, doi:10.3354/cr01125, 2012.

Dyrrdal, A. V., Saloranta, T., Skaugen, T., and Stranden, H. B.: Changes in snow depth in Norway during the period 1961–2010, Hydrol. Res., 44, 169, 169–179, doi:10.2166/nh.2012.064, 2013.

Engen-Skaugen, T., Haugen, J. E., and Tveito, O. E.: Temperature scenarios for Norway: from regional to local scale, Clim. Dynam., 29, 441–453, doi:10.1007/s00382-007-0241-1, 2007.

Fleig, A. K., Andreassen, L. M., Barfod, E., Haga, J., Haugen, L. E., Hisdal, H., Melvold, K., and Saloranta, T.: Norwegian Hydrological Reference Dataset for Climate Change Studies, Report No. 2, Norwegian Water Resources and Energy Directorate – NVE, Oslo, 2013.

Fowler, H. J. and Ekström, M.: Multi-model ensemble estimates of climate change impacts on UK seasonal precipitation extremes, Int. J. Climatol., 29, 385–416, doi:10.1002/joc.1827, 2009.

Fowler, H. J., Blenkinsop, S., and Tebaldi, C.: Linking climate change modelling to impacts studies: recent advances in downscaling techniques for hydrological modelling, Int. J. Climatol., 27, 1547–1578, doi:10.1002/joc.1556, 2007.

Gottschalk, L., Jensen Jørgen, L., Lundquist, D., Solantie, R., and Tollan, A.: Hydrologic Regions in the Nordic Countries, Nord. Hydrol., 10, 273–286, 1979.

Gudmundsson, L.: qmap: Statistical transformations for post-processing climate model output, R package version 1.0-3, http://cran.r-project.org/web/packages/qmap/citation.html (last access: February 2015), 2014.

Gudmundsson, L., Bremnes, J. B., Haugen, J. E., and Engen-Skaugen, T.: Technical Note: Downscaling RCM precipitation to the station scale using statistical transformations – a comparison of methods, Hydrol. Earth Syst. Sci., 16, 3383–3390, doi:10.5194/hess-16-3383-2012, 2012.

Hall, J., Arheimer, B., Borga, M., Brázdil, R., Claps, P., Kiss, A., Kjeldsen, T. R., Kriaučiūnienė, J., Kundzewicz, Z. W., Lang, M., Llasat, M. C., Macdonald, N., McIntyre, N., Mediero, L., Merz, B., Merz, R., Molnar, P., Montanari, A., Neuhold, C., Parajka, J., Perdigão, R. A. P., Plavcová, L., Rogger, M., Salinas, J. L., Sauquet, E., Schär, C., Szolgay, J., Viglione, A., and Blöschl, G.: Understanding flood regime changes in Europe: a state-of-the-art assessment, Hydrol. Earth Syst. Sci., 18, 2735–2772, doi:10.5194/hess-18-2735-2014, 2014.

Hanssen-Bauer, I., Achberger, C., Benestad, R. E., Chen, D., and Førland, E. J.: Statistical downscaling of climate scenarios over Scandinavia, Clim. Res., 29, 255–268, 2005.

Hanssen-Bauer, I., Drange, H., Førland, E., Roald, L. A., Børsheim, K. Y., Hisdal, H., Lawrence, D., Nesje, A., Sandven, S., Sorteberg, A., Sndby, S., Vasskog, K., and Ådlandsvik, B.: Klima i Norge 2100 Bakgrunnsmateriale til NOU Klimatilpasning – Climate in Norway 2100 background material for NOU climate adaptation, Norsk klimasenter, Oslo, 2009.

Hanssen-Bauer, I., Førland, E. J., Haugen, J. E., and Tveito, O. E.: Temperature and precipitation scenarios for Norway: comparison of results from dynamical and empirical downscaling, Clim. Res., 25, 15–27, 2003.

IPCC: Special Report on Emission Scenarios – Summary for Policymakers, Cambridge University Press, Cambridge, UK, 2000.

IPCC: Climate Change 2007: The Physical Science Basis, edited by: Solomon, S., Qin, D., Manning, M., Chen, Z., Marquis, M., Averyt, K. B., Tignor, M., and Miller, H. L., Cambridge University Press, Cambridge, UK, and New York, NY, USA, 2007.

Kay, a. L., Davies, H. N., Bell, V. A., and Jones, R. G.: Comparison of uncertainty sources for climate change impacts: flood frequency in England, Climatic Change, 92, 41–63, doi:10.1007/s10584-008-9471-4, 2008.

Klemeš, V.: Operational testing of hydrological simulation models, Hydrolog. Sci. J., 31, 13–24, doi:10.1080/02626668609491024, 1986.

Köplin, N., Schädler, B., Viviroli, D., and Weingartner, R.: Seasonality and magnitude of floods in Switzerland under future climate change, Hydrol. Process., 28, 2567–2578, doi:10.1002/hyp.9757, 2014.

Kormann, C., Francke, T., and Bronstert, A.: Detection of regional climate change effects on alpine hydrology by daily resolution trend analysis in Tyrol, Austria, J. Water Clim. Change, doi:10.2166/wcc.2014.099, in press, 2014.

Lang, M., Ouarda, T. B. M. J., and Bobée, B.: Towards operational guidelines for over-threshold modeling, J. Hydrol., 225, 103–117, doi:10.1016/S0022-1694(99)00167-5, 1999.

Lawrence, D. and Haddeland, I.: Uncertainty in hydrological modelling of climate change impacts in four Norwegian catchments, Hydrol. Res., 42, 457–471, doi:10.2166/nh.2011.010, 2011.

Lawrence, D. and Hisdal, H.: Hydrological projections for floods in Norway under a future climate, Report No. 5, Norwegian Water Resources and Energy Directorate – NVE, Oslo, 2011.

Maraun, D., Wetterhall, F., Ireson, A., Chandler, R., Kendon, E., Widmann, M., Brienen, S., Rust, H., Sauter, T., Themessl, M., Venema, V., Chun, K., Goodess, C., Jones, R., Onof, C., Vrac, M., and Thiele-Eich, I.: Precipitation Downscaling Under Climate Change: Recent Developments To Bridge the Gap Between Dynamical Models and the End User, Rev. Geophys., 48, 1–34, 2010.

Merz, R. and Blöschl, G.: A process typology of regional floods, Water Resour. Res., 39, 1340, doi:10.1029/2002WR001952, 2003.

Merz, R., Parajka, J., and Blöschl, G.: Scale effects in conceptual hydrological modeling, Water Resour. Res., 45, W09405, doi:10.1029/2009WR007872, 2009.

Merz, R., Parajka, J., and Blöschl, G.: Time stability of catchment model parameters: Implications for climate impact analyses, Water Resour. Res., 47, W02531, doi:10.1029/2010WR009505, 2011.

Midttømme, G. H., Petterson, L. E., Holmqvist, E., Nøtsund, Ø., Hisdal, H., and Sivertsgård, R.: Retningslinjer for flomberegninger – Guidelines for flood estimation, Retningslinjer nr. 5, Norwegian Water Resources and Energy Directorate – NVE, Oslo, 2011.

Parajka, J., Kohnová, S., Bálint, G., Barbuc, M., Borga, M., Claps, P., Cheval, S., Dumitrescu, A., Gaume, E., Hlavčová, K., Merz, R., Pfaundler, M., Stancalie, G., Szolgay, J., and Blöschl, G.: Seasonal characteristics of flood regimes across the Alpine–Carpathian range, J. Hydrol., 394, 78–89, doi:10.1016/j.jhydrol.2010.05.015, 2010.

Petterson, L. E.: Aktive vannføringsstasjoner i Norge – Active streamflow gauges in Norway, Rapport nr. 16, Norwegian Water Resources and Energy Directorate – NVE, Oslo, 2004.

Piani, C., Haerter, J. O., and Coppola, E.: Statistical bias correction for daily precipitation in regional climate models over Europe, Theor. Appl. Climatol., 99, 187–192, doi:10.1007/s00704-009-0134-9, 2009.

Prudhomme, C. and Davies, H.: Assessing uncertainties in climate change impact analyses on the river flow regimes in the UK, Part 2: future climate, Climatic Change, 93, 197–222, doi:10.1007/s10584-008-9461-6, 2008.

R Core Team: R: A language and environment for statistical computing, Foundation for Statistical Computing, Vienna, Austria, 2012.

Renner, M. and Bernhofer, C.: Long term variability of the annual hydrological regime and sensitivity to temperature phase shifts in Saxony/Germany, Hydrol. Earth Syst. Sci., 15, 1819–1833, doi:10.5194/hess-15-1819-2011, 2011.

Roald, L. A.: Climate change impacts on streamflow in Norway, Consultancy report A no. 1, Norwegian Water Resources and Energy Directorate – NVE, Oslo, 2006.

Sælthun, N.: The Nordic HBV model, Publication no. 7, Norweian Water Resources and Endergy Directorate – NVE, Oslo, 1996.

Seibert, J.: Reliability of Model Predictions Outside Calibration Conditions, Nord. Hydrol., 34, 477–492, 2003.

Seneviratne, S. I., Nicholls, N., Easterling, D. R., Goodess, C. M., Kanae, S., Kossin, J., Luo, Y., Marengo, J., McInnes, K., Rahimi, N., Reichstein, M., Sorteberg, A., Vera, C., and Zhang, X.: Changes in climate extremes and their impacts on the natural physical environment, in: Managing the Risks of Extreme Events and Disasters to Advance Climate Change Adaptation, edited by: Field, C. B., Barros, V., Stocker, T. F., Qin, D., Dokken, D. J., Ebi, K., Mastrandrea, D. M., Mach, K. J., Plattner, G.-K., Allen, S. K., Tignor, M., and Midgley, G. F., Cambridge University Press, Cambridge, UK, and New York, NY, USA, 109–230, 2012.

Stahl, K., Hisdal, H., Hannaford, J., Tallaksen, L. M., van Lanen, H. A. J., Sauquet, E., Demuth, S., Fendekova, M., and Jódar, J.: Streamflow trends in Europe: evidence from a dataset of near-natural catchments, Hydrol. Earth Syst. Sci., 14, 2367–2382, doi:10.5194/hess-14-2367-2010, 2010.

Stewart, I. T., Cayan, D. R., and Dettinger, M. D.: Changes toward Earlier Streamflow Timing across Western North America, J. Climate, 18, 1136–1155, doi:10.1175/JCLI3321.1, 2005.

Sunyer, M. A., Madsen, H., Rosbjerg, D., and Arnbjerg-Nielsen, K.: Regional interdependency of precipitation indices across Denmark in two ensembles of high resolution RCMs, J. Climate, 26, 7912–7928, doi:10.1175/JCLI-D-12-00707.1, 2013.

Teutschbein, C. and Seibert, J.: Bias correction of regional climate model simulations for hydrological climate-change impact studies: Review and evaluation of different methods, J. Hydrol., 456–457, 12–29, doi:10.1016/j.jhydrol.2012.05.052, 2012.

Tollan, A.: Hydrologiske regioner i Norden, Vannet i Nord., 1, 1–41, 1975.

Tolson, B. A. and Shoemaker, C. A.: Dynamically dimensioned search algorithm for computationally efficient watershed model calibration, Water Resour. Res., 43, W01413, doi:10.1029/2005WR004723, 2007.

Uppala, S. M., Kållberg, P. W., Simmons, A. J., Andrae, U., Bechtold, V. D. C., Fiorino, M., Gibson, J. K., Haseler, J., Hernandez, A., Kelly, G. A., Li, X., Onogi, K., Saarinen, S., Sokka, N., Allan, R. P., Andersson, E., Arpe, K., Balmaseda, M. A., Beljaars, A. C. M., Van De Berg, L., Bidlot, J., Bormann, N., Caires, S., Chevallier, F., Dethof, A., Dragosavac, M., Fisher, M., Fuentes, M., Hagemann, S., Hólm, E., Hoskins, B. J., Isaksen, L., Janssen, P. A. E. M., Jenne, R., Mcnally, A. P., Mahfouf, J.-F., Morcrette, J.-J., Rayner, N. A., Saunders, R. W., Simon, P., Sterl, A., Trenberth, K. E., Untch, A., Vasiljevic, D., Viterbo, P., and Woollen, J.: The ERA-40 re-analysis, Q. J. Roy. Meteorol. Soc., 131, 2961–3012, doi:10.1256/qj.04.176, 2005.

Van der Linden, P. and Mitchell, J. F. B.: ENSEMBLES: Climate Change and its Impacts: Summary of research and results from the ENSEMBLES project, M. O. H. Centre, Exeter, UK, 2009.

van Roosmalen, L., Sonnenborg, T. O., Jensen, K. H., and Christensen, J. H.: Comparison of Hydrological Simulations of Climate Change Using Perturbation of Observations and Distribution-Based Scaling, Vadose Zone J., 10, 136–150, doi:10.2136/vzj2010.0112, 2011.

Veijalainen, N., Lotsari, E., Alho, P., Vehviläinen, B., and Käyhkö, J.: National scale assessment of climate change impacts on flooding in Finland, J. Hydrol., 391, 333–350, doi:10.1016/j.jhydrol.2010.07.035, 2010.

Velázquez, J. A., Schmid, J., Ricard, S., Muerth, M. J., Gauvin St-Denis, B., Minville, M., Chaumont, D., Caya, D., Ludwig, R., and Turcotte, R.: An ensemble approach to assess hydrological models' contribution to uncertainties in the analysis of climate change impact on water resources, Hydrol. Earth Syst. Sci., 17, 565–578, doi:10.5194/hess-17-565-2013, 2013.

Vikhamar Schuler, D., Beldring, S., Førland, E. J., Roald, L. A., and Engen-Skaugen, T.: Snow cover and snow water equivalent in Norway: current conditions (1961–1990) and scenarios for the future (2071–2100), met.no report no. 1, Norwegian Meteorological Institute – met.no, Oslo, 2006.

Wilby, R. L. and Harris, I.: A framework for assessing uncertainties in climate change impacts: Low-flow scenarios for the River Thames, UK, Water Resour. Res., 42, W02419, doi:10.1029/2005WR004065, 2006.

Wilson, D., Hisdal, H., and Lawrence, D.: Has streamflow changed in the Nordic countries? – Recent trends and comparisons to hydrological projections, J. Hydrol., 394, 334–346, doi:10.1016/j.jhydrol.2010.09.010, 2010.

Permissions

List of Contributors

O. Fovet
INRA, UMR1069 SAS, 65 route de Saint Brieuc, 35042 Rennes, France
Agrocampus Ouest, UMR1069 SAS, 65 route de Saint Brieuc, 35042 Rennes, France

L. Ruiz
INRA, UMR1069 SAS, 65 route de Saint Brieuc, 35042 Rennes, France
Agrocampus Ouest, UMR1069 SAS, 65 route de Saint Brieuc, 35042 Rennes, France

M. Hrachowitz
Delft University of Technology, Water Resources Section, Faculty of Civil Engineering and Applied Geosciences, Stevinweg 1, 2600 GA Delft, the Netherlands

M. Faucheux
INRA, UMR1069 SAS, 65 route de Saint Brieuc, 35042 Rennes, France
Agrocampus Ouest, UMR1069 SAS, 65 route de Saint Brieuc, 35042 Rennes, France

C. Gascuel-Odoux
INRA, UMR1069 SAS, 65 route de Saint Brieuc, 35042 Rennes, France
Agrocampus Ouest, UMR1069 SAS, 65 route de Saint Brieuc, 35042 Rennes, France

I. Andrés-Doménech
Instituto Universitario de Investigación de Ingeniería del Agua y Medio Ambiente, Universitat Politècnica de València, Camino de Vera s/n, 46022 Valencia, Spain

R. García-Bartual
Instituto Universitario de Investigación de Ingeniería del Agua y Medio Ambiente, Universitat Politècnica de València, Camino de Vera s/n, 46022 Valencia, Spain

A. Montanari
Facoltà di Ingegneria, Università di Bologna, Via del Risorgimento 2, 40136 Bologna, Italy

J. B. Marco
Instituto Universitario de Investigación de Ingeniería del Agua y Medio Ambiente, Universitat Politècnica de València, Camino de Vera s/n, 46022 Valencia, Spain

J. M. Campbell
School of Environmental Sciences, University of Ulster, Coleraine, Northern Ireland, BT52 1SA, UK

P. Jordan
School of Environmental Sciences, University of Ulster, Coleraine, Northern Ireland, BT52 1SA, UK

J. Arnscheidt
School of Environmental Sciences, University of Ulster, Coleraine, Northern Ireland, BT52 1SA, UK

R. Hübner
Institute of Geography, Dresden University of Technology, Helmholtzstr. 10, 01069 Dresden, Germany Leibniz Institute for Applied Geophysics (LIAG), Stilleweg 2, 30655 Hanover, Germany

K. Heller
Institute of Geography, Dresden University of Technology, Helmholtzstr. 10, 01069 Dresden, Germany Leibniz Institute for Applied Geophysics (LIAG), Stilleweg 2, 30655 Hanover, Germany

T. Günther
Leibniz Institute for Applied Geophysics (LIAG), Stilleweg 2, 30655 Hanover, Germany

A. Kleber
Institute of Geography, Dresden University of Technology, Helmholtzstr. 10, 01069 Dresden, Germany Leibniz Institute for Applied Geophysics (LIAG), Stilleweg 2, 30655 Hanover, Germany

I. E. M. de Graaf
Department of Physical Geography, Faculty of Geosciences, Utrecht University, Utrecht, the Netherlands

E. H. Sutanudjaja
Department of Physical Geography, Faculty of Geosciences, Utrecht University, Utrecht, the Netherlands

L. P. H. van Beek
Department of Physical Geography, Faculty of Geosciences, Utrecht University, Utrecht, the Netherlands

M. F. P. Bierkens
Department of Physical Geography, Faculty of Geosciences, Utrecht University, Utrecht, the Netherlands Unit Soil and Groundwater Systems, Deltares, Utrecht, the Netherlands

A. Guadagnini
Department of Hydrology and Water Resources, University of Arizona, Tucson, Arizona 85721, USA
Dipartimento di Ingegneria Civile e Ambientale, Politecnico di Milano, Piazza L. Da Vinci 32, 20133 Milan, Italy

S. P. Neuman
Department of Hydrology and Water Resources, University of Arizona, Tucson, Arizona 85721, USA

T. Nan
Department of Hydrology and Water Resources, University of Arizona, Tucson, Arizona 85721, USA

M. Riva
Department of Hydrology and Water Resources, University of Arizona, Tucson, Arizona 85721, USA
Dipartimento di Ingegneria Civile e Ambientale, Politecnico di Milano, Piazza L. Da Vinci 32, 20133 Milan, Italy

C. L. Winter
Department of Hydrology and Water Resources, University of Arizona, Tucson, Arizona 85721, USA

N. Peleg
Hydrology and Water Resources Program, Hebrew University of Jerusalem, Givat Ram, Jerusalem 91904, Israel

E. Shamir
Hydrologic Research Center, San Diego, California, USA

K. P. Georgakakos
Hydrologic Research Center, San Diego, California, USA
Scripps Institution of Oceanography, University of California San Diego, California, USA

E. Morin
Department of Geography, Hebrew University of Jerusalem, Jerusalem 91905, Israel

F. Liu
Key Laboratory of Water Cycle and Related Land Surface Processes, Institute of Geographic Sciences and Natural Resources Research, Chinese Academy of Sciences, 11 A, Datun Road, Chaoyang District, Beijing, 100101, China
University of Chinese Academy of Sciences, Beijing, 100049, China

X. Song
Key Laboratory of Water Cycle and Related Land Surface Processes, Institute of Geographic Sciences and Natural Resources Research, Chinese Academy of Sciences, 11 A, Datun Road, Chaoyang District, Beijing, 100101, China

L. Yang
Key Laboratory of Water Cycle and Related Land Surface Processes, Institute of Geographic Sciences and Natural Resources Research, Chinese Academy of Sciences, 11 A, Datun Road, Chaoyang District, Beijing, 100101, China

Y. Zhang
Key Laboratory of Water Cycle and Related Land Surface Processes, Institute of Geographic Sciences and Natural Resources Research, Chinese Academy of Sciences, 11 A, Datun Road, Chaoyang District, Beijing, 100101, China

D. Han
Key Laboratory of Water Cycle and Related Land Surface Processes, Institute of Geographic Sciences and Natural Resources Research, Chinese Academy of Sciences, 11 A, Datun Road, Chaoyang District, Beijing, 100101, China

Y. Ma
Key Laboratory of Water Cycle and Related Land Surface Processes, Institute of Geographic Sciences and Natural Resources Research, Chinese Academy of Sciences, 11 A, Datun Road, Chaoyang District, Beijing, 100101, China

H. Bu
Key Laboratory of Water Cycle and Related Land Surface Processes, Institute of Geographic Sciences and Natural Resources Research, Chinese Academy of Sciences, 11 A, Datun Road, Chaoyang District, Beijing, 100101, China

S. Oehlmann
Geoscience Center, University of Göttingen, Göttingen, Germany

T. Geyer
Geoscience Center, University of Göttingen, Göttingen, Germany
Landesamt für Geologie, Rohstoffe und Bergbau, Regierungspräsidium Freiburg, Freiburg, Germany

T. Licha
Geoscience Center, University of Göttingen, Göttingen, Germany

M. Sauter
Geoscience Center, University of Göttingen, Göttingen, Germany

U. Morgenstern
GNS Science, P.O. Box 30368, Lower Hutt, New Zealand

C. J. Daughney
GNS Science, P.O. Box 30368, Lower Hutt, New Zealand

G. Leonard
GNS Science, P.O. Box 30368, Lower Hutt, New Zealand

D. Gordon
Hawke's Bay Regional Council, Private Bag 6006, Napier, New Zealand

F. M. Donath
Department of Applied Geology, Georg-August-Universität Göttingen, Goldschmidtstr. 3, 37077 Göttingen, Germany

GNS Science, Private Bag 2000, Taupo, New Zealand

R. Reeves
GNS Science, Private Bag 2000, Taupo, New Zealand

A. Vallet
UMR6249 – Chrono-Environnement, Université de Franche-Comté, 16 route de Gray, 25030 Besançon CEDEX, France

C. Bertrand
UMR6249 – Chrono-Environnement, Université de Franche-Comté, 16 route de Gray, 25030 Besançon CEDEX, France

O. Fabbri
UMR6249 – Chrono-Environnement, Université de Franche-Comté, 16 route de Gray, 25030 Besançon CEDEX, France

J. Mudry
UMR6249 – Chrono-Environnement, Université de Franche-Comté, 16 route de Gray, 25030 Besançon CEDEX, France

K. Vormoor
Institute of Earth and Environmental Science, University of Potsdam, Postdam, Germany

D. Lawrence
Norwegian Water Resources and Energy Directorate (NVE), Oslo, Norway

M. Heistermann
Institute of Earth and Environmental Science, University of Potsdam, Postdam, Germany

A. Bronstert
Institute of Earth and Environmental Science, University of Potsdam, Postdam, Germany